MECHANICS OF MULTISCALE HYBRID NANOCOMPOSITES

MECHANICS OF MULTISCALE HYBRID NANOCOMPOSITES

Farzad Ebrahimi
Department of Mechanical Engineering, Faculty of Engineering, Imam Khomeini International University, Qazvin, Iran

Ali Dabbagh
School of Mechanical Engineering, College of Engineering, University of Tehran, Tehran, Iran

Elsevier

Radarweg 29, PO Box 211, 1000 AE Amsterdam, Netherlands
The Boulevard, Langford Lane, Kidlington, Oxford OX5 1GB, United Kingdom
50 Hampshire Street, 5th Floor, Cambridge, MA 02139, United States

Copyright © 2022 Elsevier Ltd. All rights reserved.

No part of this publication may be reproduced or transmitted in any form or by any means, electronic or mechanical, including photocopying, recording, or any information storage and retrieval system, without permission in writing from the publisher. Details on how to seek permission, further information about the Publisher's permissions policies and our arrangements with organizations such as the Copyright Clearance Center and the Copyright Licensing Agency, can be found at our website: www.elsevier.com/permissions.

This book and the individual contributions contained in it are protected under copyright by the Publisher (other than as may be noted herein).

Notices
Knowledge and best practice in this field are constantly changing. As new research and experience broaden our understanding, changes in research methods, professional practices, or medical treatment may become necessary.

Practitioners and researchers must always rely on their own experience and knowledge in evaluating and using any information, methods, compounds, or experiments described herein. In using such information or methods they should be mindful of their own safety and the safety of others, including parties for whom they have a professional responsibility.

To the fullest extent of the law, neither the Publisher nor the authors, contributors, or editors, assume any liability for any injury and/or damage to persons or property as a matter of products liability, negligence or otherwise, or from any use or operation of any methods, products, instructions, or ideas contained in the material herein.

British Library Cataloguing-in-Publication Data
A catalogue record for this book is available from the British Library

Library of Congress Cataloging-in-Publication Data
A catalog record for this book is available from the Library of Congress

ISBN: 978-0-12-819614-4

For Information on all Elsevier publications visit our website at
https://www.elsevier.com/books-and-journals

Publisher: Matthew Deans
Acquisitions Editor: Dennis McGonagle
Editorial Project Manager: Alice Grant
Production Project Manager: Prasanna Kalyanaraman
Cover Designer: Victoria Pearson

Typeset by Aptara, New Delhi, India

Dedication

"To my honorable parents"
Farzard Ebrahimi

"To my dearests: my honorable parents and my beloved sisters"
Ali Dabbagh

Contents

About the Authors	ix
Preface	xi
Acknowledgments	xiii

1. Introduction to composites, nanocomposites, and hybrid nanocomposites — 1

1.1 A brief review of composites	1
1.1.1 General concepts	1
1.1.2 Applications	3
1.2 Carbon nanotube-reinforced nanocomposites	29
1.3 Fabrication methods	35
1.3.1 CNTs' synthesization	35
1.3.2 Fabrication of CNTR NCs	39
1.4 Characterization of CNTR NCs	47
1.4.1 Experimental methods	49
1.4.2 Atomistic modeling	51
1.4.3 Micromechanical methods	56
1.5 CNTR PNC structures: literature review	69
1.6 Introduction to multiscale hybrid NCs	74

2. Micromechanical homogenization and kinematic relations — 81

2.1 Micromechanical homogenization	81
2.1.1 Homogenization of CNTR PNCs	81
2.1.2 Homogenization of MSH PNCs	104
2.2 Kinematic relations	107
2.2.1 Kinematic relations of beams	108
2.2.2 Kinematic relations of plates	136
2.2.3 Kinematic relations of shells	161

3. Static analysis of multiscale hybrid nanocomposite structures — 187

3.1 Bending, buckling, and postbuckling of beams	187
3.1.1 Bending analysis of MSH PNC beams	187
3.1.2 Buckling analysis of MSH PNC beams	193
3.1.3 Postbuckling analysis of MSH PNC beams	199

3.2 Bending and buckling of plates	207
3.2.1 Bending analysis of MSH PNC plates	207
3.2.2 Buckling analysis of MSH PNC plates	215
3.3 Buckling of shells	221
3.4 Concluding remarks	230

4. Dynamic analysis of multiscale hybrid nanocomposite structures   231

4.1 Wave propagation, free vibration, and nonlinear forced vibration of beams	231
4.1.1 Free vibration analysis of MSH PNC beams	231
4.1.2 Nonlinear forced vibration analysis of MSH PNC beams	239
4.1.3 Wave propagation analysis of MSH PNC beams	247
4.2 Wave propagation, free vibration, and nonlinear forced vibration of plates	262
4.2.1 Free vibration analysis of MSH PNC plates	262
4.2.2 Nonlinear forced vibration analysis of MSH PNC plates	268
4.2.3 Wave propagation analysis of MSH PNC plates	277
4.3 Wave propagation, free vibration, and nonlinear forced vibration of shells	290
4.3.1 Free vibration analysis of MSH PNC shells	290
4.3.2 Nonlinear forced vibration analysis of MSH PNC shells	297
4.3.3 Wave propagation analysis of MSH PNC shells	303
4.4 Concluding remarks	309

References	**313**
Index	**359**

About the Authors

Farzad Ebrahimi received his B.Sc., M.Sc., and Ph.D. degrees in Mechanical Engineering from University of Tehran, Tehran, Iran. In 2012, he joined to the Department of Mechanical Engineering at Imam Khomeini International University, Qazvin, Iran as an assistant professor. Half a decade later, he was elected as an associate professor there. His research interests include mechanical behaviors of nanoengineered continuous systems, mechanics of composites and nanocomposites, functionally graded materials (FGMs), viscoelasticity, and smart materials and structures. He has authored more than 400 high-quality peer reviewed research articles in his fields of interest. Besides, he has edited and authored a few books for international publishers. His most recent book, *Mechanics of Nanocomposites: Homogenization and Analysis*, addresses static and dynamic behaviors of polymer nanocomposite structures subjected to various types of loadings.

Moreover, he is an associate editor in the journal of *Shock and Vibration* and one of the editorial board members of *Journal of Computational Applied Mechanics*. In addition to his aforementioned memberships, he is a distinguished peer reviewer whose expertise helps the editors of prestigious journals to judge research articles. He reviews a large number of manuscripts every year for reputable journals like *Composites Part B: Engineering, Materials & Design, Composites Science and Technology, Composite Structures, Thin-Walled Structures, European Journal of Mechanics – A/Solids, Mechanical Systems and Signal Processing, Engineering with Computers, Journal of Sound and Vibration, Journal of Vibration and Control*, etc.

Ali Dabbagh entered Imam Khomeini International University, Qazvin, Iran in 2012 to start his Bachelor studies in Mechanical Engineering. After graduation in 2017, he continued his graduate studies in the School of Mechanical Engineering, College of Engineering, University of Tehran, Tehran, Iran. In 2020, he received his M.Sc. diploma; his thesis was titled "On Mechanical Analysis of Multi-Scale Nanocomposite Structures." His research atmosphere lies in the field of solid mechanics with emphasis on the mechanics of composites and polymer nanocomposites, micromechanical homogenization techniques, dispersion behaviors of waves scattered in FGM nanostructures, and dynamics of smart continuous systems. The outcome of his research activities can be summarized in more than 55 research articles and two books authored by him. Due to his contribution to some original researches, he was awarded a membership in the Iran's National Elites Foundation in 2018.

In his first book, *Wave Propagation Analysis of Smart Nanostructures*, dispersion curves of nanosize continuous systems made from smart electro-/magnetoresponsive materials were analyzed. His second book, *Mechanics of Nanocomposites: Homogenization and Analysis,* deals with the static and dynamic responses of polymer nanocomposite structures in different working conditions. He is presently seeking an appropriate Ph.D. position to march on with his research interests.

Preface

About three decades ago, nanotechnology was born by invention of carbon nanotube (CNT). From then on, this newborn field has exponentially affected our world. Nowadays, a wide range of industries can be mentioned which are in close contact with nanotechnology and it is simply because of the enhanced mechanical, thermal, electrical, and optical properties of nanosize elements compared to their bulk specimen. Thanks to all of these superiorities, nanoengineered design has attracted the attention of both scientists and engineers. Relying on the mechanical merits of nanosize elements, they have been broadly hired as reinforcements in the fabrication of composites. In the present millennium, CNT-reinforced polymer nanocomposites (CNTR PNCs) promoted extraordinary stiffness, strength, and fracture toughness from themselves which made them excellent candidates for the design of structural components in different engineering disciplines such as mechanics, aerospace, civil, marine, automotive, etc. Although PNCs can be named as one of the best alternatives for the aforesaid applications, it must be kept in mind that they need to be well characterized prior to utilization. CNTs, for example, are generally synthesized in a nonstraight shape due to the existence of topological defects in their microstructure. Also, they are always found in the form of aggregated bundles because of the interatomic van der Waals attractive potentials between their individuals. Hence, the properties of PNCs may be influenced by such phenomena and this issue needs to be well captured in the engineering designs.

PNCs introduced in the previous paragraph can be assumed as two-phase materials if we dismiss the very small interphase between CNTs and matrix. Such nanomaterials can be strengthened even more if they are treated as a matrix which is aimed to host macrosize fibers as reinforcements. In such three-phase materials, multiscale reinforcements are implemented to improve the material properties of polymers. Therefore, they are known as multiscale hybrid (MSH) nanocomposites. Thanks to their dual reinforcing mechanism in both macro- and nanoscales, these hybrid nanomaterials can be well utilized in all of engineering applications that two-phase PNCs were. Due to their complicated composition, it is hard to characterize the properties of MSH nanomaterials via experimental measurement or atomic simulations. However, there exist many techniques for characterization of such three-phase nanomaterials in the framework of micromechanics with respect to the destroying impacts of

CNTs agglomerates and their wavy shape on the mechanical behavior of hybrid nanocomposites.

This book instructs how to extract the properties of MSH nanocomposites through a hierarchical micromechanical homogenization while the effects of waviness phenomenon, agglomeration of the nanofillers in the matrix, and thin/thick being of the CNTs on the effective modulus of the nanomaterial are captured. In addition, a variational-based method will be employed to describe the motion equations of structures of beam-, plate-, and shell-type subjected to both static and dynamic excitations. To this end, classical and shear deformable kinematic theories are used in order to make it possible to analyze both thin and thick continuous systems manufactured from MSH nanocomposites. It is noteworthy that within this text, the stability of the dynamic systems will be monitored, too, as well as that of the static systems. Plus, it tries to express the mathematical formulations so that the readers need not to refer to any other reference. Besides, all of the provided formulation can be derived by any individual with a primary knowledge from the fundamentals of engineering mathematics. The data presented in this book can be implemented by the designers of structural elements. Also, this book can be useful for both graduate and postgraduate researchers interested in analysis of mechanical behaviors of nanocomposite structures.

The authors did their best to present a general and complete book in order to make it easier for the readers to comprehend static and dynamic characteristics of hybrid nanocomposite structures. However, they want to sincerely ask all of the readers not to spare the authors their thoughts and comments. For sure, critical comments of any respectful reader can help the authors to strengthen the quality of this text in future editions.

Farzard Ebrahimi
Imam Khomeini International University, Qazvin, Iran

Ali Dabbagh
University of Tehran, Tehran, Iran

Acknowledgments

The authors are utterly certain that this book could not have been born had they only relied on their own abilities. So, they want to express their deepest gratitude to those who contributed to this project with all they had. First, the authors would like to appreciate their own family for their sacrifice during the long period of the book preparation. There is no doubt that this book could not be brought to life without their patience and mental assistance. Second, the authors want to cordially thank their colleagues for the time they put in proof reading the book. They want to especially extend thanks to Prof. Timon Rabczuk from Bauhaus University Weimar, Germany and Prof. Ömer Civalek from Akdeniz University, Turkey for their constructive comments. Needless to say, they led the authors to write a better book with their critical points of view. We are also grateful to Prof. Francesco Tornabene from University of Salento, Italy for his positive feedback on this work. In addition to the aforementioned international specialists, the authors would like to thank Mr. Saman Golpaygani Sani from K.N. Toosi University of Technology, Iran who selflessly helped the authors with his expertise in control engineering. Last but not least, the authors feel themselves in debt to the Elsevier team for their professional attitude during this project. The deepest appreciation goes to the Editorial Project Manager, Alice Grant. Her sense of responsibility, flexibility, and kindness made it possible for the authors to concentrate on their job free from any concern.

CHAPTER 1

Introduction to composites, nanocomposites, and hybrid nanocomposites

1.1 A brief review of composites

1.1.1 General concepts

From several years to now, people are familiar with the word "*composite.*" The widespread applications of composites in various fields have made them one of the most famous types of engineering materials in the community's point of view. Now it is not strange to see composites in the superficies of buildings. Also, many of the people know composites from their referral to the dentists. As you know, composite materials are the best-suit materials for tooth implants which are going to be implemented instead of the original damaged tooth. The main reason of this growing tendency in the real-world toward using composites is the superior features of this type of materials compared with the conventional types of homogeneous materials. In other words, composites' invention was due to the fact that such materials could enable us to gain from the benefits of the constituents as well as covering their weak points. As a very simple example of this potential, one can recall the concept of the manufacturing of reinforced concrete (RC). Remembering from the Mechanics of Materials course, steel as a ductile solid has proper tension capability whereas it cannot endure compressive loading as well. By contrast, concrete, that is, a brittle material, can exhibit excellent performance from itself once subjected to compressive loading; however, it cannot show reliable tension resistance. Therefore, thin steel bars were dispersed in concrete in order to reach a new material whose behavior in both tension and compression modes is improved [1]. As same as this example one can understand the general concept of the composites. In the composites' literature it is usual to call the main constituent material, whose volume fraction is greater, *matrix*.

Besides, usual composites possess a great strength-to-weight ratio which is of great importance in the engineering designs. In fact, you as a designer will be allowed to have elements with a lowered weight able to endure huge loadings. In practice, relying on the designers' understanding of the required material behavior, the volume fraction of the host matrix and the reinforcing element will be tailored [1].

Commonly used composites can be categorized to three main groups of *fiber-reinforced, particulate,* and *laminated* composites. In fiber-reinforced ones a finite content of fiber-like elements of the reinforcing agent will be dispersed in the matrix. In these composites, the fibers' orientation angle in the matrix determines the direction which has the best strengthened feature. As a very global attitude, the reason for selecting fibers-like reinforcing agents is the direct relationship between the size-lowering of the materials and their properties. In other words, small pieces of any material can promote better mechanical, thermal, and electrical execution compared with the bulk material [1]. It is reported that the matrix in a fiber-reinforced composite is responsible for saving the alignment of the fibers and avoid from their deviation in various directions as well as being in charge of transferring the load from the environment to the fibers [1,2]. The second type of composites, that is, particulate ones, can be enriched by mixing a group of particles from the reinforcing elements with the host material. As the final type of composites, laminates can be manufactured by arranging a set of stacking plies on the top of each other with desired orientation angles. This category can cover two previous types of composites in itself. Indeed, some of the plies can be made from particulate composites and some other can be selected from fiber-reinforced ones depending on the design obligations. Obviously, the mechanical reaction of the laminated composites to the external excitations can be varied if the orientation angles of the constituent plies are modulated [1]. It is worth mentioning that both short and long fibers can be selected in fiber-reinforced composite materials. As a whole recommendation it is better to utilize short-type fibers due to their better material properties compared with long ones. Also, it is notable that four major types of fiber-reinforced composites can be achieved in general. These types are *unidirectional, bidirectional, discontinuous,* and *woven* fiber-reinforced composites. In this book, we will not enter in the detail of these four types of composites reinforced with fibers and the volunteers are highly advised to explore about the aforementioned data in the composites' literature [2,3]. At the end, it must be noticed that utilization of composites has some troubles in addition to its advantages. Actually, the above text concerned with the appropriate features of composites and there was no data about the destructive regimes in the composites. As one of the most crucial and common phenomena in the laminated composites the separation of the stacking plies from each other, also known as *delamination* phenomenon, can be mentioned. The delamination in the laminates takes place due to the instantaneous change in the profile of shear stress at the intersection of the adjacent plies [1]. For the purpose of covering the delamination phenomenon in the case of using laminated composites it is better to use the concept of layer-wise theory instead of the so-called equivalent single-layer method [4]. Another issue that must be in mind is the possibility of the availability of pores and voids in the composite because of the usual problems in the fabrication procedure. Once fiber-reinforced type composites are selected, the real bonding circumstance between the fiber and the matrix must be included in the investigations. Indeed, the assumption of having perfect bonding between fiber and matrix is not accurate in all cases and sometimes the debonding will occur and affects the promotion of the composite material in a negative way [1].

1.1.2 Applications

In the beginning of the previous subsection some explanations were depicted about the popular applications of the composites in the daily life. Now, authors are aimed to discuss about some of the most magnificent applications of the different types of composites in a wide range of industries. Therefore, in this subsection the applications of composite materials in different aspects of engineering realm such as automotive, naval, space and aerospace, thermal, electronics, civil, and fusion will be discussed. In addition to the aforementioned applications, we will explain about the role of the composites in the medical and bioenvironmental cases. In what follows, each of the mentioned applications will be investigated within the framework of a separate section for the sake of simplicity.

1.1.2.1 *Composites in the automotive industry*

Maybe everyone is aware of the widespread application of the composites in the production of the automobiles. In this part of the book we will go through understanding the conceptual reason of this trend. As we know from our elementary knowledge of material science, general fibers commonly used in the fabrication of the composites possess a very high stiffness-to-weight ratio. For instance, consider carbon fiber (CF) and steel to be compared with each other. Low-stiffness CFs are able to provide Young's modulus of higher than steel whereas the mass density of such fibers is about one-fourth of the steel. Therefore, implementation of CF-reinforced polymer (CFRP) in the automotive industry can be an excellent choice because the utilized composite can be at least as efficient as the conventionally used steel alloys. On the other hand, a remarkable reduction will appear in the total weight of the designed segment once metallic alloys are replaced by CFRPs. Motivated by this primary reason, Ford Corporation used CFRPs in the prototype of one of its cars at the end of the 1970s decade [5]. They used CFRP in some parts of the car instead of the previously used steel. Some of these parts were front end, frame, doors, bumpers, driveshaft, and body-in-white. This material substitution resulted in an approximately 33% decrease in the total weight of the car. However, the replacement did not affect the performance of the automobile in a negative manner and the manufactured car's performance was at the worst situation identical with the formerly fully metal one [5].

Although the aforementioned example was a proof for the suitability of the CFRPs in the automotive industry from the performance viewpoint, such composites cannot be widely implemented in the fabrication of the cars. As you might have predicted, the main problem in this way is the high cost of the utilization of the CFRPs in the manufacturing procedure. Several attempts were made by researchers to lessen the cost of using such materials. However, the final cost for the fibers was nearby 8$ per lb, which is not appropriate for this industry because of the fact that the final price of the fabricated automobile will be unaffordable for the customer. Due to this limitation, glass fibers (GFs) were selected to be involved in the automotive industry for the goal of lessening the costs. As stated in Section 1.1, the glass fiber-reinforced (GFR) composites can be from various types. In fact, it is possible to use each of the fiber-reinforced or particulate composites in the automotive industry. Also, it is worth mentioning that in common applications thermoset polymers such as polyester and vinyl-ester are utilized as the matrix for the hosting of the dispersed particles or fibers of the reinforcing glasses. However, thermoplastics can be used in such composites with lower efficiency. Indeed, implementation of thermoplastics as the resin will result in a weakened thermal resistance which is not a desirable design outcome. Therefore, it is better to use thermosets in the composites reinforced with glass to be used in the production of the automobiles [5].

Another issue that is of high importance in the design of automobile devices is that they should be able to show a perfect energy absorption from themselves. In the previously widely utilized metallic materials like steel alloys or aluminum, the energy absorption will be carried out by the element through the plastic deformation that appears in the material. Indeed, the absorption procedure will be satisfied due to the ductile behavior of such metallic materials that can endure a certain value of plastic strain before their collapse. So, the absorption phenomenon can be related to the stress–strain behavior of such materials. However, no similarity can be found in the stress–strain behavior of the FRPs with that of the metals [5]. These materials have a stress–strain curve as same as that of the brittle materials in common. Although it may seem that this difference makes it impossible for the FRPs to be in the role of an excellent energy absorber, it must be noted that such materials can be better choices for the goal of absorbing the energy once utilized instead of metallic ones.

Indeed, the energy absorption mechanism in FRPs is not similar with that of the metallic materials. It is proven that brittle materials can absorb the high energy of a desired projectile by dispersing its initial energy in a conical shape fragmentation of the specimen's surface [5]. Hence, only the approach of energy absorption in FRPs differs with its mechanism in the metals. Also, it is shown that low-quality composites possess a greater specific energy absorption in comparison with the metals. It in natural that this specific value can be improved in the case of using high-quality manufactured FRPs instead of the low-quality ones [5].

On the other hand, stiffness of the implemented materials in the automotive industry is one of the key factors affecting the designers' mind. The general thought of the designer is that the best material is that with the greatest possible stiffness. Also, it is evident that the stiffness of the FRPs is very smaller than that of the metals. However, the equivalent stiffness-to-weight ratio of such composites is at least identical with that of the metals. Also, it must be taken in mind that because of the lower weight of the FRPs in comparison with metals, their stiffness can be enhanced by adding the thickness of the material [5]. Furthermore, it is worth mentioning that utilization of the composites can be resulted in the improvement of the structural stiffness of the elements via omitting the flexibility-making joints that cannot be ignored in the case of using metals. In other words, some of the joints that increase the flexibility of the system (i.e., corresponding with the reduction of the structural stiffness) will disappear due to the parts integration alternative that can be applied once composites are used in the design of the parts in an automobile [5]. In addition to the stiffness issue, composites can lead to a better damping mechanism originated from the fact that FRPs have greater internal damping compared with the metallic materials. As you know, the comfortability of the customers in the automotive industry can be better satisfied once the car has a better suspension system and this issue can be directly related to the damping ability of the implemented material. So as a brief conclusion, utilization of composites can be resulted in a better stiffness-to-weight ratio for the designed elements of a cat followed by a better oscillation dissipation [5].

It is also proven that the implementation of light-weight textile-reinforced composites hosted by a thermoplastic polymer can be resulted in observing marvelous mechanical performance even better than formerly known composites using a lower content of the reinforcing phase [6]. Indeed, researchers showed that such composites are able to exhibit an improved impact resistance, enhanced rheological behavior, and also acceptable stiffness caused by particular architecture of the textile's lattice. The aforementioned architecture was proven to be resulted in a remarkable enhancement in the energy absorption capability of the composite which is of great importance in the automotive industry as expressed in Ref. [5].

Although utilization of the GFs in the framework of the FRPs in the automotive industry has brought several benefits for the manufacturers and customers, there exists some deficiencies in the implementation of such composites which cannot be dismissed easily. For instance, the poor machinability potential, low recyclability of the FRPs, and their dangers for the living environment motivated the engineers and scientists to seek for another type of composites to be utilized instead of the so-called FRPs [7]. To this purpose, researchers found composites manufactured from natural fibers an appropriate choice. Natural fiber-reinforced composites can provide lowered weight, higher recycling probability, and lesser dependence on the oil resources. On the basis of the European Union's approval, the car manufacturers in the Europe were compelled to produce cars that a certain amount of them be recyclable. This amount which was determined as a fraction of the car's total weight was about to 80% in 2006 and it was risen to 85% in 2015. This decision resulted in the fabrication of less-pollutant cars via a similar quality by substitution of the GFs with natural fibers like kenaf, hemp, flax, jute, and sisal [7,8]. Toward these benefits, many of the world-class car manufacturers such as Mercedes-Benz, Ford, Audi, and Toyota started to use natural fibers in their productions. From then on, we will use the *green composites* terminology for these composites because of their help the living environment by saving oil resources and decreasing the pollutants. Now, we will go through a detailed data about the constituents of the green composites in the present paragraph. Among all of the natural fibers used in the production of the green composites, jute is the most known one due to its aboundness all over the world. Also, flax and ramie are popular green reinforcements in the producers' viewpoint. These fibers are capable enough to exhibit an outstanding mechanical performance from themselves [8]. Researches have shown that among all types of natural fibers the greatest stiffness-to-weight ratio belongs to ramie followed by hemp, kenaf, flax, abaca, curaua, jute, and sisal (refer to Fig. 1.1 for more data). It is interesting to point out that the stiffness-to-weight ratio of ramie, hemp, and kenaf is bigger than that of the E-GF which proves the acceptable mechanical performance of the green composites [8].

In addition to the polymer-based composites, metal-matrix ones possess a giant role in the automotive industry. These composites can be fabricated from dispersion of various fibers or particles in a metallic matrix. Depending on the selected constituents and their utilized content, metal-based composites can possess great strength-to-weight ratio, high stiffness-to-weight ratio, excellent damping performance, machinability, improved wear, corrosion, and friction resistance, and lessened coefficient of thermal expansion (CTE) [9]. Each of these features can be applicable for one of the parts of a car. For example, the wear resistance and thermal performance of metallic composites as well as their excellent fatigue life can be

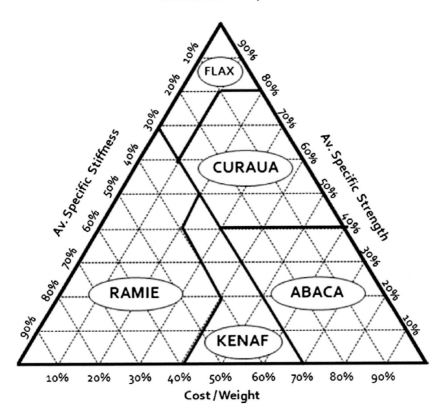

FIGURE 1.1 Ternary diagram for some of the natural fibers utilized in the green composites for reinforcement. It can be figured out that ramie, kenaf, and abaca possess great specific stiffnesses; while flax and curaua are known for their great specific strength. That is why many of the famous manufacturers in the field of automotive are aimed at hiring such fibers as reinforcing elements in their products. From G. Koronis, et al., Green composites: a review of adequate materials for automotive applications, Compos. Part B: Eng., 44, 1, 125, 2013, reprinted with permission from Elsevier [8].

logical justifications of their broad use in the piston crown and valves of the cars' engine. The low weight of such composites and their resistance against friction make these materials suitable choices for the fabrication of the bearings in the cars' engine. As another application, metal-based composites are widely used in the manufacturing of the cylinder block of the cars' engine because such composites are able to promote fantastic corrosion and wear resistance from themselves as well as their low friction and weight. In addition to the application of metal-based composites in the engine's parts, the low weight of such composites and their wear resistance was appealing enough for the designers to implement such materials in the cars' brakes, gearbox and differential bearings, and the gears and shift forks of the driveline. For more information about the vast applications of metallic composites in the automobiles, please take a look at the Ref. [9].

1.1.2.2 Naval and marine applications

Another engineering field which is in a very close touch with the application of the composites is the ship engineering and naval area. From historical viewpoint, composites were only implemented in noncritical ingredients of naval devices due to the fact that their material properties were not proven to be efficient enough to be utilized as critical elements, too. However, from the Second World War on, composites are broadly used in the fabrication of naval and also marine elements. For instance, the use of composites in the patrol boats and corvettes of US navy is reported in the open literature [10]. It is noteworthy that since 1960s, composites as light-weight materials have been utilized in the procedure of fabricating patrol boats. It must be mentioned that these boats are generally limited in length up to 10 m only because of the confined stiffness of the hull girder. In such boats, composites reinforced with GFs are used widely [10]. After successful production of patrol boats by US navy, other navies all over the world began to employ composites for the purpose of manufacturing patrol boats with various configurations. More historical data in this area can be found taking a brief look at the review published in the early 2000s [10]. Furthermore, composites have attracted the attention of the mine-countermeasure vessels' designers, too. Conventionally, such vessels were manufactured from wood because of its independency on the magnetic signals which made it possible for the ships to move toward directions armed with magnetic mines. However, infrequent distribution of wood in the world after the Second World War pursued the engineers to go through finding another alternative for such an application [10]. On the other hand, replacing wood with composites resulted in a remarkable reduction in the through-life costs of the mine-countermeasure vessels caused by the need for the maintenance in woody mine-countermeasure vessels. Use of composites in the fabrication of mine-countermeasures seemed appealing enough to the designers to produce fully composite ones in the 1970s [10] (refer to Fig. 1.2A). Another naval application of the composites can be found in the corvettes. Such instruments are giant ships designed to satisfy a group of purposes such as surveillance, combat, mine laying, mine-countermeasures, and antisubmarine warfare operations. In the production of such huge naval elements, composites possess a very critical role. The leading country in this area was Sweden who started to fabricate composite corvettes before any other country all over the world under the project YS-2000 [10]. To meet the designed properties in practice, hybrid composites reinforced with fibers of carbon and glass were utilized by the engineers according to those reported in Ref. [10]. It must be declared that composites are not implemented in the naval industry only

(A)

(B)

FIGURE 1.2 (A) The Alkmaar-class mine-countermeasure vessel manufactured in the 1974 by the Dutch navy. The hull of this mine-countermeasure vessel is designed and fabricated via GFRP. (B) GFRP air ducts utilized in the Karel Doorman ship of the Dutch navy to be responsible for both air-conditioning and heating (Color figures available in eBook only). From D. Mathijsen, Now is the time to make the change from metal to composites in naval shipbuilding, Reinforced Plast., 60, 5, 290 and 291, 2016, reprinted with permission from Elsevier [11].

for their improved stiffness/strength or their mine scaping capability. Such materials can be used as air ducts in the military ships to be in charge of air-conditioning and heating purposes (see Fig. 1.2B). In addition to the above examples, composites are broadly used in other parts of naval and marine instruments like superstructures, masts, propulsion, and so on whose detailed explanation can be found in complementary references in the open literature [10–12].

Polymeric composites can show remarkable resistance from themselves against corrosion and also are able to provide the light-weight criterion which is a key factor in the designing of naval instruments [13]. Also, utilization of this class of advanced materials can be lead to omit the magnetic-pulse dependency of the previously used metallic materials, too [13]. However, the issue of using ideal thermal barriers in the naval facilities is one of the most important design criteria in the procedure of their design. In fact, great thermo-mechanical working conditions and also possibility of occurrence of conflagration in the large ships makes it necessary to use fire-resistant ingredients in the fabrication of naval devices. In addition to the aforementioned requirements, elements of a huge ship must be powerful enough to provide acceptable strength and stiffness combined with low weight. Hence, fire-retardant composites are one of the most potential materials whose selection in this field can be resulted in interesting performance. In order to satisfy the above requirements, various types of composites were tested in the framework of an experimental study to examine their thermal behavior to understand if such materials can be implemented in naval applications. It was figured out that using intumescent coatings individually or in combination with ceramic coatings can satisfy the design criterion released by American Society for Testing and Materials [14]. It was shown that the flame spread index and heat release of such composites seemed appealing enough to be employed in the naval applications. In another attempt, it was experimentally found that utilization of a modified type of halogenated composites will be resulted in observing an improved thermal performance [13]. It is also proven that using matrices modified with aluminum tri-hydrate in the manufacturing process can be resulted in an enhanced ignition time incorporated with lower heat release rate [13]. At the end of the 1990s, researchers found it interesting to utilize brominated fire-resistant composites in the marine applications in order to provide better thermal barriers in the naval instruments [15]. It was reported that a maximum of 20%–25% reduction in the peak of the release rate can be met using aluminum tri-hydrate composites under cone calorimeter testing.

1.1.2.3 Space and aerospace applications

It is known from several years ago that the local temperature in some of the space activities around Earth and Mars exceeds nearby 1800 K which is corresponding with a huge thermal loading applied on the device responsible for working under aforementioned working conditions. Moreover, many parts of structural elements in aerospace applications must possess high specific stiffness and strength and therefore, fibrous composites are widely used in such applications [16]. Hence, the structural elements used in the fabrication of such devices are forced to survive under such severe thermal loading which cannot be satisfied unless a material with nearly zero CTE is implemented as the constituent material of the mentioned devices. Motivated by this reality, metal-matrix composites (MMCs) achieved a magnificent role in the field of aerospace engineering. In the above applications, it is common to employ MMC structures of type tube, plate, and panel in critical conditions [16]. Although MMCs are proven to be efficient choices in space and aerospace applications, it must be regarded that their production faces some glass ceilings like forming, manufacturing costs and so on. US's National Aeronautics and Space Administration (NASA) was one of the leading organizations could successfully use boron-aluminum (B/Al) MMC as the frames of truss lattices in the mid-fuselage section of the space shuttle orbiter [16]. It is worth mentioning that this attempt was resulted in a 45% weight-saving in comparison with the case of employing full aluminum bars as the struts of the underobservation truss lattice. Within another successful endeavor, graphite-reinforced aluminum-based (Gr/Al) MMC was implemented as the antenna boom of Hubble Space Telescope [16]. The later was resulted in certifying the required stiffness incorporated with a very low CTE, that is, one of the most crucial design criteria in this field of interest. Another achievement of such a material selection was the possibility of receiving electrical signals efficiently due to the outstanding electrical conductivity of the manufactured MMC [16]. Furthermore, other types of MMCs can be found whose multitask characteristic is close to Gr/Al one. For instance, graphite-reinforced copper (Gr/Cu) are able to provide enhanced stiffness in conjunction with high thermo-electrical conductivity which suits requirements of aerospace applications as well [16].

On the other hand, discontinuous reinforced aluminum (DRA) MMCs are appropriate candidates for the purpose of being used in a wide range of aerospace applications containing joints, attachment fittings of truss lattices, longerons, electronic packaging, thermal planes, bushings, and so on. In such applications, either particulate- or whisker-type reinforcing agents will be utilized in order to manufacture the composite [16]. To attain more data about various types of composites, referring to the authors' former book seems to be beneficial [1]. Getting back to our main discussion, it must be considered that among all of the potential applications of DRAs, such MMCs' best functionality can be observed once they are chosen to be used in electronic packaging and thermal-management because of their remarkable thermo-electrical conductivity in addition to their tunable CTE [16]. For example, particulate DRAs are broadly employed in either communication or global positioning system satellites. It is also reported that in the case of using DRAs for thermal-management in spacecraft power semiconductor module, more than 80% weight-saving will be enriched [16].

In addition to MMCs, polymer-matrix composites (PMCs) are able to provide many of the design criteria in the aerospace industry. For instance, PMCs can manifest suitable specific stiffness incorporated with an acceptable interlaminar shear strength (ILSS). Also, the proper bonding between the fiber and polymeric matrix is generally a powerful one which avoids from the occurrence of failure caused by the debonding phenomenon. Even though such composites possess lots of merits which can be useful in the aerospace applications, specimens manufactured from some resins like epoxy suffer from high possibility of ignition which is not a negotiable issue in the aforementioned application at all. So, such composites are required to be modified from flame-retardancy viewpoint before being utilized in the critical positions [17]. An especial category of PMCs whose exclusive material properties are highly demanded in the space and aerospace industry is PMC hosted by a shape memory polymer (SMP) matrix, also known as SMP composite (SMPC). Prior to state the potential applications of SMPCs, it is not bad to be familiar with some of the general characteristics of SMPs. Such polymers are a unique group of polymeric materials that are able to react against external excitations by manipulating their shape and color followed by a recovery procedure. SMPs are lightweight, low-cost, easy-to-manufacture materials with astonishing biodegradability and tunable glass transition temperature (T_g). However, these polymers cannot show an appropriate stiff behavior from themselves. To solve this shortcoming in such materials, they will be used as host matrices of composite materials reinforced with high-stiffness fibers whose stiffness is many times greater than that of the SMPs [18]. Therefore, the obtained SMPCs possess wide range of merits which make them applicable for space and aerospace applications. Meanwhile, it must be considered that the mechanical behaviors of SMPCs can be affected by some determining factors like high vacuum, severe thermal loading, and ultraviolet radiation whose influences in the space and aerospace engineering cannot be ignored [18]. As a simple proof of this claim, it is reported that the tensile characteristics of SMPCs will be dramatically affected negatively once the aforementioned material is assumed to be attacked by γ-radiation [19]. Also, it is noteworthy that the SMPCs' T_g might be stimulated in the enormous thermal conditions which take place in the practical applications in this area [20].

The shape memory feature of SMPCs has made it possible for such advanced composites to enter into the aerospace industry as hinges. Implementation of rubber-based SMPCs in the tape hinges can be resulted in a better shock control and intensifies the stored strain energy. In the late 2000s, it is reported that CFR SMPCs can be efficient candidates for the goal of being used as tape hinges whose recovery is of great importance in the aerospace applications [21] (see Fig. 1.3A for illustrative data). On the other hand, SMPCs can be utilized as booms in the satellites, too. Implementation of SMPCs as booms of such advanced instruments will be resulted in a mentionable decrease in the weight of the boom compared with the case of which a metallic material is employed instead. It is reported that accurate acquisition from the marvelous characteristics of SMPCs leads to fabrication of different types of booms for the space satellites. Moreover, as you might be aware, conventional metallic booms are consisted of tubular, extendible booms, and collapsible truss booms whose manipulation must be governed by a motor. However, once the smart behaviors of SMPCs are tailored, the desired elongation and compression are reachable without using extra mechanical elements [18]. According to the advanced characteristics of SMPCs, different types of booms can be produced including foldable truss boom, coilable truss boom, and storable tubular extendible member boom [18]. As one of the vanguard activities in the present era about SMPC booms, a funding was devoted by US Air Force Academy (USAFA) for the goal of manufacturing foldable SMPC truss booms with either two or three longerons, as illustrated in Fig. 1.3B, for the FalconSat-3 mission [22]. In the case of fabricating coilable booms, it is common to utilize SMPCs in the longerons of the boom in order to make the extension procedure simpler. Obviously, foldable and coilable booms are not the only aspects in which SMPCs can be implemented. In addition to the positions discussed above, SMPCs are going to be more and more used in various parts of space and aerospace devices containing deployable reflector, morphing structures, and expandable lunar habitats, etc. (see Fig. 1.3C–E for illustrations and complementary explanation) [18].

1.1.2.4 Composites in thermal environments

Without any doubt, one of the most crucial engineering concern is to dominate the huge thermal gradient which takes place in various industrial working conditions. Although many attempts have been carried out by several researchers to develop metallic superalloys with acceptable thermal stability, some inherent limitations make it impossible to improve the thermal characteristics of such alloys more than a certain bound. In many high-temperature circumstances, the local temperature exceeds nearby 1500 K, that is, roughly corresponding with 80% of the melting temperature of most of the metallic alloys [26]. Therefore, in such severe thermal conditions, metallic elements cannot exhibit proper stiffness and creep resistance from themselves. Motivated by above realities, ceramic-matrix composites (CMCs) have attracted the engineers' attention and seemed interesting enough in their opinion to be utilized in such working

14 1. Introduction to composites, nanocomposites, and hybrid nanocomposites

(A)

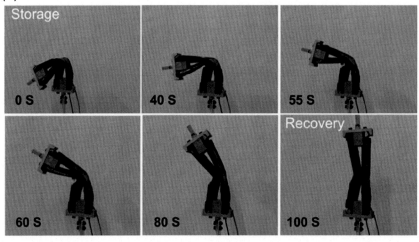

FIGURE 1.3 (A) Experimental illustration of both storage and shape recovery processes in a CFR SMPC in a time period of 100 seconds with regard to aerospace application of this smart device as deployable hinges. (B) SMPC booms with two and three longerons shown at the left and right, respectively. The two-longeron SMPC boom was claimed to be a better choice due to the smoother motion of its longerons along with the central guideline. However, the diagonal members utilized in the three-longeron SMPC were reported to face with difficulties during their bending motion to be tied down. Based on the design criteria, the two-longeron SMPC boom was utilized in the FalconSat-3 [22]. (C) The illustration of the stages passed for the deployment of the SMPC breadboard utilized in flexible precision reflectors (FPRs) developed to improve the satellite communication. By this way, the antenna of the FPR will be able to operate under radio frequencies even higher than 40 GHz. (D) Experimental measurement of the sensitivity of the deflection of the SMPC-assisted morphing skins to the increment of the local temperature applied on the SMPC tube. It is clear that change of the temperature, that is, corresponding with variation of the stiffness of the SMPC, leads to changes in the flexural deflection of the manufactured morphing structure. This behavior can be well matched to the tunable behavior which is demanded in the modern aerospace structures. (E) The consequent steps passed to carry out the deployment test of a mid-expandable composite lunar habitat, ordered by NASA, under low pressure at ILC Dover. This lunar habitat is 3 m large in diameter and possesses a length of 10 m. Color figures available in eBook only. Part A from X. Lan, et al., Fiber reinforced shape-memory polymer composite and its application in a deployable hinge, Smart Mater. Struct. 18, 2, 5, 2009, reprinted with permission from IOP Publishing [21]. Part B from P. Keller, et al., Development of a deployable boom for microsatellites using elastic memory composite material, in: 45th AIAA/ASME/ASCE/AHS/ASC Structures, Structural Dynamics & Materials Conference, p. 7, 2004, reprinted with permission from Composite Technology Development, Inc. [22]. Part C from P. Keller, et al., Development of elastic memory composite stiffeners for a flexible precision reflector, in: 47th AIAA/ASME/ASCE/AHS/ASC Structures, Structural Dynamics, and Materials Conference, p. 10, 2006, reprinted with permission from Composite Technology Development, Inc. [23]. Part D from S. Chen, et al., Experiment and analysis of morphing skin embedded with shape memory polymer composite tube, J. Intell. Mater. Syst. Struct., 25, 16, 2057, 2014, reprinted with permission from the Authors [24]. Part E from J. Hinkle, et al., Design and materials study on secondary structures in deployable planetary and space habitats, in: 52nd AIAA/ASME/ASCE/AHS/ASC Structures, Structural Dynamics and Materials Conference, p. 4, 2011, reprinted with permission from Composite Technology Development, Inc. [25].

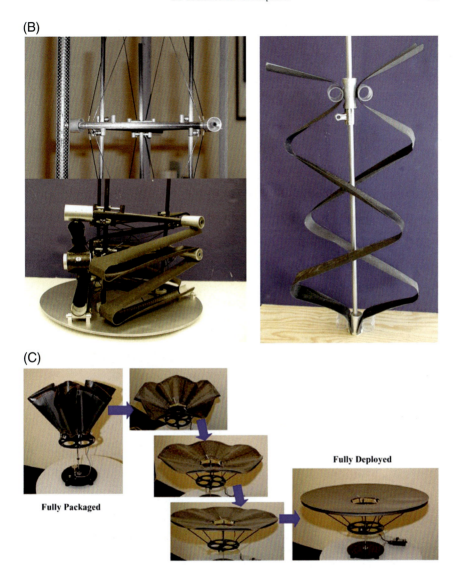

FIGURE 1.3 Continued

conditions, for example, in engines. In fact, implementation of CMCs guarantees the thermal durability for temperatures up to about 1800 K. This excellent manifest is not strange at all due to the enhanced thermal stability of CMCs. According to the literature, even temperatures in the range of 2300 K are supported by the aforementioned type of composite materials [27]. Such an achievement enables the designers to use these

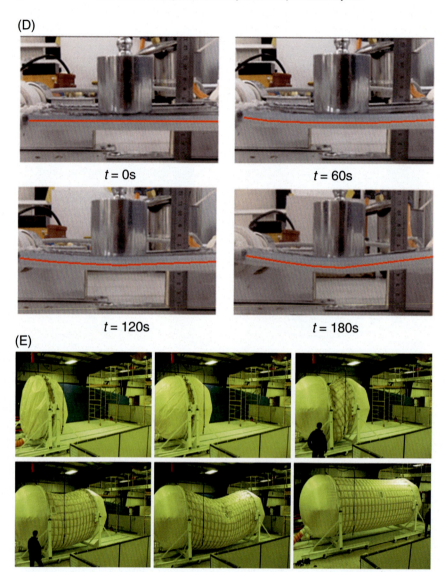

FIGURE 1.3 Continued

composites in aircraft engine and gas turbines. It is worth mentioning that individual ceramics cannot be employed in the abovementioned critical applications because of their brittle nature which leads to weak tensile properties [26].

The only use of composites in the engineering fields concerning high-temperature working conditions cannot be summarized as those discussed in the previous paragraph. In other words, in some other circumstances the important issue is to provide a situation in which heat transfer is speeded up [28]. One of the obvious examples of the abovementioned cases is the light-emitting diode (LED) industry. LEDs are used in a vast variety of conditions due to their longer lifetime in comparison with conventional lightings. But, the performance of these elements can be affected in a negative way because of the high heat flux generated inside them. Hence, it is of high importance to facilitate the heat transfer from the LEDs' inside to their outer atmosphere. To this purpose, materials with high thermal conductivity are required and this feature can be simply satisfied by PMCs in general [28]. Also, once the electronic packaging of electrical elements is aimed, the critical role of high-range thermal conductivity composites can be sensed again. In other words, the importance of fast cooling in the ingredients of electrical devices has made it obligatory to take as much gain as possible from the capabilities of highly thermal conductive materials. For instance, the cooling procedure in single chip packages must be performed using thermally conductive composites to provide an acceptable adhesion to do the cooling task in an efficient way [28]. Furthermore, PMCs well suit electrical control unit applications in the automotive industry, too. In these cases, thermally conductive, low-stiffness, viscoelastic materials are required and all of these design requirements together can be achieved using PMCs. Based on the literature, silicone PMCs are one of the most efficient candidates for the aforementioned situation [28]. Another field of interest involved with thermally conductive PMCs for the purpose of satisfying the process of rapid heat transfer is the battery industry. If high charge/discharge working rate of the batteries is met, the rate of heat generation will be greater than that of its attenuation. So, composites with remarkable thermal conductivity are deeply demanded in such cases to avoid from the onset of burning phenomenon in the battery [28]. Finally, it is noteworthy that another application of composites is in the solar cell industry. The increase of the working temperature can dramatically lessen the energy conversion efficiency of solar cells and this negative impact can be covered by providing a heat dissipation mechanism in such systems. However, the photovoltaic materials used in the fabrication of solar cells cannot be cooled efficiently and this drawback is usually compensated with the aid of electrical insulating fillers whose constituents are PMCs in general [28].

1.1.2.5 *Composites in transducers, sensors, and actuators*

In many of the modern multitask advanced devices, the concept of energy conversion, especially from electrical-type to mechanical one vice versa, is used. In these high-technology applications, smart piezoelectric transducers are increasingly utilized as well. Generally, two major types of transducers are employed, namely sensors and actuators. In the first one, an input of mechanical-type (i.e., strain or stress) must be converted to an electrical measurable variant with the aid of piezoelectricity phenomenon. However, the later type of transducers is asked to show the influence of an initial electrical signal in the context of a mechanical sign. It is noteworthy that sensors and actuators are respectively known as direct (passive sound receiver or hydrophone) and indirect (active sound transmitter or loudspeaker) transducers [29]. For example, hydrophones are generally expected to sense the underwater noises with relative low frequency (i.e., waves with frequencies inside 40 kHz). Obviously, a hydrostatic stress state exists in such cases due to the greater length of the sound wave in comparison with the sensor's dimensions. In the above circumstances, the sensitivity of the hydrophone will be judged by measuring the output voltage generated by the hydrostatic stress. Also, to measure the suitability of the piezoelectric for such applications, the voltage coefficient (g_h) will be monitored which relates the hydrostatic strain coefficient (d_h) to the permittivity of the utilized material. In addition to the electric-based viewpoint, the mechanical flexibility, shock resistance, and easy-to-form being of the sensor is of high importance in the mentioned application [29].

Another key parameter in the transducers is their thickness electromechanical coupling (k_t) which reveals the efficiency of the transducer in converting mechanical and electrical energies to each other. Besides, the mechanical loss (i.e., from acoustic type) in the transducers will be evaluated by the mechanical quality factor (Q_m). It is obvious that maximizing this factor can be led to a mentionable reduction in the acoustic losses. However, this parameter cannot be aggrandized limitless because its increase corresponds with low-resolution images in the applications like smart medical imaging. Hence, a mid-range Q_m must be selected to obtain a combination of desired design requirements all together. It is reported that quality factors from 2 to 10 can meet the desired goal [29]. Regarding the aforementioned design criteria, smart piezocomposites seem to be efficient enough to be used in the transducers industry. Using such smart composites can satisfy the discussed material properties in addition to the fact that enables the manufacturer to form the transducer in any arbitrary shape which is required for any particular case [29].

One of the materials which was previously used in the transducers is $Pb(Zr_{1-x}Ti_x)O_3$, also known as PZT. However, this group of piezoelectric materials are not proper choices for the case of being used as hydrostatic transducers mentioned in the previous paragraphs because their low hydrostatic piezoelectric coefficient will be resulted in a low voltage coefficient which is not a desirable feature in such applications. It is reported that the above phenomenon is caused by the coupling between d_{31} and d_{33} coefficients of PZT. To handle this problem, smart PMCs with piezoelectric properties can be employed efficiently. In the later, the aforementioned coefficients are decoupled from each other and also the permittivity of the enriched piezocomposite will be lessened to. The secondary effect of the mentioned material selection amplifies the voltage coefficient (g_h) due to the inverse relation between the permittivity and voltage coefficient. Experimental investigation of silicone rubber piezocomposites revealed that they are powerful enough to be employed as transducers because they exhibited remarkable sensitivity to low-frequency sound in addition to their lightweight property and easy-to-form feature [30]. Combination of polymeric matrices with piezoelectric ceramics can be resulted in generation of smart piezocomposites whose applications as electrical elements varies as the ceramic-polymer connectivity changes. For instance, 1–3 piezocomposites can be utilized as transducers in some applications of which both sensing and actuation procedures must be happened in conjunction to each other. For the sake of clarity, assume a pulse-echo medical ultrasonic imaging instrument in which both passive and active conversions must be happened in, followed by each other [29,31]. The design requirements expressed in former paragraphs (e.g., high thickness electromechanical coupling, low acoustic impedance, and supporting a wide range of dielectric constants) can all be met in the case of using the aforementioned type of piezocomposites [31].

Although the aforementioned type of piezocomposites has lower acoustic impedance in comparison with conventional ones, their acoustic impedance is many times greater than that of the human tissue and this reality affects the medical ultrasound imaging negatively. This issue is reported to be compensable using an interlayer connective piezoelectric between the original one and tissue. But, there still exists a problem and it is the dependency of the system's sensitivity to the thickness of the layers. To this purpose, a better alternative is reported, that is, implementation of the single crystals of the solid solution of $Pb(Zn_{1/3}Nb_{2/3})O_3$, also known as PZN, and $PbTiO_3$, also known as PT [32]. Via this approach, the required electromechanical coupling incorporated with low-range acoustic impedance can be met easily. In another similar attempt, the same behavior was monitored from 1 to 3 PMCs reinforced with $Pb(Mg_{1/3}Nb_{2/3})O_3-PbTiO_3$, also known as PMN-PT [33]. To conclude, PZN-PT and PMN-PT piezocomposites can be proper choices for the purpose of being employed in the ultrasonic transducers [32,33].

1.1.2.6 Composites in civil structures

FRPs possess magnificent features which can satisfy a wide range of material demands in civil engineering. Among all of their features, their eye-catching specific stiffness and strength, light weight, and durability are the most crucial ones. Therefore, conventional metal- and concrete-based elements have recently been replaced with FRPs. In general, applications of composites in the civil structures fall into two major groups of rehabilitation and new construction. FRPs will be implemented to repair an old structure; either strengthening it or retrofitting it against external environmental excitations like seismic in the first group. In the later, however, FRPs in either individual form or in combination with concrete will be used for the purpose of making new structures. For example, hybrid FRP/concrete composites showed excellent performance from themselves once implemented in the fabrication of bridges. In the mentioned application, a remarkable durability was observed from such lightweight hybrid composites [34]. In an experiment performed by researchers of University of California, San Diego, a five-story masonry building was strengthened with FRPs containing the fibers of carbon. Once the pseudodynamic seismic excitation is applied on the structure, an enhanced retrofitting is observed, that is, attributed to the material properties of the FRP [35].

Another issue which is covered in civil structures by the means of composites is the improvement of shear strength of walls. To this end, FRP wall overlays will be coated on the original wall. Due to the fact that the failure mode in this condition corresponds with the diagonal direction, the reinforcing fibers are orientated in direction with 45° incline along with longitudinal direction [34]. In addition, FRPs can be implemented for the purpose of satisfying better flexural resistance of civil structures. Because the main purpose in such cases is to enrich better buckling performance, it must be considered that the mentioned demand cannot be satisfied unless an ideal load transfer from the original structure to the overlay is happened [34]. Moreover, CFRPs are proven to be efficient choices in order to strengthen the slabs. In this regard, an experimental investigation was carried out to compare the results of both strengthened and individual slabs subjected to three- and four-point bending tests. It is observed that the deflection of slab will be dramatically lessened if CFRP strengtheners are utilized. According to this study, a minimum of twice greater ultimate strength for the strengthened slabs was observed in comparison with the state of which individual ones are used [34].

On the other hand, the concept of retrofitting columns of suspended structures like bridges was paid more attention since earthquakes of 1980s and 1990s in the California state. To this purpose, steel and FRP jackets were employed to retrofit the columns against seismic stimulations. It is reported that a mentionable improvement in the dynamic resistance of the structure was observed. Although both steel and FRP were able to improve the dynamic resistance of the underobservation structures, however, it was conceived that FRP jackets were better choices to this end. In fact, a larger enhancement was observed in the case of using steel jackets which is proven to be unnecessary in the aforementioned working conditions

[34]. Besides, lightweight, low-maintenance cost FRPs are reported to be suitable materials to be used as the constituents of bridge decks. Implementation of FRPs in the bridge decks makes it possible to construct long, lightweight bridges powerful enough to support enormous live loadings [34]. Even though FRPs possess features which sound interesting for civil applications, it is known that their manufacturing and curing process restricts their applications. It is reported that one of the best ways to overcome this shortcoming is to employ microwave-assisted curing routes. Following this method, the mechanical strength and storage modulus of the FRPs will be increased significantly [36].

In another type of applications, hybrid FRP-concrete composites will be used as an efficient material in civil structures. The major reason of such material selection is to cover the weakness of concrete's tensile performance. Actually, a novel type of RC will be generated in this way which has a very lighter weight compared with the conventional RCs [34]. One key factor which must be taken into consideration during implementation of FRP/concrete composite structures is the thermal durability of the obtained hybrid composite. For instance, phenomena such as freezing and thawing of composite can take place because of the thermally graded working conditions which must be endured by the composite structure. It is mentioned that the failure possibility is caused by the difference between properties of concrete and FRP [37].

On the other hand, nobody doubts that the issue of environment maintenance is one of the hot topics all over the world. Of course, one of the engineering aspects which is in close touch with this issue is the field of civil. In fact, the compatibility of the civil structures with the living environment is one of the most crucial subjects. Based on this logical background, a revolution in the constituent materials of civil structures has been begun since the last years of 1970s. In this regard, civil engineers did their best to find green materials to minimize the destroying effect of civilization on the environment as well as constructing modern, efficient structures. One of the green composites which seemed appealing in the engineers' mind is cellulosic fibers-reinforced one. Such natural fibers exhibited an acceptable mechanical performance from themselves once they were used as reinforcing agents in cement- or concrete-based composites. Actually, cellulosic cementitious composites were proven to have remarkable specific stiffness and strength as well as improved tensile, flexural, and impact properties never been observed in individual concrete at all [38]. In addition, such composites are environmentally friendly materials with mentionable biodegradability, renewability, cost-efficiency, and light weight features. It is reported that such materials are able to compensate for the old weaknesses of both cement and concrete. For example, cellulosic composites hosted by concrete can show an improved crack propagation resistance because of the fact that cellulosic fibers act as a bond trying to retain the matrix's cracks. The most famous cellulosic fibers which can be used in such composites are flax, sisal, jute, hemp, coir, hibiscus, cannabinus, eucalyptus, malva, ramie, pineapple leaf, kenaf bast, sansevieria leaf, abaca leaf, vakka, bamboo, banana, and palm [38]. Even though cellulosic cement- and concrete-based composites are excellent choices to be implemented in civil structures, it is reported that three major issues must be considered prior to start their industrial fabrication. The first concern is due to the fact that there exists no reliable data in the open literature about the durability of such composites. Moreover, the possibility of existence of pores in the composite and weak distribution of reinforcing natural fibers in different segments of the composite are reported to be the possible outcomes of using high content of reinforcing fibers in such composites. Final issue is attributed to the changes that exist in the chemical composition, geometrical dimensions, and surface roughness of cellulosic fibers. Such variable properties are reported to be key factors affecting the performance of the natural composite in a negative manner [38]. However, it is demonstrated that all of the above possible shortcomings can be compensated in the case of using woven composites of such cellulosic fibers as the reinforcing agent of the concrete in civil structures [38].

In addition to the broad applications of composites in civil structures, they are appropriate choices for the goal of being utilized as constituent material of infrastructures. Actually, the advanced properties of FRPs cannot be summarized in their specific strength and stiffness. Such materials are able to manifest outstanding resistance against corrosion, humidity, ultraviolet radiation and are able to promote suitable reaction to fatigue loading from themselves. Because of the aforementioned abilities, FRPs are implemented as connective joints, bridge piers, bumper systems, frame structures, truss bridges, and supports of solar panels [39] (see Fig. 1.4).

1.1.2.7 Fusion applications of composites

CMCs of silicone carbide-matrix/silicon carbide-fiber (SiC/SiC$_f$) possess an incredible role in the fusion reactors [40,41]. The main reason of this application is the ability of such composites to tolerate huge temperature gradient as well as their capability of showing excellent irradiation performance. Besides, such composites are good in some other aspects such as providing low chemical sputtering, high oxygen gettering, and negligible activation at both short and medium terms [40]. However, they suffer from existence of pores. Also, they cannot be shaped into large-scale samples, too. Therefore, giant ingredients of fusion reactors cannot be manufactured from SiC/SiC$_f$ composites, unless a connective joint is used (refer to Fig. 1.5A for an example of the jointed element in the fusion industry). In nuclear reactors, materials with high thermal conductivity and low tritium retention are asked. Hence, some of the famous composites may be unsuitable for such applications. For example, CFRPs cannot be

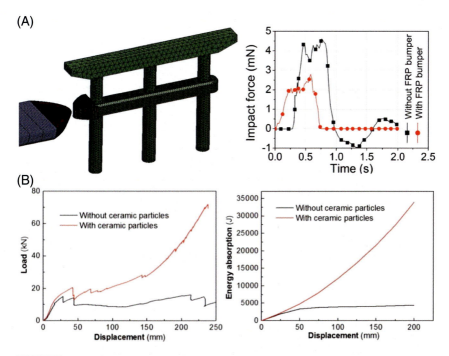

FIGURE 1.4 (A) Numerical simulation of an accident between a ship and FRP bumper installed on the piles of a bridge (left) and a comparison between the mechanism of the impact loading's damping in the cases with and without bumper (right). (B) Illustration of the effect of existence of ceramic particles in the FRP bumpers on the load–displacement and absorbed energy of the bumpers aimed to be utilized in the bridges for instance. It is observed that implementation of ceramic fillers can be resulted in a much better bumper performance. (C) Manufactured samples of FRP lattice sheets aimed to be implemented in the waterways. (D) A sample FRP short span truss bridge (left) and its flexural loading examination (right). In this small bridge, GFs were employed to fabricate the constituent FRP. Color figures available in eBook only. From H. Fang, et al., Connections and structural applications of fibre reinforced polymer composites for civil infrastructure in aggressive environments, Compos. Part B: Eng., 164, 134, 136, 138, and 139, 2019, reprinted with permission from Elsevier [39].

implemented in fusion applications due to the remarkable dependency of their thermal conductivity on the temperature. In the early 2000s, graphite fiber/SiC composites were experimentally tested and it was proven that they have a mentionable thermal conductivity incorporated with small tritium retention. So, such composites are excellent candidate for the fusion reactor applications [42].

Commonly, chemical vapor infiltration method is employed to fabricate SiC/SiC$_f$ composites. In an experimental attempt concerning this approach, it was understood that mechanical, thermal, and electrical properties of such composite vary gradually as temperature changes and therefore, a tunable feature can be attributed to the material properties of these composites [43]. However, this tunable characteristic does not affect the

(C)

(D)

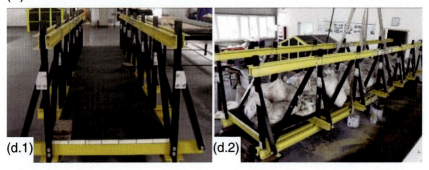

(d.1) (d.2)

FIGURE 1.4   Continued

efficiency of such composites in fusion applications at all. It is reported that SiC/SiC$_f$ composites are able to show a remarkable specific stiffness incorporated with a very low CTE at temperatures about 1800 K [44]. In addition to the above feature, SiC/SiC$_f$ composites are proven to be a proper choice for fusion applications due to their marvelous radiation resistance [44]. According to the advanced characteristics of such composites, they were implemented in power reactor blankets, flow channel inserts, and also applications concerning interaction with fission materials [44,45]. Within the framework of an experimental study, SiC fibers synthetized with chemical vapor deposition (CVD) were used to produce high-purity SiC/SiC$_f$ composites. The SiC matrix was fabricated through chemical vapor

1.1 A brief review of composites 25

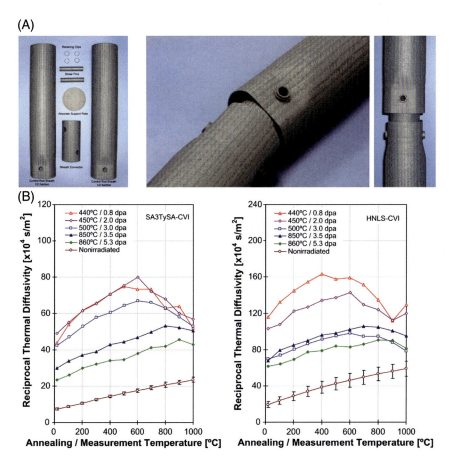

FIGURE 1.5 (A) The SiC/SiC$_f$ ingredients of a nuclear grade control rod sheath (left) and assembled view of the mentioned jointed device (right). (B) The variation of the reciprocal thermal diffusivity of the CVI-synthesized SiC/SiCf composites versus annealing/test temperature for both SA3 and HNLS products at left and right subplots, respectively. It can be observed that neutron radiation affects the HNLS specimen more than the SA3 one. It must be noticed that the thermal diffusivities at any desired temperature were reported 30 minutes after that the system touches into equilibrium state. (C) The variation of the defect thermal resistance of the CVI-synthesized SiC/SiC$_f$ (irradiated) composites versus annealing/test temperature for both SA3 and HNLS products at left and right subplots, respectively. It can be observed that for the temperatures inside 600°C, the defect thermal resistance of the SA3 specimen remains almost consistent; however, those of the HNLS products experience a decreasing trend in the mentioned temperature range. In other words, this figure denotes that the effect of various defects on the thermal conductivity of the SA3 and HNLS composites is not similar thanks to the varieties existing in the microstructure and impurities of such samples. Color figures available in eBook only. Part A from L.L. Snead, et al., Silicon carbide composites as fusion power reactor structural materials, J. Nucl. Mater., 417, 1, 332, 2011, reprinted with permission from Elsevier [47]. Parts B and C from Y. Katoh, et al., Thermophysical and mechanical properties of near-stoichiometric fiber CVI SiC/SiC composites after neutron irradiation at elevated temperatures, J. Nucl. Mater., 403, 1, 52 and 57, 2010, reprinted with permission from Elsevier [46].

(C)

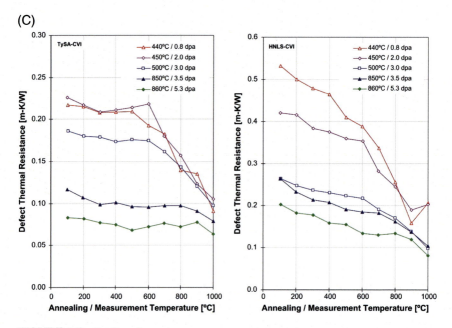

FIGURE 1.5 Continued

infiltration procedure. It was shown that thermal conductivity, diffusivity, and stiffness of the composite vary under action of the radiation [46] (see Fig. 1.5B, C). In another investigation in this domain, it is demonstrated that the resistance against neutron radiations is the most crucial feature which is needed to be satisfied in fusion power reactors. It is mentioned that thermo-electro-magnetic loadings can be better endured by the CMCs and the critical issue which must be considered is the radiation phenomenon [47]. In addition to the aforementioned applications, SiC and its composites are one of the most popular materials in the core of the light water reactors. In such cases, SiC-based composites are required to be employed as fuel cladding and fuel channel boxes [48].

In the previous paragraphs, the general concepts of utilization of composite materials in the fusion and fission applications were explained briefly. As the objective of this text is not related to such issues, more information about this part will not be presented and the volunteers are deeply advised to refer to complementary Refs. [49–52].

1.1.2.8 *Biomedical applications*

In the recent years, advanced characteristics of composites are going to be hired in biomedical applications. For example, such materials are used for the purpose of manufacturing artificial ventricular assist-type muscles. These artificial composite muscles can be implanted into heart patients' bodies to strengthen them against abnormalities taken place in heart due to the weakness of the cardiac muscles. It is noteworthy that implementation of composites in the aforementioned applications will be resulted in arrhythmia control. In such cases, ionic polymeric metal composite sensors and actuators will be implanted in the patients' bodies without affecting the blood circulation procedure. Because the artificial muscle is in touch with the heart ventricle, it must be soft enough to avoid from damaging the patient's heart during the compression. Furthermore, the composite must be a smart one with electronic robustness. It is worth mentioning that the recharging of the artificial muscle's battery will be carried out through the contact between battery and patient's skin [53] (refer to Fig. 1.6A to see a sample of the discussed device). In another biomedical application, ionic polymeric metal composites are utilized in order to produce artificial smooth muscle actuators which can be widely used as the constituents of the artificial parts of human bodies. Such elements can be produced by arranging a large number of segments attached together. Once again, the smart nature of composite materials will be employed in this case to activate any segment followed by another one via a controlled framework. After activation, a wave-based motion starts and due to the smart electromechanical coupling available in the composite, the artificial smooth muscle will be enabled to move and do its predefined task as well. These smart artificial devices can be used for the goal of amplifying the human bodies' motional sections. To this end, various types of artificial elements including wearable, electrically self-powered, exoskeletal prostheses, orthoses, and integrated muscle fabric system ingredients like jackets, trousers, gloves, and boots can be implemented [53]. Another biomedical application which can be addressed to the aforesaid type of composites is correction of refractive errors of the human eyes. In this critical case, many problems about human's vision, for example, myopia, can be overcame in the context of utilizing smart composites [53]. It must be mentioned that the aforementioned cases are not the only cases of which ionic polymeric metal composites can be used. In other words, such smart composites can be also employed in incontinence assist devices, peristaltic pumps, surgical instruments, and so on [53].

Besides, polymeric composites reinforced with natural fibers are of great magnificence in biomedical applications, too. For instance, PMCs reinforced with silk fibers can be used for production of wound sutures. In such circumstance, the knot strength and handling feature of the silk fibers sound interesting for the surgeons to suture the tissues, for example, cardiovascular tissues, with. The same composites are also useful in the case of being employed as scaffolds of injured tissues. The main reason of the suitability of these composites in the mentioned situations is the slow degradation rate of such materials that makes it possible for the injured tissue to be repaired during a relatively long period of time. Also, silk

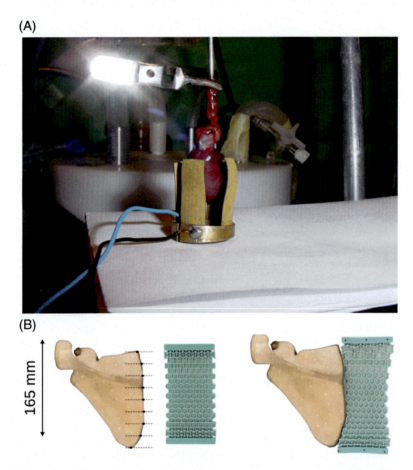

FIGURE 1.6 (A) A mini heart compression system manufactured with ionic polymer metal composite. This device is designed in a way that be in an excellent compatibility with the heart tissue and able to be charged by itself only due to the existing touch with the heart. (B) An additively manufactured sample of a soft mechanical metamaterial (i.e., consisted of three auxetic, transitional, and conventional lattices in the longitudinal direction) utilized for the purpose of matching the shape of the anatomical model of the scapula. (C) Application of viscoelastic/hyperelastic bibeams for the goal of organizing metamaterials with strain rate-dependent behaviors to be implemented as the constituents of the wearable devices used by human. Because of the difference between the properties of the selected hyperelastic and viscoelastic materials employed to manufacture the bibeams, such elements will bend in specific side. Hence, appropriate arrangement of such strain rate-dependent elements can be resulted in generating lattices similar to auxetic metamaterials. Color figures available in eBook only. Part A from M. Shahinpoor, K.J. Kim, Ionic polymer–metal composites: IV. Industrial and medical applications, Smart Mater. Struct., 14, 1, 211, 2004, reprinted with permission from IOP Publishing [53]. Part B from M.J. Mirzaali, et al., Shape-matching soft mechanical metamaterials, Sci. Rep., 8, 1, 4, 2018, reprinted with permission from the Authors [57]. Part C from S. Janbaz, et al., Strain rate-dependent mechanical metamaterials, Sci. Adv., 6, 25, 7, 2020, reprinted with permission from the Authors [59].

(C)

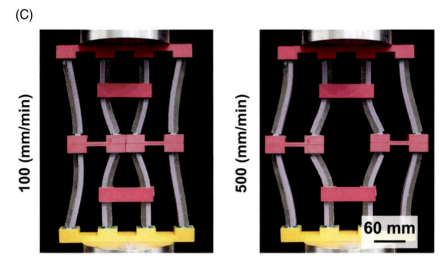

FIGURE 1.6 Continued

fiber-reinforced composites allow the designer to adjust the degradation time with the approximate time, that is, needed for the neotissue to be formed. One of the most famous parts of human body which can be repaired by composites reinforced with silk fibers is the damaged tissue of anterior cruciate ligament, also known as ACL. Plus, it is reported that the combination of natural fibers and biopolymers like polyvinyl alcohol can be resulted in production of biocomposites with a wide range of applications [54]. On the other hand, other types of composites can be found which can be broadly used as biomedical artificial instruments. For instance, functionally graded composites consisted of at least two different phases are of major importance in biomedical applications. Also, metamaterials with specific geometry-dependent mechanical behaviors are finding an incredible role in the biomedical engineering applications more and more, as shown in Fig. 1.6A, B. Among all types of metamaterials, those with auxetic lattice has attracted the attention of the research community more than the others. For more data about other cases of which composites are involved in, please refer to complementary Refs. [55–59].

1.2 Carbon nanotube-reinforced nanocomposites

It is obvious that small pieces of any arbitrary material possess properties superior to those of the same material's bulk. This phenomenon, which can be even observed in microscale, will be sensed clearer once a nanosize specimen of a material is compared with its bulk [1]. For example, carbon-based nanomaterials like graphite, graphene, carbon nanotube (CNT), graphene oxide, and graphene platelet (GPL) provide Young's modules many times greater than carbon itself in macroscopic scale. The credit of this nowadays obvious reality can be attributed to Iijima, the Japanese physicist who may be was the first one to discover CNT about three decades ago [60,61]. The outstanding material properties of nanomaterials, however, cannot be only summarized in their marvelous stiffness. In fact, the aforesaid nanomaterials are capable of promoting improved thermal and electrical properties in addition to their exclusive mechanical ones. Therefore, these nanomaterials seemed interesting enough to the research community to use them as reinforcing agents in composite materials hosted by different types of matrices (e.g., PMCs, CMCs, and MMCs) [1,62]. Because of the implementation of nanomaterials in such composites, they are usually called *nanocomposites*. It is worth mentioning that due to the fact that the main focus in this text book will be on the *CNT-reinforced* (CNTR) nanocomposites, the discussions will be orientated to be about such nanocomposites. However, the concepts are generalizable and readers can use them to understand what happens in other types of carbonic nanocomposites.

Concentrating on the possibly most famous type of nanocomposites, that is, CNTR polymer nanocomposite (CNTR PNC), they can be defined as multifunctional advanced materials with several thermo-electro-mechanical superiorities compared to other materials ever made. The mentioned potential of CNTR PNCs originates from the exceptional characteristics of CNTs. For instance, the chemical bond between the carbon atoms in the CNT's lattice is σ, that is, the most powerful bond. This powerful bonding between the atoms will show its influence in the framework of CNTs' marvelous stiffness. CNTs are as stiff as the stiffest material ever made, that is, diamond. In addition to their ability to provide Young's modulus of TPa-order, they are able to withstand 15% normal strain at the onset of their fracture. The aforementioned properties are reported to be caused by the sp^2 rehybridization of the valance orbitals of such nanomaterials [62]. CNTs are ranked as one of the best thermal and electrical conductors, which lead to their widespread applications in tremendous thermo-electric working conditions. Although CNTs are capable of showing a wide range of enhanced material properties, their addition to a matrix does not guarantee an improvement in the properties of the obtained NC. In other words, there exist some physical phenomena acting in a way leading to a reduction in the efficiency of the CNTR PNCs. Among all of these phenomena, the tendency of CNTs to be entangled by each other [63,64], their *wavy* shape [65,66], and *interfacial bonding* between matrix and CNT [67–69] are the most crucial ones which must be regarded once analyzing the behaviors of CNTR NCs.

To discuss the later, it must be recalled that CNTs are known as nanostructured materials whose diameter are in the nanometer range. Based on the type of the CNT (e.g., single-walled CNT [SWCNT], double-walled CNT [DWCNT], or multi-walled CNT [MWCNT]) and also its chirality, the diameter of such nanomaterial can be varied [62]. According to the inverse relation between the CNTs' diameter and their surface-to-volume ratio [70], such nanofillers have a very big surface-to-volume ratio corresponding with a great interfacial region between matrix and CNT. The load (stress)-transfer mechanism between matrix and CNT (i.e., required to be in its best ever possible state to enrich an efficient CNTR NC) is deeply affected by the type of the interfacial bonding between the constituents [71]. Although the interfacial bonding is of great importance in CNTR NCs, such bonding in the CNTR PNCs is weak because the CNTs cannot be connected to the polymer chains as well [72–74] and this bonding in such materials is a weak van der Waals (vdW)-type one [75,76]. Hence, the interfacial debonding takes place in CNTR PNCs once the as-received CNTs are utilized in the manufacturing procedure. To compensate this shortcoming, using functionalized CNTs is a recommended alternative. With the aid of this treatment, the revealed CNTs will be able to join the polymer chains easier than before caused by the fact that the compatibility of CNTs with the polymeric matrix will be increased in such circumstance [71].

On the other hand, the existence of groups of entangled CNTs in the PNC affects the mechanical performance of CNTR PNCs in an awful manner. In such situation, the CNTs are distributed in a nonuniform pattern within the NC which leads to the reduction of the materials' performance. Actually, any CNT will be entangled in the interatomic attractive vdW potential of the adjacent ones leading to construction of a number of randomly distributed clusters. So, many of the CNTs will be came together in a limited space of a cluster; whereas, some other regions can be found suffering from the lack of the reinforcing CNTs. The aforementioned phenomenon, also known as *agglomeration*, plays a crucial negative role in the determination of the material properties of the CNTR NC [77]. Existence of agglomerated CNTs in the NCs results in stress concentration which leads to appearance of microcracks and finally the failure of the material [71]. However, there exists beneficial approaches to overcome the aforesaid destroying influence. For example, one useful method is to employ CNTs modified within the framework of surface functionalization instead of the as-received ones in the fabrication of PNCs [78]. Functionalized CNTs have a better resistance against construction of the possible local agglomerates. In other words, not only functionalized CNTs increase the interfacial bonding between polymer and CNTs, but, they are also proven to be appropriate candidates for the goal of lessening the agglomeration's possibility through cutting the surface interactions of CNTs with each other [71]. This modification can be resulted to achieve CNTR PNCs exhibiting improved Young's modulus as well as better tensile strength. Researches have shown that both carboxyl (acid)- and amine-functionalized CNTs play crucial roles in aggrandizing the stiffness of PNCs hosted by thermoplastics [71]. Among two aforementioned types of surface modifications, amine functionalization is reported to be more efficient due to the fact that CNTs treated via this approach possess more compatibility with the polymeric matrix. This compatibility results in generation of a bridge between the opposite edges of matrix's internal cracks. The same improvement can be also observed in PNCs made from thermoset polymers [71]. Within the framework of an experimental study, it is shown that using plasma-polymerization procedure incorporated with utilizing functionalized CNTs reveals a CNTR PNC with material properties much better than those of PNCs manufactured via raw CNTs [79]. Another key issue which must be taken into consideration is the size of the clusters containing aggregated CNTs. This parameter is of high magnificence in the determination of the mechanical properties of CNTR NCs. Actually, better properties can be achieved once the dimensions of the clusters are minimized. Simply, a large number of small clusters affects the reinforcing purpose lower than a confined number of big ones [80]. In some other endeavors, bioinspired techniques like using acetone solvent [81] and employment of functionalized MWCNTs for producing enzymatic glucose biosensors [82] were suggested whose similarity was in remarkable reduction of the agglomeration possibility.

The inherent wavy shape of the CNTs (i.e., known as *waviness* phenomenon) is of course another issue which must be taken in mind during any study concerned with the material properties of CNTR NCs. As an opening to understand his issue, imagine a CNT as a cylindrical rod whose length is many times greater than its diameter. The above geometrical dimensions make the CNT a flexible nanostructure which has high potential to be bent due to its very big length-to-diameter ratio. According to this fact, CNTs with straight shape are rare and most of them have several curvatures in their shape. This curvy shape shows itself in the framework of reducing the reinforcing efficiency of such nanofillers in the NCs. To be more specifically involved in the physics of this issue, it must be expressed that CNTs will bent in the synthesizing procedure even if no external excitation is applied on them. The reason of this phenomenon is the available topological defects caused by the existence of pentagon–heptagon pairs in the hexagonal lattice of the CNT [77,83]. Another issue, that is, attributed to be another possible reason for the wavy shape of the CNT is the local deformations which may appear in an individual CNT. In fact, the CNTs' sensitivity to any excitation is high enough to make bending in this nanostructure easily. The resource of the bending stress can be even the weight of the CNT, its interaction with other adjacent CNTs, or the limited space of which this nanostructure is being synthesized in [83]. It is also reported that the nonstraight CNTs include a wider range. Some other types of CNTs (e.g., coiled CNT [CCNT], branched CNT, regularly bent CNT) can be found whose shape is nonstraight [83]. It is noteworthy that such CNTs cannot improve the material properties of polymers as well as straight ones. Back to the main discussion about wavy CNTs, it is worth mentioning that functionalization process can decrease the possibility of waviness phenomenon by lessening the attractive force between the CNTs, which was previously mentioned as one of the sources of the aforesaid phenomenon. It is shown that the effect of the waviness phenomenon on the mechanical properties of CNTR PNCs is at least identical with that of the agglomeration phenomenon [84].

After above discussions about various phenomena whose impact on the performance of the PNCs should be considered, the exceptional thermo-electro-mechanical properties of such materials will be reviewed in the following paragraphs. To start with the electrical superiorities of CNTR PNC in comparison with other materials, it is interesting to point out that polymers are individually one of the weakest types of materials if the electrical conductivity criterion is emphasized. Whereas, CNTs are able to support high-range electrical conductivities because of the existence of remarkable interatomic interactions between their atoms. Actually, the σ-π bond rehybridization which happens in the CNTs is the reason of the metallic-like electrical conductivity of such nanostructures [62]. So, addition of a limited content of CNTs to the polymer results in an electrical percolation in the obtained PNC. Such NCs are able to provide the high electrical conductivity as well as low percolation threshold, that is, desirable in the electronic applications, if long CNTs are selected to manufacture the PNC [71]. Details about the necessity of utilization of long CNTs to satisfy the aforesaid criteria is provided in complementary Refs. [85,86]. Unlike the crucial impact of surface functionalization of CNTs on the mechanical properties of CNTR PNCs, this treatment is reported not to be an efficient way to aggrandize the electrical conductivity of such NCs. Using acid-functionalized CNTs decreases the electrical conductivity of the CNTR PNCs; whereas, this feature can be intensified by using polyaniline-modified CNTs [71]. Another approach, that is, reported to be beneficial for the enhancement of the electrical conductivity of CNTR PNCs is to add a limited content of nanoparticles of highly electrical-conductive iron or silver to the composition of such NCs [71]. In addition to their high electrical conductivity, CNTR PNCs are smart materials with mentionable electromechanical coupling. The dependency of strain sensitivity of such materials to their electrical resistance makes them potential materials to be employed for fabrication of smart strain sensors [71].

Polymers are also unable to exhibit mentionable thermal conductivity. Among various types of polymers, thermal conductivities greater than 0.44 W/m K are not observed [71]; while, CNTs can provide thermal conductivities as big as 6000 W/m K [62]. Thus, adding a low content of CNTs to the polymers can be resulted in appearance of a great thermal conductivity which can be useful in the cases that thermal management through heat transfer is purposed. It is also revealed that the thermal conductivity of the CNTs grows proportional to the square of the local temperature in $T < 100$ K [62]. This temperature dependency of the CNTs' thermal conductivity acts positively in the cases that heat transfer is required. Although CNTR PNC possesses better thermal conductivity compared to the polymer itself, this enhancement is not as efficient as the improvement observed in the electrical conductivity of the same material. The major reason of this difference is that the mechanisms of which thermal and electrical conductivities will be increased are not the same. Indeed, the thermal conductivity of the CNTR PNC is restricted due to the high thermal boundary resistance of the interfacial region, also known as *Kapitza resistance* [71]. In other words, the controlling parameter of thermal conductivity is transportation of the phonons; while, electron transportation is in charge of the same role for electrical conductivity [71]. This phenomenon was previously mentioned in the context of the dependency of the thermal conductivity on the temperature change in the CNTs. Meanwhile, it is worth mentioning that the thermal conductivity of the CNTR PNCs can be aggrandized once functionalized CNTs are used in the fabrication procedure [71,87]. Amine functionalization, however, is shown to be a better way to enhance the thermal conductivity of CNTR PNCs compared with acid functionalization. In addition to their improved thermal conductivity, CNTR PNCs can show excellent thermal stability from themselves. Recalling our knowledge about the thermal stability of polymers, the credit of such an improvement can be attributed to the CNTs. The thermal durability of the CNTs is reported to be better than that of fullerene, graphite, and even diamond [62]. Moreover, the T_g of the polymers are proven to be increased whenever they are mixed with a small amount of the CNTs. The aforementioned positive effect of the CNTs on the thermal durability of the polymer originates from the interactions between the polymer chains and CNTs' surface which leads to remarkable reduction in the mobility of the polymeric matrix [71]. Obviously, the abovementioned interactions can be increased by adding the compatibility of the polymer and CNT with the aid of surface functionalization [88]. Another way to improve the thermal durability of such NCs is reported to be nitrogen plasma treatment. A recent research has proven the positive effect of implementation of nitrogen plasma treated CNTs instead of as-received raw ones on the thermal durability of CNTR PNCs through reduction of the amount of the aggregated CNTs in the NC [89]. Furthermore, the CNTR PNCs manufactured from silane-functionalized CNTs are capable of promoting an acceptable fire retardancy [71,88]. As the final clue about the thermal characteristics of CNTR PNCs, it must be considered that if the polymers' cross linking is influenced in a negative manner, the decomposition temperature of the achieved PNC will be lessened dramatically [71].

At the end of this section, it is interesting to mention the fascinating viscoelastic properties of the CNTR PNCs. Such materials are excellent candidates for the purpose of being used as structural damping elements. The results of dynamic mechanical analysis of such NCs revealed that CNTR polymers are capable of showing a 1400% improved loss factor in comparison with the individual polymer. The reason of the aforesaid enhanced loss factor is reported to be the slippage between the reinforcing CNTs [90,91]. Based on their improved viscoelastic properties and recalling their remarkable thermal durability and conductivity, lightweight CNTR PNCs have achieved a mentionable role in the applications concerned with the damping phenomenon [71]. On the other hand, it is interestingly shown that the reinforcing efficiency of the CNTs is in its maximum amount once soft polymers are chosen to the hosting matrix of the PNC. Indeed, the ductility, fracture toughness and strain, and Young's modulus of soft polymers are reported to be increased by addition of CNTs; however, those of harder polymers are shown to be decreased in the same position. Besides, the crack bridging phenomenon cannot be observed in the CNTR PNCs made from a hard polymeric matrix [71].

1.3 Fabrication methods

Obviously, achieving the aforesaid extraordinary material properties cannot be satisfied unless a homogeneous distribution of the CNTs in the NC is provided and the possibility of existence of agglomerates of CNTs is lessened to reach an efficient bonding in the interfacial region. According to the above reasons, it is very important to employ a suitable fabrication procedure for the goal of manufacturing CNTR NCs. In this section of this text, a brief insight about the frequently used methods of production of the abovementioned type of materials will be provided for the sake of clarity. In this regard, a primary discussion about the methods of synthesization of the CNTs will be presented as the opening.

1.3.1 CNTs' synthesization

Here, the most crucial methods of synthesization of the CNTs will be explained. Commonly, SWCNTs and MWCNTs will be synthesized via electric arc discharge (EAD) method [60,61,92,93], laser ablation method [94,95], gas-phase catalytic growth method [96], and CVD method [97]. In this subsection, general concepts of the aforementioned methods will be presented briefly. Afterward, the most magnificent ones will be surveyed in detail.

In the EAD method, a direct current (DC) will be used to connect a pair of graphite electrodes in the presence of an inert gas at high pressures of about 500 torr [98]. Once the above approach is undertaken in the presence of catalyst (e.g., Ni, Fe, and Co), SWCNTs are obtained; else, MWCNTs will be achieved at the end. Once the primarily known EAD method is utilized, CNTs with a mentionable structural perfection [99] can be synthesized whose size and structure may be negatively affected by a group of variants containing chamber's temperature, catalyst's composition, and existence of hydrogen for example [100,101]. Similar to CNTs synthesized through EAD method, those produced by laser ablation possess high structural perfection [101]. Besides, independent production of the SWCNTs can be met in the framework of this method. However, the content of the CNTs produced by this method is restrained and not as large as that produced via EAD method [102]. On the other hand, a high-range power is needed to make the CNT via this method [102]. CNTs synthesized with this method can be of variable quality depending on the chemical composition of the target material, laser's power, and wavelength, etc. [101,103].

The above methods suffer from some mutual shortcomings which makes them unsuitable choices if it is aimed to produce CNTs for industrial uses. For instance, generation of the arc needed in the EAD and laser required for the ablation are expensive enough make these methods not only unreachable but also unreasonable ones [62]. In both of the aforesaid methods, a large content of carbon and graphite is needed for the start of the synthesization. Providing such great amount of graphite is not easy and this reality makes it hard to utilize these methods in mass production of the CNTs [62]. In addition, nanotubes synthesized via EAD or laser ablation are always found in entangled and unaligned state and they need to be processed and purified before being used in critical applications [62]. In another famous method of synthesization of the CNTs (i.e., CVD method), a gas-phase hydrocarbon resource (e.g., ethanol, acetylene, propylene, methane, ethylene, and so on) will be subjected to a thermal loading caused by local temperatures of nearby 1000–1400 K. The aforesaid process will be handled in the presence of metallic catalysts like Ni, Fe, Co, etc. for the goal of growing CNTs. Implementation of the CVD method brings properly aligned [97], controllable grown [104] CNTs to life in large-scale dozes. Based on these reasons, CVD is the most popular synthesization method which is broadly utilized for the mass production of different types of CNTs. With the aid of this method, even the number of the walls of MWCNTs and the CNTs' diameter can be tailored [105]. It must be considered that although CVD possesses many positive features, CNTs synthesized via this method suffer from negatively influenced thermal [101], mechanical [106], and electrical [107] properties due to the high defect density of the produced CNTs [108]. In addition to the above characteristics, the CVD method is known as an efficient method for the purpose of fabricating ultralong CNTs with acceptable material properties [101]. A

TABLE 1.1 A brief comparison between different methods of synthesization of the CNTs [109].

Method	Advantage(s)	Disadvantage(s)
Electric arc discharge	High structural perfection	Impurity of carbon atoms, synthesization of short CNTs
Laser ablation	High structural perfection, high purity CNTs	High production cost, low yield
Chemical vapor deposition	Possibility of massive production, acceptable alignment of CNTs, controllability over CNTs' diameter, and number of the walls	High defect density

compact conclusion about what stated above can be found by referring to Table 1.1.

In the previous paragraph, it was tried to give a general insight about the differences of the most famous methods which can be used for the purpose of synthesizing CNTs. In what follows, a more detailed explanation will be presented for each of the aforementioned methods to be familiar with the mechanisms among which the CNTs can be produced.

1.3.1.1 Electric arc discharge

In the EAD, a pair of rod-like electrodes of circular cross-section will be used inside a chamber filled with either gas or liquid. Generally, He will be used if it is aimed to utilized gas-type inert. A segment of the anode will be in the shape of a hollow cylinder which will be filled with pure or hybrid metallic catalyst. The constituent material of the electrodes can be either metallic or carbonic. Positive and negative poles of a DC electricity resource will be subjected to the electrodes in order to start the procedure of the CNTs' formation. Before complete connection of the electric current, the cathode and anode will be kept in vacuum condition. Once the electric current is applied, the distance between electrodes must be tuned for the goal of generating the spark required to form the plasma which leads to the deposition of the CNTs from the anode. Indeed, the high temperature which appears in such conditions will be resulted in the vaporization of the anode's material and formation of the CNT. The above process can be speeded up in the case of using inert gas in the chamber. It is noteworthy that the DC will be applied only if the pressure of the inert gas is fixed on a certain value. Critically, the arcing current and pressure of the inert gas must be tracked instantaneously because any variation in the above parameters can affect the quality and features of the synthesized CNTs.

As mentioned before, both SWCNT and MWCNT are easy to be produced within the framework of EAD by changing the catalyst [62,101].

1.3.1.2 Laser ablation

The setup required to use this method is consisted of a cylindrical chamber which contains a target rod as well as a trap which is responsible for hosting the synthesized CNTs. The chamber itself is covered by a furnace to provide the high temperature required for the construction of the CNTs. The target rod is commonly consisted of combination of graphite and catalyst. The formation mechanism begins with the radiation of the laser beam from a source. Once the laser's emission starts, it will follow its predefined path to coincide with the target material and vaporize it accordingly [62,101]. This process will be progressed in the presence of an inert gas inside the chamber. Meanwhile, the chamber's working temperature will be increased by the furnace. Therefore, a cloud will appear in the neighborhood of the incidence point which is formed by ions of carbon and an inclusion of the catalyst's atoms. The above process will be resulted in the construction of the walls of the CNTs around the previously introduced trap. In general, CNTs synthesized via this method are single-walled ones whose diameter, length, and chirality are easy to control [101]. It must be declared that the SWCNTs achieved through this method are capable of exhibiting low defects. SWCNTs with purities up to 90% can be achieved by the means of laser ablation which could not be obtained via EAD. It is interesting to point out that the total elapsed time in the laser ablation will be in the range of some milliseconds [62].

As same as EAD, herein, the material properties of the synthesized CNTs can be varied if the process parameters are changed. For example, it must be noticed that the characteristics of the synthesized CNTs can be manipulated if the temperature and pressure of the inert gas are changed or a different catalyst is utilized. Obviously, the laser beam can affect the synthesization procedure because of the variable nature of the laser's source which may be resulted in emission of beams with different wavelength, pulse width, and intensity [62,101]. Also, it must be mentioned that a wide range of temperatures can be employed depending on the type of the source, that is, chosen to emit the laser beam. However, it is better to use high temperatures. As a comparison, lessening the temperature from about 1500 K to 1200 K results in a reciprocal reduction in both length and diameter of the synthesized CNTs [110,111]. So, the working temperature of the inert gas is the most crucial controlling parameter. Although implementation of high-range temperatures is better for the synthesization of the CNTs, this statement is not identical with the boundless increase of the working temperature. In other words, CNT has been rarely synthesized neither in 800°C nor in 1300°C. Hence, the formerly mentioned range of temperature must be used to manufacture CNT via laser ablation [62].

1.3.1.3 Chemical vapor deposition

Similar with laser ablation, CVD method requires a cylindrical chamber covered by a furnace. There exists a substrate in the chamber and this substrate is in charge of hosting the catalyst. In this method, a precursor will be sprayed in the chamber and heated enough to be deposited on the substrate. So, the CNTs will be shaped perpendicular to the substrate in two major ways. If the connection between substrate and catalyst is a powerful one, the CNT will be formed in a way that the primarily made carbon atoms are at the top part of the synthesized CNT and those that have been formed later are closer to the catalyst. In contrast, the location of the previous groups of carbon atoms in the longitudinal direction of the CNT will be changed if the interaction of the substrate and catalyst is a weak one. It is worth mentioning that the most common catalysts which are used in the CVD process are Fe, Ni, and Co. Implementation of the CVD technique enables the user to synthesize both SWCNT and MWCNT whose lengths are the maximum possible length ever made. Besides, this method possesses the advantage of low-cost being in comparison with previously discussed ones (EAD and laser ablation) [101].

All of the above explanations were concerned with the thermally influenced CVD. Other types of CVD algorithms can be found in the open literature that do not undergo with the aforementioned original format of the CVD. In one of these methods, also known as plasma-enhanced CVD or PECVD, the CNTs can be synthesized without using elevated temperatures. This type of CVD can be employed in the microelectronics-involved applications where some processes can be found which cannot be managed in the cases that the local temperature is high [62]. Lately, a roll-to-roll CVD (R2R CVD) framework was employed by researchers of the Massachusetts Institute of Technology that makes it possible to synthesize acceptable carbon nanomaterials with better material properties compared with those synthesized via conventional methods [112].

1.3.2 Fabrication of CNTR NCs

As explained in the previous section, the quality of the CNTs can be varied if the processing parameters are changed. The same trend can be observed once the compound of CNTs and a desired matrix is aimed to be surveyed. In other words, the production of CNTR NCs can be deeply influenced by the interfacial region between CNT and matrix, formation of local agglomerates leading to appearance of clusters of CNTs, the topological defects like wavy shape of the CNTs, and so on. Although all of the above problems are probable to be appeared in the CNTR NCs, a finite number of famous fabrication techniques are utilized by the producers of CNTR NCs in common for industrial production of high-quality NCs.

These methods are: solution mixing, in situ polymerization, melt mixing, layer-by-layer (LBL), and bucky paper. In the following subsections, it will be tried to give a conceptual explanation about each of the aforesaid methods for the sake of clarity.

1.3.2.1 Method of solution mixing

Based on this method, a third-party solvent is used in order to manufacture CNTR NCs. To this end, the CNT powder will be solved in the solvent at first. A proper solution can be enriched by using the ultrasound-assisted baths for this purpose. However, in the circumstances that ultrasonic bath is not in hand, one is allowed to use a powerful stirring to mix the CNTs with the liquid solvent. Then, it will be the time to prepare the mixture of CNT/solvent/polymer via sonication. Afterward, the solvent material will be evaporated and the final NC will be formed as a thin film [113]. The benefit of using an ultrasonic bath in this method is that the solvent can be easily evaporated by heating the bath with the aid of the piezoelectric-assisted controller provided in this device. This method can be applied for the fabrication of CNTR NCs hosted by either thermosetting or thermoplastic polymers. For example, MWCNTs synthesized through EAD method were employed to be solved in ethanol by the means of sonication in the middle 1990s [114]. Thereafter, the obtained mixture was stirred with epoxy matrix (i.e., a thermoset) to produce CNTR NCs. It is noteworthy that CNT and polymer can be mixed by shear mixing, too. However, this technique will bring some shortcomings with itself. For example, implementation of the shear mixing can be resulted in the reduction of the length of the CNTs which leads to a decrease in the material properties of the obtained PNC. To avoid from this deficiency, it is recommended to use surface-functionalized CNTs in the fabrication procedure. Such a selection will be resulted in improved material properties as well as avoiding from the reduction of the CNTs' length [115] (see Fig. 1.7 for detailed data).

In another production approach, two other steps were added to the original solution mixing for the goal of making CNTR NCs consisted of thermoplastic polymers. In this method, the CNTs achieved from EAD were solved in chloroform via sonication. Next, the CNTs were mixed with polyhydroxyaminoether and the outcome solution was splashed in a Teflon mold. Thereafter, this solution was cured in the ambient temperature to be dried in a finite period of time [116]. Furthermore, there exists another efficient approach to fabricate CNTR NCs on the basis of the solution mixing. In one of these methods, once the conventional solution mixing was fulfilled, the suspension of CNT and polymer will be used to be leaked and dropped on a roller which is rotating via a certain frequency. Afterward, another roller touches the first one (the wet roller) and a thin film of high-quality CNTR NC will be observed. This high quality of the

1.3 Fabrication methods 41

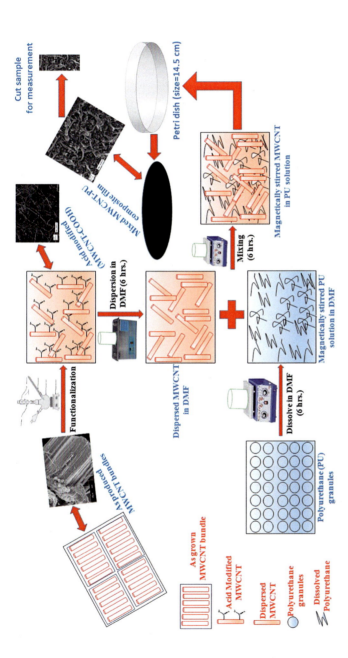

FIGURE 1.7 Schematic of the process required for fabrication of PNCs reinforced with CNTs in the context of the solution mixing method (Color figure available in eBook only). In the method described above, the as-produced MWCNTs were functionalized and then dispersed in the dimethylformamide (DMF) solvent with the aid of an ultrasonic mixer for 6 hours. At the same time, the granules of the polymer matrix (e.g., polyurethane in the above illustration) will be dissolved in the same solvent by the means of magnetically assisted stirring. Then, another magnetic stirring step for 6 hours will be passed to combine the processed nanofillers and polymer. At the end, the CNTR PNC solution will be poured into a desired mold (e.g., a petri dish in this case). Once the mixture is treated in an oven with temperature fixed on 80°C for 12 hours, the solid thin film of the PNC will be attained. From T.K. Gupta, et al., Improved nanoindentation and microwave shielding properties of modified MWCNT reinforced polyurethane composites, J. Mater. Chem. A, 1, 32, 9140, 2013, reprinted with permission from The Royal Society of Chemistry [115].

manufactured NCs is reported to be attributed to the proper load (stress)-transfer mechanism between the polymer and the functionalized CNTs [117].

In addition to the above types of solution mixing and their individual merits, it must be pointed out that the coupling casting/evaporation algorithm results in the reagglomeration of the CNTs. With regard to the destroying effect of the existence of agglomerates in the NCs, another method is suggested to compensate for this shortcoming. To this end, the suspension of the CNT and polymer will be dripped into water for the goal of generating a precipitation in the polymer chains. The outcome of this procedure will be the prevention from the construction of CNT agglomerates by entrapping them via polymer chains. The final high-quality CNTR NC will be attained by drying the obtained material in the vacuum condition. This method is also known as coagulation method [118]. It is worth mentioning that this method will be utilized for the goal of functionalizing the CNTs' sidewalls in high temperatures. The main concept of this approach is to use the controlled crystallization process in temperatures close to the polymer's crystallization one (T_c) [119,120].

1.3.2.2 Method of melt mixing

Although the method of solution mixing is a user-friendly efficient method for the production of the PNCs, some insoluble thermoplastic-based PNCs cannot be manufactured in the framework of this method [62]. In such cases, the polymeric matrix cannot be solved in the common solvents and therefore, another method must be employed for the purpose of producing NCs hosted by them. One of the best low-cost methods to this end is definitely the melt mixing method. According to this method, a combination of the polymer and CNT will be provided by applying a huge intense shear loading in high working temperatures. Based on this procedure, the polymeric matrix will be melted in order to pull the debundled CNTs into itself for the goal of manufacturing PNC [62,113]. Commonly, two major types of this method are known. In the first type, a shear mixer will be utilized to combine the polymer and CNT. Based on this, the master batches of the PNC will be produced. In the second type, however, large-scale production of continuous PNC will be met by the means of an extruder as a mixing tool [121,122].

To be more specifically familiar with the latter, it must be mentioned that two separate funnels will be used for the purpose of guiding both polymer grains and CNTs into the extruder. The polymer grains will be transferred to liquid phase by passing through the heated zone. Meanwhile, CNTs will be guided to the same zone and a melt mixture will be provided by the rotation of the extruder around its axis [123,124]. In the next step, the mixture will be cooled and dried by the means of a water bath for example in order to be shaped as desired. To this end, injection molding

1.3 Fabrication methods 43

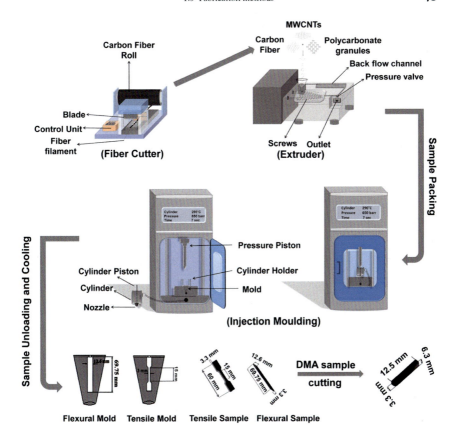

FIGURE 1.8 The schematic illustration of the procedure passed to manufacture PNC material with the aid of the extrusion-assisted melt mixing (Color figure available in eBook only). The polymer granules were dried for 24 hours at temperature of 100°C prior to start the fabrication. In this process, high temperatures of 250°C and 290°C were applied in the extruder and injection cylinder, respectively, for the purpose of satisfying the homogeneous distribution of the reinforcing elements in the microstructure of the PNC. At the end, various types of molds can be implemented to prepare samples which can be used for various experimental examinations. For instance, tensile and flexural molds are shown in the above illustration. From A.S. Babal, et al., Synergistic effect on static and dynamic mechanical properties of carbon fiber-multiwalled carbon nanotube hybrid polycarbonate composites, RSC Adv., 6, 72, 67956, 2016, reprinted with permission from The Royal Society of Chemistry [123].

machines are utilized accordingly (refer to Fig. 1.8 for schematic data). According to this method, environmentally friendly PNCs can be rapidly produced in industrial scopes without using extra solvents. However, it must be considered that the bundles of CNTs may exist in the PNC which was not observed in those synthesized via solution mixing [125]. It is noteworthy that the thermal durability of the constituents is a key factor in the determination of the quality of the PNCs made by this method

FIGURE 1.9 Illustration of the general algorithm of the in situ polymerization method for the production of the CNTR polymers (Color figure available in eBook only). From T.K. Gupta, S. Kumar, Fabrication of carbon nanotube/polymer nanocomposites, in R. Rafiee (Ed.), Carbon Nanotube-Reinforced Polymers: From Nanoscale to Macroscale, p. 67, Elsevier, 2018, reprinted with permission from Elsevier [62].

[126]. On the other hand, experimental observations have shown that PNCs manufactured via melt mixing exhibit higher internal viscosity from themselves with the enhancement of the CNTs' loading. This improved viscosity is reported to be more eye catching once the volume fraction of the utilized reinforcing phase exceeds 2% [127]. This feature can be well suited to the cases that an extraordinary viscose damping is needed. Besides, electron microscopy of the PNCs indicates on the proper distribution of the CNTs in the polymer which is possibly originated from the inverse relationship between increment of the temperature and appearance of the agglomeration phenomenon [128].

1.3.2.3 Method of in situ polymerization

At the late 1990s, the in situ polymerization method seemed appealing enough to be employed for the purpose of fabricating PNCs. One of the superiorities of this method in comparison with previously discussed ones such as solution mixing and melt blending is the higher homogeneity of the product due to the tiny dimensions of the monomeric molecules utilized instead of polymeric chains [113]. On the other hand, this method can be used efficiently in the cases that the host polymer is either insoluble or thermally unstable. Within the framework of this method, the manufactured PNCs will be able to promote better interfacial bonding between the constituents [62,113]. Also, the dispersion of the CNTs in the matrix can be improved using this method. To be more involved in the detail of fabrication of the PNCs via this method it must be considered that the CNTs need to be solved in a solution of the monomers; followed by the beginning of the polymerization in the presence of the initiator [62]. The schematic of the steps needed to be passed to manufacture PNCs by the means of in situ polymerization can be observed in Fig. 1.9.

In one of the earliest endeavors, radical initiators were implemented for the purpose of starting the reaction between the constituents [129]. It was proven that this type of in situ polymerization makes it possible to overcome the π-bonds of the CNTs' lattice. This phenomenon is the outcome of the fragmentation of the radical initiator. So, the nanotubes can attend in the chemical polymerization procedure more easily. It is worth mentioning that the length of the polymerized chains in this method can be tailored by generating a delay in the time of which the CNT material must be added to the mixture of initiator and monomer. This change is reported to be resulted in an enhancement in the mechanical properties of the PNC [129,130]. Parallel with and independent from Ref. [129], two separate groups tried to examine the radical-type in situ polymerization of polypyrrole-based [131] and phenylacetylene-based [132] PNCs. Experimental characterization based on the spectroscopy revealed that the chemical bonding between the polymer and sidewalls of the CNT are not proper ones. In another approach, researchers tried to use wet CNTs, whose sidewalls are wetted via conducting polymers, in the framework of electrochemical in situ polymerization [133,134]. Data provided by electron microscopy revealed that the CNTs in PNCs manufactured with this method are entirely covered by the molecules of the polymer. This reality guarantees the possibility of the surface functionalization of via a vast range of functions. To take a detailed industrial look at the product, it must be declared that such PNCs are able to manifest excellent optoelectrical features due to the perfect bonding between the polymer chains and the CNTs [133,134].

During the efforts for the production of PNCs possessing anisotropic behavior, a group of Japanese researchers came to the decision of using magnetically affected polymerization technique. To this end, they provided a compound of styrene monomer with CNT and applied a magnetic field of 10 T in magnitude on it [135]. Afterward, they accomplished the polymerization and observed that the alignment of the CNTs in the final PNC has not changed at all. The above results were regenerated by other researchers working on the similar issue. In these attempts [136,137], epoxy matrix was chosen and the efficiency of this method was proven in a way that they reported an enhancement in the thermoelectrical properties of the PNC due to the existence of the magnetic field [136]. In addition to the above works, smart silicone hosted PNCs were fabricated based on combining the graphitic lattice of the CNT with viscose monomers followed by mechanical grinding of the components [138]. Interestingly, the high amount of thermal energy flow of the grinding causes an improvement in the thermal conductivity of the achieved PNC. Furthermore, the calendering method (i.e., one of the simplest types of shear mixing) can be selected for the goal of fabricating PNCs from epoxy monomers and nanotubes in the framework of the in situ polymerization. In this method [139], the first step is allocated to the coarse combination of the CNT and matrix. In the next step, it will be tried to implement three-roll calender. Within the latter step, a relatively suitable dispersion of the agglomerated CNTs in the PNC can be achieved with the aid of the knead-vortex available between the rolls. This process will be finalized in the area between the rolls [139].

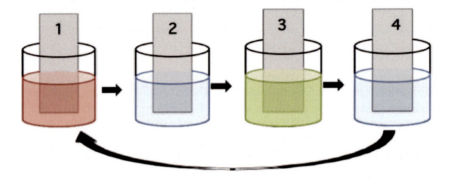

FIGURE 1.10 The PNC film-deposition in the context of the LBL method (Color figure available in eBook only). The adsorption of polyanion and polycation will be accomplished in steps one and three, respectively. Meanwhile, steps two and four are related to the washing procedure. It is noteworthy that the above cycle will be passed gradually till a particular number of the bilayers are generated in the LBL algorithm. One must pay attention that the thickness of the produced layers can be manipulated by changing the charge density, molecular weight, temperature, time of deposition, concentration, etc. of the ingredients aimed to be deposited. From S. Srivastava, N.A. Kotov, Composite layer-by-layer (LBL) assembly with inorganic nanoparticles and nanowires, Acc. Chem. Res., 41, 12, 1832, 2008, reprinted with permission from American Chemical Society [141].

Although PNCs produced via in situ polymerization are high-quality ones, it must be considered that this method is not a proper way for the goal of manufacturing PNCs in industrial scopes. The most important reason is of course the high amount of the solvent, that is, required in the processing of the aforesaid method [62].

1.3.2.4 Layer-by-layer method

In addition to the previously mentioned methods, another procedure for the production of the PNCs was developed in the early 2000s, called LBL method. According to this method [140–142], something like a sandwich film of the PNC will be produced by manipulating the quality of the adsorption of the individual sheets of the CNTs and polyelectrolyte chains, bonded together via electrostatic and vdW forces. You might be curious about the way of which this process is going to be exceeded. A substrate consisted of glass, for example, will be fallen into the solution of the CNT and polymer. Just a try is adequate for the assembly of layers as many as some tens. The schematic of the steps of the LBL algorithm can be observed in Fig. 1.10. It is noteworthy that the total structural integrity of the PNC film can be improved by using poly(acrylic acid) instead of the CNT after five working cycle, periodically [140]. This results in better cross-linking between the polyelectrolytes due to the carboxyl functionalization. In the next step, the prototype will be heated up to

120°C in order to strengthen the cross-linking through construction of amide bonds between poly(acrylic acid) and polyelectrolyte chains. The secondary impact of this heating is the formation of the covalent bonds between polyelectrolyte chain and CNT [140]. It is worth regarding that utilization of the LBL method enables the producer to manufacture PNCs possessing CNT weight fractions as large as 50% [140,142].

1.3.2.5 Bucky paper-assisted method

Another approach is known for the goal of fabricating PNC which employs thin porous sheets consisted of a large number of CNTs. The aforesaid sheets, also known as bucky papers, can be prepared by filtration of CNT dispersion in the water by the means of surfactant [143,144]. Such sheets possess a porosity volume fraction of 70% and a thickness varying from 50 to 200 micron. According to this method, the PNC can be easily produced by soaking the bucky papers in the polymer solution [143]. In a different bucky paper-assisted investigation, the mixture of epoxy-hardener was utilized via Buchner filtration through the thickness direction. In this attempt [145], the epoxy-hardener mixture was combined with acetone in order to lessen the viscosity of the mentioned blend. In this method, a high content of the CNTs can be used for the goal of manufacturing the PNC. It must be considered that within the framework of this method, a high temperature press molding is needed to form the PNC in its final shape [145]. Implementing the strategy introduced in [145], PNCs containing high CNT loading with mass fractions varying from 22% to 82% were produced following a vacuum filtration process [146]. In another study, phenolic matrix PNCs were manufactured by dispersing CNTs into solvent to generate the bucky paper which is aimed to be added to the phenolic resin [147]. Thereafter, the achieved suspension was mixed with rotation frequency of about 30,000 rpm. The rest of the fabrication detail such as filtration, drying, and press molding are similar with those reported in the above references and they will not be explained anymore. The schematic of the abovementioned approach can be observed in Fig. 1.11.

The above discussions were all about the methods of manufacturing PNCs with the aid of the bucky papers. However, it is worth mentioning that such thin-walled porous sheets can be employed as the thin layers of laminated composites in order to amplify the mechanical properties of the laminate [142].

1.4 Characterization of CNTR NCs

This section is allocated to the explanation of the different types of techniques commonly used to characterize the CNTR NC materials. Obviously, different methods are known up to now for the goal of

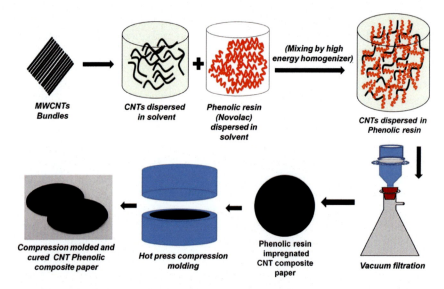

FIGURE 1.11 The schematic view of the bucky paper-involved production of the PNCs hosted by a phenolic matrix (Color figure available in eBook only). Once both nanofiller and matrix were solved in the solvent (e.g., acetone solvent was implemented in Ref. [147]), the mixture will be homogenized using a high-frequency mixer for a particular time period. Thereafter, the filtration followed by a drying process will be carried out. At the end, the PNC will be formed as thin-walled disks with the aid of the compression molding technique. The produced PNC films will be then carbonized in a high temperature furnace in the presence of Nitrogen gas to attain the PNC which can be used for various purposes. From S. Teotia, et al., Multifunctional, robust, light-weight, free-standing MWCNT/phenolic composite paper as anodes for lithium ion batteries and EMI shielding material, RSC Adv., 4, 63, 33170, 2014, reprinted with permission from The Royal Society of Chemistry [147].

deriving the material properties of the CNTR nanomaterials containing their mechanical, thermal, and electrical features. However, all of the usual methods can be fallen in the scope of one of the experimental, atomistic, and micromechanical viewpoints. In the upcoming subsection (Section 1.4.1), a brief explanation on the way of obtaining the material properties of the CNTR NC materials through experimental measurements will be presented. For sure, such measurements are the most reliable choice for accurate extraction of the material properties at the nanoscale. However, difficulties of these methods like calibration problems, measuring errors, and special examination circumstances makes it a little hard to analyze the nanomaterials via them. Hence, two other categories are to some extent preferred by a large number of research scholars due to their simplicity.

Although many of the researchers call atomistic and micromechanical methods simpler alternatives, this statement cannot be accepted completely. Indeed, the aforementioned methods are simpler due to their experiment-free nature; but they are involved with other types of complexities instead. For example, atomistic modeling of the NC materials undergoes with models suffering from extreme amount of computational costs. Also, a small change in an input variant may result in strange response which might be in some cases never justified by the user. Furthermore, micromechanical techniques contain other type of complexities caused by their close touch with the linear and nonlinear theories of solid mechanics. Also, it will be very important to choose an appropriate model in the micromechanics-based attempts to obtain a reliable understanding from the characteristics of the modeled NC material. In the next two subsections (Sections 1.4.2 and 1.4.3), detailed discussion about atomistic and micromechanical methods will be presented, respectively.

1.4.1 Experimental methods

There exist several methods in order to realize the behaviors of CNTR NCs. In one of the most famous methods, vibrational spectroscopy will be utilized to this end. The vibrational spectroscopy can be accomplished using infrared and Raman tests. Herein, the focus is on the Raman spectroscopy. In this method, the oscillations of any desired molecule will be tracked to identify the underobservation sample. To this end, a photon will be emitted from a laser source in order to interact with the sample's molecules. The energy of this photon is many times higher than the vibrational quantum energy. Once the incident between the photon and sample is happened, a portion of the photon's energy will be dissipated and the Raman spectra of the molecule can be reached. Afterward, the spectra can be interpreted based on three main features, namely frequency, intensity, and band shape. Besides, with respect to the fact that any molecule possesses its unique energy level, a particular fingerprint of each molecule can be obtained from its Raman spectrum. In the interpretation procedure, it is usual to employ either empirical or group theory approach for the goal of finding more details about the molecular structure. According to the latter viewpoint, molecules of the same functional groups will be treated similarly. In other words, the fingerprint of any desired group of molecules will be presumed to be independent from the individual molecules of that category. With the aid of the aforementioned method, the availability of different functional groups in a sample can be investigated according to the frequencies and their corresponding wave numbers in the spectra of the examined specimen [148].

In an experimental investigation, the Raman spectra of CNTR PNCs with different CNT loadings were obtained in order to probe the crystallization in such materials [149]. The Raman spectroscopy of CNTR NCs revealed that the peak frequencies of the system's spectra can be shifted due to the heat generated by the existence of laser beam whenever the NC is subjected to a four-point bending test [150]. In another study, the Raman spectra of CNTR NCs showed that implementation of thermally precured CNTs in the composition of the NCs can be resulted in an enhancement of the CNTs' uniform distribution in the polymeric matrix [151]. The variation of the G band in the Raman spectra of NCs reinforced with either as-received or functionalized CNTs in various temperatures proved the positive impact of the surface functionalization on the better transfer of the loading from matrix to the CNT, vice versa [152]. Moreover, a multiscale experimental characterization of the CNTR NCs was conducted on the basis of monitoring the Raman spectra of the aforesaid material [153]. The study of the residual stresses in the cured CNTR NCs was accomplished on the basis of interpreting the frequency shifts appearing in the NC's Raman spectra [154]. In another attempt, it was shown that a shift in the D* band of the Raman spectra of CNTR NCs can be resulted in a change in the strain of the specimen and this feature seemed to be useful for the act of strain sensing [155]. The measurements of the frequency peaks in the Raman spectra of CNTR PNCs showed that addition of bioinspired functional groups to the CNTs leads to a remarkable enhancement in the interfacial shear strength of the NC [156]. A nearly similar spectroscopy-assisted research was accomplished to show the dependency of the residual strains in the CNTR NCs in different temperatures to a mismatch between the CTE of the reinforcing phase with that of the polymer [157]. In addition, a combined investigation was carried out concerning the electro-mechanical features of the CNTR NCs such as resistance and strain variations by the means of the well-known Raman spectroscopy [158]. On the basis of the Raman spectroscopy, the losses in the interface between polymeric matrix and CNT were shown as well as highlighting the relation between the G band shift of the NC's spectra on the polymer's strain [159]. Experimental analysis of the interfacial bonding between CNT and matrix was procured in the framework of Raman spectroscopy [160]. It was shown that the addition of the CNTs in the polymer can be resulted in a positive shift in the wave numbers corresponding with both D and G bands; however, the interfacial bonding is not a powerful one. The resistance of CNTR NCs against ballistic impact loading was surveyed in the framework of an experimental study including a spectroscopy-assisted characterization [161]. On the other hand, spectroscopy of CNTR NCs revealed that cryomilling processing of the CNTs followed by their chemical functionalization can improve the fracture resistance of the manufactured NC [162]. The tensile strength of the CNTR NCs containing 3 wt.% reinforcing phase reported to be enhanced by about 30% in comparison with the pure matrix and the results were validated with those achieved from experimental characterization including spectroscopy [163].

1.4.2 Atomistic modeling

This part will concentrate on the algorithms which are commonly used in order to attain a rough estimation of the material properties of the CNTR PNCs. Many methods can be found in the literature to this purpose. However, in what follows, three most famous methods will be mentioned and discussed. The first and most crucial one which will be discussed is the molecular dynamics (MD) simulation. This method can be employed in many interdisciplinary fields like nanotechnology, biotechnology, nanomaterials, etc. Afterward, a modified branch of the MD, also known as coarse grained (CG) modeling, will be explained in detail. With the aid of this method, an answer similar to that predicted via original MD can be reached with a lessened computational cost. At the end, the ab initio simulation which is founded on the basis of the density functional theory (DFT) will be discussed briefly. The reason of the brief investigation in the last method is the lesser application of this method in the investigation of the CNTR PNCs due to its very high computational cost.

1.4.2.1 Molecular dynamics simulation

Within this part of the book, it is aimed to give the reader a general insight about application of the MD in characterization of CNTR PNCs. However, it must be pointed out that this section is written briefly and cannot cover all of the issues in the MD literature. This is logical because we are not about to hire MD in the homogenization procedure in the present text. Thus, only general issues in this area will be discussed herein as well as reviewing some of the crucial attempts made by researchers in this field. Before starting to discuss about the investigations fulfilled to reach the material properties of CNTR NCs via MD, it is interesting to present brief explanations about the framework of the MD. MD is an efficient method for the goal of analyzing phenomena concerned with material science, physics, chemistry, and other related aspects of science and technology by considering the interactions between the atoms and molecules of the underobservation object. This method is founded on the basis of the second law of the Newton [164]. According to this law, the interactions between various parts of any desired object will be covered. Any MD investigation consists of the below definite steps:

1. Generation of the appropriate atomic geometry
2. Defining the proper atomic interactions
3. Determination of the governing equations
4. Initialization and minimization of the total potential energy
5. Assigning the simulation circumstances
6. Implementation of an integration method
7. Computation of the required unknowns

Within the first step, an appropriate atomistic shape must be implemented which can best describe the underobservation object. The properties of the mentioned object such as its mass, the chemical bonding between this object and the adjacent ones, atom's charge, and boundary conditions (BCs) must be determined. Thereafter, proper interatomic potential must be employed to make it possible to compute the total energy of the system. It is worth mentioning that many potentials are available in the library of MD simulation. However, the experience of the user and his/her understanding of the nature of the designed problem can be useful means to find the suitable potential. Once the bonding conditions and proper potential are determined, the entire value of the potential energy will be calculated by individual computation of both bonded and nonbonded interactions followed by summing all of them up [164]. It must be kept in mind that the total energy of any particle comes from the interactions of the mentioned particle with other adjacent ones. Next, the initial state of any particle at the beginning of the modeling procedure will be presumed to be the initial condition of the problem. Afterward, the total energy will be minimized to attain the most stable situation from thermodynamic viewpoint. Once this process in completely done, it will be the time to capture the constraints that must be utilized to reach the desired variants. These constraints can be assumed to be responsible for a task which can be presumed to be similar with the role of the essential BCs in the FEM. To this purpose, *"ensemble"* will be determined by the user. In one sentence, ensemble is something like a package including all of the limitations. Three major ensembles of microcanonical (NVE), canonical (NVT), and Gibb's (NPT) are utilized in the MD simulations. In the first one, number of the atoms (N), entire volume (V), and total energy (E) will be kept changeless. In the second ensemble, temperature (T) will be controlled instead of energy. At the end, pressure (P), temperature, and number of the atoms (or molecules) will be fixed in the third ensemble [164].

Herein, some of the most important attempts toward analyzing the mechanical behaviors of CNTR NCs with the aid of the MD simulation will be reviewed. For instance, the estimation of both longitudinal and transverse Young's modules of the CNTR PNCs was the main objective of one of the initial attempts in this field [165]. In this study, the CNT is treated similar to a transversely isotropic linearly elastic solid. The mechanism of pull out of the SWCNTs from PNC was modeled by the means of MD and with respect to the vdW interactions [166]. In this study, the impacts of both length and diameter of the CNT on the pull-out characteristics of the PNC were covered. The findings of this simulation indicate on the inverse relation between CNTs' diameter and the IFSS of the PNC [166]. The destroying influence of existence of topological defects in the lattice of the CNTs on the equivalent modulus of the PNC is included in the

MD simulation accomplished with the aid of the well-known COMPASS[1] potential [167]. According to a combined semiatomistic study, the elasto-plastic constitutive analysis of CNTR PNCs was performed by the means of the MD incorporated with the nonlinear relations of the micromechanics [168]. The results generated in this semiatomistic simulation were in a remarkable agreement with those achieved from the experimental examinations. Moreover, a different atomistic point of view dealing with the approximation of the material properties of the interfacial region in the CNTR PNCs was introduced in [169]. In this investigation, the signs of local fracture at the simulation cell of the tensioned PNC were employed to estimate the material properties in the interfacial area. By considering the crucial role of the interfacial region on the determination of the material properties of CNTR PNCs, an investigation was conducted with the aid of the MD atomistic method for the goal of monitoring the stress–strain curve of the aforesaid type of the PNCs [170]. In another atomistic simulation [171], it was observed that zigzag (9,0) CNTs can best improve both longitudinal and transverse modules of the CNTR PNCs. Consideration of the impact of the low length-to-diameter ratio of the short CNTs on the equivalent stiffness of the CNTR PNCs hosted by epoxy was the issue of another MD-based atomistic modeling fulfilled in [172]. Furthermore, the type of the bonding between CNT and matrix in a PNC was simulated by the means of the MD [173]. In the mentioned study, an annular region was assumed to be in the role of the interphase between CNT and polymer. By choosing a thermoset polymer as the matrix of the NC, CNTs with inherent wavy shape were utilized in the framework of MD simulations for the goal of analyzing the impact of waviness phenomenon on the IFSS of the PNCs [174]. In addition, this investigation includes the effect of the unavoidable phenomenon of aggregation of the CNTs on the material properties, too. The pull-out of the CNTs from the polymer matrix in a CNTR PNC was simulated in a commercial MD-based software in order to determine the IFSS of this type of the PNCs [175]. In the context of a hybrid simulation study, researchers utilized the MD for the purpose of showing the improved IFSS of the PNCs reinforced with boron nitride-modified CNTs [176]. It was revealed that such boron nitride-CNT reinforced nanomaterials can provide a 90% improvement in the IFSS compared with the bare CNTR PNCs. By the means of the constant strain method of minimizing the total energy in the framework of MD, the components of the elasticity tensor of CNTR PNCs were derived in [177]. Moreover, the influence of nitrogen-doping process on the variation of the thermomechanical properties of the CNTR PNCs were tracked in [178] with the aid of the so-called MD. It was demonstrated that implementation of

---

1 Condensed-phase Optimized Molecular Potential for Atomistic Simulation Studies.

the pyridinic-type nitrogen-doped CNTs can be resulted in a noticeable improve in thermal conductivity, modulus, and IFSS of the CNTR PNC. In another effort, a combination of MD and FEM was utilized for the goal of probing the constitutive behaviors of PNCs. Comparison of the results with the experiments showed that the implemented atomistic simulation can estimate the material properties with an acceptable precision in low CNT loadings [179]. The MD simulation of the creep behaviors of CNTR PNCs revealed that the geometrical features of the reinforcing CNTs like their length and diameter can affect the relaxation time, creep strain, and strain rate of the CNTR PNCs in a remarkable manner [180]. On the other hand, MD simulations proved that employing CNTs as reinforcing elements in the PNCs can be resulted in an approximate 25% reduction in the friction coefficient in comparison with the bare polymer [181]. According to the MD and by capturing the variations of the density with temperature, both T_g and CTE of the CNTR PNCs were characterized [182]. Besides, MD simulations certified the efficiency of the plasma treating of the CNTs in improving the thermal conductivity, storage, and loss modules of the CNTR PNCs [89]. With emphasis on the role of the chemical bonding on the constitutive characteristics of CNTR PNCs hosted by geopolymers, MD simulations were accomplished in [183] for the purpose of showing the impact of bonding condition on the yield strength, Young's modulus, and IFSS of the PNCs. It was reported that chemical bonding can be resulted in a mentionable enhancement in each of the abovementioned features. More data concerned with the MD simulations can be found referring to complementary Refs. [184,185].

1.4.2.2 Coarse grained simulation

Even though the previously discussed full-atomistic MD simulation reveals the reaction of the PNCs to various excitations with high precision, in some cases such a precision is not required. In other words, the bulk features of the system are aimed to be known instead of a deep detailed insight into the molecular behaviors of the underobservation object in the above cases. In this regard, a mesoscale investigation, which falls in the scope of neither atomistic nor continuum mechanics studies, can satisfy the desired goals [186]. In these cases, CG simulation matches the desired purposes. So, in one sentence, CG modeling can be assumed as a leveled down form of the MD simulations. According to this method, a group of atoms in any desired object will be treated as a unit, named CG beads. Hence, lower run-time will be required for the purpose of extracting a comprehension from the behaviors of the object, definitely in conjunction with a decreased precision [62]. In this type of simulations, it is common to utilize the center of mass of the CG beads as the point that must be considered for construction of the potential energy of the object which is going to be monitored. Obviously, this type of mapping algorithm reduces the precision of the gathered response in comparison with fully atomistic MD modeling. However, suitable selection of the atomic potential can be resulted in minimizing the difference between CG and original MD modeling [62].

From the middle 2010s to now, some of the researchers tried to present CG-assisted investigations addressing the material properties of CNTR PNCs. In one of the initial attempts in this area, the CG method is employed in order to approximate the Young's modulus of the CNTR PNCs [187]. In this study, the concept of the conservation of strain energy between CG and atomistic MD simulations was implemented in order to lessen the difference between the responses achieved from these methods. It was observed that the modulus of the PNCs can be derived with a remarkable precision on the basis of the CG modeling, in addition to a mentionable reduction in the computational cost. In another CG-based simulation, the Young's modulus, tensile strength, and stress–strain curves of CNTR PNCs were obtained with respect to the crucial impacts of the geometrical properties of short CNTs and different amounts of the volume fraction of the CNTs [188]. Although there exist differences between the predicted properties with those reported in the experimental examinations, the provided data are of great importance because of the fact that a large number of atoms (approximately 170,000 atoms) were categorized in only 11,400 CG beads. This leads to a critically lessened computational cost. The same researchers could also simulate the fracture mechanics behaviors of CNTR PNC materials by the means of the CG MD simulation [189]. In this work, the fracture characteristics of the PNC were determined free from the precise position of the impactor only by monitoring the load–displacement curve of the nanomaterial. It is noteworthy that the presented CG simulation was able to predict the crack growth in a CNTR specimen with a good precision. By the means of the same CG algorithm, the tensile properties and Young's modulus of the PNCs reinforced with cross-linked CNTs were characterized with a very low computational cost [190]. Based on this simulation study, the material properties of the CNTR PNCs can be attained for high CNT loadings by reserving the accuracy of the results. In another CG simulation concerned with the interfacial load transfer in the CNTR PNCs, it was shown that applying reasonable changes in the CG modeling can lead to have a model which enables the user to provide data in complete agreement with those achieved from the MD [191]. In addition to the aforementioned investigations, CG modeling can be utilized for the goal of analyzing bare polymer [192] and nonpolymeric NCs strengthened via CNTs [193].

1.4.2.3 Density functional theory

Another group of atomistic simulations is reported in the open literature which estimates the material properties of nanomaterials through a quantum mechanics-based viewpoint. According to this approach, also known as DFT, the electronic density at any desired part of the object will be approximated. One of the most famous DFT-based methods is the well-known ab initio simulation [186]. Although ab initio simulation can provide exact responses which are in agreement with the real response, this method is not a reasonable choice for the goal of analyzing the nanomaterials. The reason of this claim is the growing rate of the computational costs in this approach. To make sense, assume the computational cost to be a polynomial function of the system's inputs with powers between 3 and 7 [186]. By the means of the DFT, the intrinsic binding energy available between the polymer matrix and reinforcing nanoparticles can be covered. Also, it will be possible to capture the adhesion energy between the abovementioned parts of an arbitrary PNC [194]. In 2017, a group of researchers carried out an atomistic investigation for the goal of probing the effectiveness of either silanization or functionalization of the CNTs on tensile and bending strength, adsorption energy, and equilibrium distance of the CNTR PNCs implementing a coupled DFT-assisted MD simulation [195]. It was realized that silanization can be a better type of modification in comparison with functionalization with the aid of OH groups. The impact of existence of Thrower–Stone–Wales topological defect in the CNTs' lattice on the structural integrity and interfacial bonding strength of the PNCs was included in the context of a hybrid simulation undergoing with both MD and DFT [196]. It is revealed that the stress–strain curve of the CNTR PNC shifts downward as the number of the Thrower–Stone–Wales grows in the reinforcing nanofillers.

It must be noticed that although exact response of the underobservation NC can be obtained via DFT-based simulations, it is not usual to employ this quantum mechanics-based method for the goal of monitoring the constitutive, tensile, impact, and fracture behaviors of the CNTR PNCs.

1.4.3 Micromechanical methods

Micromechanical methods are beneficial tools for the goal of finding a reliable comprehension of the material properties of composites and NCs free from experimental measurements. According to such methods, either elasticity or compliance tensor and other thermo-electro-mechanical characteristics of heterogeneous solids can be obtained. From fundamental point of view, these algorithms are founded based on the assumptions introduced by Hill [197]. Hill's conditions are related to the axial displacement and traction applied on an arbitrary representative volume element (RVE) inside the heterogeneous material. Afterward, the classical stress–strain relationship in its general form will be written in order to reach either elasticity or compliance tensor of the underobservation material.

What stated above is in general the whole concept of the micromechanics-assisted methods of homogenization. It must be regarded that the differences between diffcrent types of micromechanical methods will be summarized in the way that the RVE and physical phenomena will be captured.

In this section and what follows, the concentration will be on three main types of micromechanical homogenization methods. At first, analytical attempts in this area will be reviewed. In such methods, the effective material properties will be extracted with the aid of linear/nonlinear continuum mechanics. Thereafter, numerical approaches like for example those in close touch with the so-called FEM will be explained. At the end of this part of the text, micromechanical multiscale modeling will be paid attention. In the latter, some probabilistic issues might be involved, too. However, it must be noticed that there exist a large number of studies concerned with a combination of the aforesaid approaches. So, one will definitely face with references in each of the triple subsections which could be appeared in any other one, too. This is on the basis of the authors' categorization and might be against one's opinion.

1.4.3.1 Analytical methods

Many studies can be found in the literature concerning the analytical micromechanics-based investigation of PNCs. In a large number of them, well-known micromechanical schemes like Eshelby, Mori–Tanaka, self-consistent, etc. are employed. So, it will be useful to have a brief review on the main idea of each of these methods. In the final years of 1950s, the Eshelby method was first introduced in order to homogenize infinite solids containing ellipsoidal heterogeneity within themselves [198]. In this method, the impact of existence of single elliptical inclusion in an inhomogeneous material on the tensor of elasticity of the same material was taken into consideration. It is worth mentioning that this method can be utilized for the situations of having inclusions in other shapes in any arbitrary material. In other words, the geometrical parameters introduced in this method can be manipulated for the purpose of modeling platelet-, sphere-, and cylinder-like inclusions in any material [198]. Implementation of the Eshelby method makes it possible to compute the tensor of elasticity in the cases that the media is subjected to infinitesimal strains. This method claims that whenever the entire continua is subjected to a uniform far stress/strain field, the stress/strain in the ellipsoidal inclusion can be assumed to be constant. Moreover, the elasticity tensor suggested by this method does not depend on the material properties of the underobservation solid and the only controlling parameters in this homogenization approach are the geometrical features of the inclusion [198]. Moreover, a particular form of the Eshelby method, called Dilute method, aims at analyzing materials with spherical heterogeneities while there exists no interaction between inclusions. This method goes through deriving the elasticity tensor of an inhomogeneous material including a group of spherical inclusions by decomposing the Eshelby tensor to two deviatoric and dilatational parts. It must be regarded that the accuracy of Dilute model is acceptable only at low volume fractions of the inclusions [199]. However, the assumption employed in the Dilute method is very simplifying and disregards real cases. Therefore, remarkable deviation between the elasticity tensor obtained from Dilute method and experimental measurements can be found. To compensate this shortcoming, modifications were applied on the Dilute model in order to extract a more realistic model. Among all of the suggested modifications, the best one, called Mori–Tanaka method, was introduced in [200]. According to this method, a tensor which can be determined by the concentration tensor, suggested in the Dilute model, with respect to two BCs will be utilized to modify the relation used to derive the tensor of elasticity. Based on these BCs, the modification tensor must be equal with concentration and identity tensors at the inclusions' volume fractions of zero and one, respectively [200]. Besides, one other approach is addressed in the literature for the goal of finding the properties of nonhomogeneous materials. Based on this method, also known as self-consistent method, the formerly mentioned framework of Dilute model will be used by applying minor modifications. In this method, the concentration tensor will be constructed by employing the Eshelby tensor instead of the matrix. It is noteworthy that if values greater than 0.5 are assigned to the volume fraction of the inclusions, negative values will be estimated for both bulk and shear modules of the material [199].

In the early 2000s, the framework of Mori–Tanaka method was employed by a group of researchers to attain an analytical model for the prediction of the Young's modulus of the CNTR PNCs. The comparison of the results achieved from the mentioned modeling with those gathered from experimental examinations revealed that the introduced model was able to estimate the modulus of the PNCs at low CNT loadings (i.e., volume fractions smaller than 2%) [201]. Based on the Eshelby method and its combination with the Mori–Tanaka approach, the effects of aggregation of the CNTs in separate inclusions, wavy shape of the CNTs, and orientation of them on the variation of the modulus of the CNTR PNCs were included in the context of an analytical micromechanical modeling in [202]. Furthermore, the study of distribution of stress along with the axial direction of the CNT in a CNTR PNC was performed in [203] with the aid of an analytical method verified via numerical FE-assisted simulation. It is illustrated that the distribution of stress in the PNC depends on the aspect ratio of the reinforcing CNTs. In another study, a continuum mechanics-assisted framework was introduced toward calculating the thermal conductivity of the CNTR PNCs whenever the CNTs are presumed to be aligned inside the PNC [204]. Furthermore, the arrays of the creep compliance tensor of the CNTR PNCs were computed based on analytical extension of the micromechanical Mori–Tanaka method incorporated with the well-known Maxwell's correspondence principle in the viscoelasticity [205]. It was shown that the presented modeling was able to provide compliances close to those achieved from the experiments. At the same year, both Mori–Tanaka and self-consistent algorithms were utilized for the goal of illustrating the variation of both longitudinal and transverse modules of the CNTR PNCs against volume fraction of the nanoparticles [206]. According to this analytical study, both of the above methods estimated similar properties at low CNT loadings; whereas, this agreement never appeared in the cases that high values were assigned to the volume fraction of the CNTs. These trends are in complete accordance with the nature of the utilized theories, as formerly discussed in Refs. [199,200] for accuracy of self-consistent and Mori–Tanaka methods, respectively. The stress transfer problem of CNTR PNCs was analytically solved in [207] with the aid of a micromechanical theorem regarding for the continuity conditions in the continua. In this study, the variation of normal stress and IFSS in the nanomaterial versus geometrical parameters were provided. On the other hand, the influence of Kapitza resistance on the evaluation of the thermal conductivity of the PNCs reinforced with highly conductive CNTs was covered in the framework of an analytical micromechanical study for the aerospace applications [208]. Moreover, the concept of micromechanics was utilized by a group of researchers in order to extract an analytical relation for the approximation of a relation for the constitutive behaviors of CNTR PNCs in the plastic domain [209]. The stress–strain curves achieved from this investigation were compared with those obtained from experimental measurements and it was found that the proposed methodology is able to predict relatively accurate results. In another analytical study, the Cholesky conjugated gradient method was utilized by a group of Portuguese researchers for the purpose of presenting a model to simulate the variation of the electrical conductivity of the PNCs reinforced via either randomly oriented or aligned CNTs with respect to the slenderness ratio of the CNTs and their volume fraction by putting emphasis on the role of the hopping between the nanofillers on the electrical response [210]. Similarly, another study addressing the electrical conductivity of the CNTR PNCs was fulfilled in [211] according to both electron hopping and conductive network frameworks for PNCs reinforced via either SWCNTs or MWCNTs.

Back to the characterization of the Young's modulus of the CNTR PNCs, both Eshelby–Mori–Tanaka and elasticity-based equivalent fiber method were utilized in [212] to derive analytical relations for the estimation of the modulus of such NCs. In this investigation, the equivalent fiber was assumed to be a combination of the nanofiller, matrix, and interphase zone. Also, comparison of the results achieved from the Eshelby–Mori–Tanaka method demonstrated that using from two specific values for the volume fraction of the inclusions and volume fraction of the entangled CNTs inside the inclusions can be resulted in having an approximation from the modulus of the PNCs that is very close to the data reported in the experimental examinations for the CNT loadings smaller than 1% [212]. Furthermore, the impacts of waviness phenomenon and cumulation of the CNTs in certain inclusions on the Young's modulus of the CNTR NCs hosted by SMPs were included in another analytical study carried out within the framework of the micromechanics [213]. In this paper, a consistent thermo-mechanical constitutive relation for such smart CNTR PNCs was derived which is able to capture the destroying effects of agglomeration and waviness phenomena. The effect of the nanofiller debonding on the stress–strain curve of the CNTR PNCs was included in the context of an analytical survey arranged in [214]. In this model, the concept of equivalent fiber was utilized by assuming the material properties in the interphase to vary gradually in the radial direction according to a power-law method. In another attempt, an analytical framework for the goal of deriving a relation for the stress–strain behaviors of the CNTR PNCs containing wavy CNTs was gathered by considering the important impact of the bonding condition on the constitutive equation of the NC system [215]. Based on this study, it was interestingly revealed that the wavy shape of the CNTs can be even resulted in an improvement in the modulus of the PNC if a weak bonding exists between the nanofillers and matrix in the underobservation nanomaterial. Furthermore, by the means of an agglomeration-dependent micromechanical homogenization scheme, both Young's and shear modules, thermal and moisture expansion coefficients, and both thermal and electrical conductivities of the PNCs reinforced with CNTs were derived in [216]. The viscoelastic analysis of CNTR PNCs under both uni- and biaxial loadings in order to achieve the creep compliance of the mentioned nanomaterial with regard to the influence of existence of an interphase region covering the CNTs was fulfilled in Ref. [217]. In this work, the simplified unit cell (SUC) micromechanical method was implemented to derive a new model for the calculation of the creep compliance of PNCs. The same authors utilized the three-dimensional (3D) form of the SUC model for the goal of deriving the creep-recovery characteristics of CNTR PNC materials in the framework of the nonlinear viscoelastic model of Schapery by considering the important influence of the existence of an interphase zone in the composition of the NC [218]. As same as their former work (i.e., Ref. [217]), the results of this study were in an excellent agreement with the experimental datasets reported in the open literature. In another study, the destroying influence of the aggregation of the CNTs on the equivalent Young's modulus of the PNCs was tracked using two well-known analytical methods of Mori-Tanaka and self-consistent [219]. It was shown that implementation of such theorems can be led to an accurate prediction of the modulus if appropriate parameters are chosen as the inputs. On the basis of the self-consistent micromechanics framework, the piezoresistivity of the PNCs reinforced with MWCNTs was simulated and monitored in Ref. [220]. It is noteworthy that the influences of interface and tunneling on the piezoresistive response of the NC material are covered, too. It is reported that the piezoresistive feature can be dramatically influenced by changes in the interfacial condition. The micromechanical Halpin–Tsai method was utilized in an analytical study [221] in order to analyze the influences of IFSS, nonideal interfacial bonding, and geometrical features of the CNTs on the evaluation of the Young's modulus and tensile strength of the CNTR PNC materials. With the aid of the effective medium framework, the equivalent thermal conductivity of the CNTR PNCs was computed in [222]. In this micromechanical investigation, the coupled influences of the curvy shape of the CNTs and their orientation in the PNC material are covered together. On the other hand, the influences of interphase thickness, wavy shape of the CNTs, and aggregation of such nanofillers on both Young's modulus and electrical conductivity of the CNTR PNCs were included in an analytical micromechanical study accomplished in Ref. [223]. A precise approximation of the thermal conductivity of the CNTR PNCs consisted of a smart SMP matrix was presented in an analytical investigation by considering the magnificent role of the CNTs' agglomeration and waviness on the thermal properties of the system [224]. In this research, the thermal conductivity was achieved using the SUC micromechanical approach. According to the aforesaid method and incorporating it with the well-known correspondence principle of Maxwell in the viscoelasticity, the creep compliance of the CNTR PNCs was predicted with respect to the agglomeration phenomenon as well as capturing the interphase's effect [225]. In another recently carried out investigation, the modulus of the CNTR PNCs was evaluated on the basis of the well-known Mori–Tanaka scheme with respect to the existence of the CNT agglomerates containing curvy CNTs. Also, the imperfect being of the interfacial bonding between the CNTs and matrix is taken into consideration, too [226]. In another effort, the electrical conductivity of the PNC materials reinforced with CNTs was simulated in the context of an analytical micromechanics-based algorithm which is able to capture the influences of the interphase's thickness and tunneling resistance on the electrical properties of the nanomaterial [227]. The presented model was able to estimate the electrical conductivity of the PNC in a way that its difference from the experimental data could be manipulated by changing the polymer matrix.

1.4.3.2 Numerical methods

In this part, the numerical modeling-based studies dealing with the mechanical properties of the CNTR PNCs will be reviewed. It must be declared that in these works, the word "numerical" generally stands for FEM. In other words, these attempts undergo with implementation of the concept of the RVE in the micromechanics in addition to the well-known FEM to estimate the properties of the PNCs in the framework of an iterative numerical method. For example, the ANSYS commercial FE-based software was utilized in a numerical study for the goal of surveying the failure mechanism and damage initiation of the CNTR PNCs while the CNTs are assumed to be distributed randomly in the matrix and their curved structure is taken into consideration [228]. In another study, an approximation of the Young's modulus of the PNCs reinforced with SWCNTs was attained by the means of the powerful FEM [229]. In this work, the influence of the interfacial bonding between the CNTs and matrix on the mechanical properties of the PNC was considered with the aid of a cylindrical RVE. Based on a hybrid elastic–viscoplastic FE investigation, the influence of the cohesion energy in the interphase region in a CNTR PNC on the debonding characteristics of the aforesaid material was monitored regarding for the variation of the interphase's properties as a function of the spatial parameters [230]. Moreover, the stress–strain curve and Young's modulus of the CNTR PNCs were extracted in the context of an FEM-assisted micromechanical analysis organized in [231] with regard to the impact of existence of nonbonded interphase zone as well as considering short CNTs as the reinforcing elements in the composition of the nanomaterial. In this research, the well-known Lennard–Jones (LJ) potential was employed to apply the influence of vdW interfacial bonding on the prediction of the tensile characteristics of the PNC. The same researchers adopted another numerical framework for the purpose of simulating the constitutive equation of the CNTR PNC materials with the aid of the well-known ANSYS commercial software by considering the effect of weakly bonded interfacial bonding using the concept of equivalent fiber [232]. According to this FE analysis, it was revealed that the simple form of the rule of the mixture provides greater modulus for the PNC which can be led to unexpected failures in either static or dynamic stimulations. Furthermore, a 3D FEM was implemented in a fracture mechanics analysis to determine the mode I fracture behaviors of the CNTR PNCs in the presence of a half-circular crack in the simulated unite cell [233]. It is noteworthy that the effect of existence of a nonbonded interphase region located between the CNT and polymer on the fracture behaviors of the system was covered, too. Another FE simulation was fulfilled by a group of researchers aiming at taking the effect of the probability distribution functions of the characteristics of the CNTs on the Young's modulus of the PNC materials into consideration [234]. Comparison of the results with those obtained from the experimental measurements certified the accuracy of this probabilistic 3D simulation. On the other hand, both isotropic and orthotropic MWCNTs were utilized in a micromechanical FE study to extract the modulus of the PNCs reinforced with such nanofillers in both axial and transverse directions with regard to the impact of the interphase region on the stiffness of the nanomaterial [235]. In a different FE modeling concerned with the investigation of the tensile behaviors of CNTR PNCs, two types of RVE were considered. In the first type, single CNT was assumed to be located at the center of the RVE; whereas, the second type of RVE contained four symmetrically located CNTs in itself [236]. It was illustrated that both approaches result in almost the same modules for the PNC. Moreover, a mixture of both analytical and FE micromechanical techniques was utilized in [237] to reach an appropriate estimation of the Young's modulus of the CNTR PNC materials once three various patterns for the wavy shape of the CNTs are considered. It was demonstrated that the negative influence of the waviness phenomenon on the modulus of the PNC can be better observed in the case of using long CNTs for reinforcement. Motivated by the crucial effects of agglomeration of the CNTs and their curved shape on the equivalent properties of the PNCs reinforced via CNTs, the FEM was implemented in Ref. [238] to predict the Young's modulus of the PNC materials. It is obvious that the embedding length of the CNTs in the matrix can play a magnificent role in the occurrence of the pull-out phenomenon. By considering this issue and with respect to the state of the interfacial bonding between CNT and polymer, the crack bridging phenomenon in the CNTR PNCs was probed on the basis of the incorporation of the FEM with the modeling of the cohesive zone in the NC's interface [239]. Iranian researchers manufactured, modeled, and analyzed the stress–strain constitutive equation of the PNCs strengthened via MWCNTs in both tension and compression modes to provide a general insight into variations of the modulus of PNCs. In this study, the FE simulation was employed to obtain the stress–strain curve of the three-walled CNTR PNCs numerically [240]. In this research, the modeling was performed with the aid of ABAQUS commercial software to simulate a small portion of the nanomaterial in a square-shaped RVE.

In another effort, FE calculations were hired to survey the Young's modulus of the CNTR PNCs whenever the modulus and thickness of the interphase zone is presumed to be varied continuously [241]. In this work, SWCNTs were chosen to reinforce the polymeric matrix and the modeling was accomplished in ANSYS software. On the other hand, the same commercial software was selected in [242] for the goal of monitoring the Young's modulus of the PNC materials reinforced via CCNTs regarding for the impacts of geometrical features of the CCNTs such as number of the coils, helix angle, and pitch of the coils as well as the content of the utilized CCNTs and interphase region on the modulus of the CCNTR PNC. In the middle of the 2016, a complicated FE simulation in close touch with the object-oriented coding expertise was introduced in [243] to automatically generate the unite cell required for the investigation of the fracture mechanics and initiation of failure in CNTR PNC materials in the framework of the ABAQUS software. Consideration of the coupled influences of local temperature, thickness of the interphase zone, and random distribution or aligned positioning of the CNTs in the polymer on the evaluation of the effective Young's modulus of the PNC materials reinforced via CNTs was carried out in the framework of an FEM-based study in [244]. Based on the 2D and 3D FE modeling of the CNTR PNCs and by considering the influence of random orientation of the nanofillers in the PNC material, the stress–strain curve of the mentioned nanomaterial in both elastic and plastic regions was derived in Ref. [245] by the means of the equivalent fiber method. In this work, the whole RVE containing the CNT, interface, and matrix was assumed to be a fiber with transversely isotropic behavior. Extending the previous research, the same authors investigated the elastoplastic constitutive behaviors of the CNTR PNCs while the effect of imperfect bonding in the interfacial zone on the variation of the Young's modulus of the system is included [246]. On the other hand, a hybrid framework between FEM and genetic algorithm was presented in a micromechanical study for the purpose of reducing the stress concentration in the PNC materials reinforced with CNTs in an indirect manner undergoing with the modification of the stress field rather than lessening the stress concentration directly [247]. Applying modifications on the embedded FEM, the stress–strain curves of the PNC materials were illustrated for a very general condition in which CNTs with various length-to-diameter ratios, curvatures, orientations, etc. could be employed as reinforcing elements in the nanomaterial [248]. It was revealed that the presented FE-based formulation was able to predict the constitutive behaviors of the CNTR PNCs accurately. Based on an experimentally verified FE micromechanical simulation, the stress–strain curves of CNTR PNCs were plotted in [249] once the influences of aggregation phenomenon, wavy shape of the CNTs, and the interphase zone on the evaluation of the modulus of the PNC were included. Furthermore, a numerical analysis was fulfilled in Ref. [250] in the context of the extended FEM (XFEM) in order to track the fracture toughness and stress intensity factor of the CNTR PNCs with several discontinuities. In this work, the impacts of debonding phenomenon and CNTs' pull-out on the fracture characteristics of the PNC material were captured, too. In a different atmosphere, the FEM was chosen in order to achieve the electrical conductivity and dielectric properties of the PNCs reinforced with CNTs [251]. In this study, the CNTs were simulated as straight wires in the RVE. Although such a simplifying assumption was employed in the numerical homogenization, it was demonstrated that the approximated properties were in an acceptable agreement with those achieved by the same researcher through experimental measurements.

1.4.3.3 *Multiscale modeling*

Evidently, the behaviors of the nanomaterials depend on a large number of scale parameters of both temporal and spatial type. There exist, for example, dimensions from nano-, micro-, macro-, and mesoscales in a PNC material which is a kind of nanomaterial. Also, several time scales can affect the dynamic behaviors of such nanomaterials. Therefore, implementation of multiscale modeling is an efficient means for the purpose of analyzing the characteristics of these materials. Multiscale modeling of PNC materials can be classified in one way to hierarchical, semiconcurrent, and concurrent groups. With the aid of hierarchical multiscale modeling, the beginning of simulation will be with finer scales and it ends in coarser ones. This order cannot be changed in such modeling approach. Due to this gradual transfer from fine to coarse scales, this approach is one of the best methods to compute the properties of the simulated nanomaterial. In the concurrent modeling, however, there exists no interaction between fine and coarse scales. In such multiscale modeling, the problem will be solved in both of the mentioned scales at the same time. In the semiconcurrent method, the user is allowed to move from fine scales to the coarser ones, vice versa. According to this flexibility between different scales, semiconcurrent modeling is known as a user-friendly multiscale modeling with an efficiency similar to that of concurrent one.

Up to now, a large number of researches have been accomplished for finding the thermo-electro-mechanical properties of CNTR PNC materials. In 2005, a multiscale modeling was developed to monitor the stress transfer in the CNTR PNCs using the shear-lag method as an elasticity-based method in combination with the atomistic concept of the molecular structural mechanics (MSM) [252]. According to a multiscale framework, the buckling endurance of the CNTR PNCs was tracked and discussed whenever atomistic and FE methodologies are employed for the purpose of deriving the material properties of the CNT and polymer, respectively [253]. It is worth mentioning that this study was accomplished with the assumption of existence of weak vdW interfacial bonding in the NC. By the means of the powerful FEM, a multiscale modeling was carried out in Ref. [254] to study the tensile strength and stress–strain behaviors of CNTR PNCs. In this investigation, the Morse potential was implemented for the atomistic-based simulation of the CNT and the final RVE was constructed using the FEM. According to this work, the interfacial bonding was assumed to be perfect at the beginning of the analysis till the time that the IFSS will be greater than the shear strength. Afterward, the debonding phenomenon was covered. Moreover, the FE-assisted multiscale investigation of the constitutive behaviors of the CNTR PNC materials was covered in [255] by considering the fact that the polymer matrix possesses a viscoelastic behavior. In this study, the MSM framework was utilized to model the CNT and the well-known Kelvin–Voigt method was employed for the goal of considering the rheological behaviors of the polymeric matrix. In addition, the LJ interatomic potential was selected to describe the interaction between the nanofillers and polymer. In another study, the stress–strain curve of the CNTR PNC materials was illustrated based on the random selection of parameters like shape of the CNTs, the status of the CNT agglomerates, orientation of the CNTs, and CNTs' distribution in the framework of a multiscale modeling [256]. The same authors could capture the impact of the interphase zone around the CNT on the effective modulus of the CNTR PNCs by choosing the LJ potential to account for the vdW-type attractive force between CNT and polymer within the framework of a multiscale model founded on the basis of the FE analysis [257].

In another effort, an FE-assisted multiscale model was developed to track the stress–strain curve and stiffness characteristics of CNTR PNCs whenever the utilized RVE of the PNC is assumed to be subjected to different types of loadings [258]. Comparison of the stress–strain curves achieved from this model and experimental measurements, procured before, proved the accuracy of the presented multiscale modeling. The evaluation of the Young's modulus of the CNTR PNCs with regard to the influence of the temperature on the stiffness was performed on the basis of a multiscale modeling involved with MD, MSM, and FEM in Ref. [259]. Furthermore, the effects of entanglement of the CNTs in the bundles, the interphase region, and random pattern of the CNTs' orientation in the polymer matrix on the equivalent electrical conductivity of the CNTR PNCs were studied based on a multiscale modeling [260]. The influence of existence of a break in the CNTs utilized for the purpose of reinforcing PNCs hosted by epoxy on the distribution of the axial stress along with the longitudinal direction of the CNT was simulated on the basis of FE multiscale analysis in Ref. [261]. In another multiscale study, the pull-out force and IFSS of the CNTR PNCs were calculated by considering an atomistic-type nonbonded interfacial condition according to the well-known LJ interatomic potential [262]. On the other hand, a group of Korean researchers implemented a multiscale framework to make it possible to predict both tensile and shear modules of the PNC materials reinforced via dispersion of the CNTs in the matrix by focusing on the size dependency of the CNTs' behavior [263]. In this work, the MD simulations and analytical micromechanical methods are incorporated together and the interfacial bonding between the CNT and matrix is simulated using stiff spring elements. The destroying influence of the curved shape of the CNTs on the Young's modulus of the CNTR PNC materials was captured in the context of a multiscale modeling based on the bottom-up approach [264]. It is worth regarding that the waviness phenomenon was taken into consideration by the means of stochastic modeling which considers for the random existence of either straight or wavy CNTs in the PNC. Although a good agreement between the Young's modules of this modeling and experiments were reported, implementation of the aforesaid modeling in the PNCs with high CNT loadings can be resulted in greater errors compared to the case of using low-content CNTR PNCs. Moreover, a nonlinear multiscale approach was outlined in [168] to predict the stress–strain curves of the CNTR PNCs in both elastic and plastic zones while the important effect of imperfect bonding in the interface and interphase region on the modulus of the PNC is covered. In this study, the well-known Mori–Tanaka method was modified and incorporated with the method of secant moduli to analyze the problem. It is noteworthy that the atomistic characteristics of the CNT were simulated in the framework of the ab initio MD. By the means of a multiscale 3D-FE simulation algorithm, the influence of the entanglement of the CNTs in the inclusions on the variation of the electrical conductivity and piezoresistivity of the CNTR PNC materials was explored in [265]. Gaining from the multiscale analytical modeling approach, the fracture behaviors of the PNCs consisted of a thermosetting matrix filled with the dispersion of the CNTs was explored to show the improvement which can be induced in the fracture toughness of the polymer by adding a limited amount of the CNTs to it [266]. Comparison of the proposed multiscale model with the experimental data showed that the presented methodology is valid for the PNCs with low CNT loadings. In another multiscale model, several unavoidable phenomena like agglomeration of the CNTs in the polymer, the tendency of the CNTs to be in wavy shape, and random distribution of such nanofillers in the matrix were captured and their impacts on the prediction of the Young's modulus of the CNTR PNCs were illustrated within the framework of the irregular tessellation algorithm [267]. It was demonstrated that the presented model is powerful enough to provide data close to those obtained from the experiments reported in the open literature. Focusing on the variations of the material properties at the interface between CNT and polymer and with the aid of a multiscale 3D-FE model, the stress transfer problem of CNTR PNCs was answered in Ref. [268].

On the other hand, another FE-assisted multiscale study concentrating on the crucial influence of the functionalization of the reinforcing CNTs in a bio-NC material hosted by a polymeric matrix was organized by Iranian researchers [269]. It was shown that the Young's modulus and tensile strength of the abovementioned PNC can be enhanced while the CNTs are covered by hydrogen atoms. The distribution of stress at the interface between CNT and polymer matrix in the CNTR PNCs with respect to the existence of the CNT agglomerates was illustrated by the means of the multiscale modeling accomplished by the means of the FEM as a powerful numerical tool [270]. Moreover, the agglomeration of the CNTs was included in another multiscale modeling for the purpose of deriving accurate values for the Young's modulus of the CNTR PNC materials while the aggregated CNTs were simulated as curvy nanofillers instead of treating them as straight nanofillers [271]. The results of this multiscale study were similar to those achieved from the experimental measurements. In another attempt, a 3D multiscale method was developed on the basis of the fundamentals of the FEM to gather the Young's modulus of the PNC materials reinforced via both armchair and zigzag CNTs having pinhole-type topological defects in their lattice [272]. This modeling was accomplished according to a square-shaped RVE by considering the influence of the covalent bonding between the CNTs and matrix. According to the combination of the three-parameter viscoelastic model for the linearly solids and Mori–Tanaka micromechanical approach, a multiscale model was introduced for the purpose of predicting the viscoelastic properties of the CNTR PNCs by considering the influence of the time-varying properties of the interface [273]. The mode I fracture characteristics of the CNTR PNCs under action of the debonding of the reinforcing nanofillers from the matrix were tracked in the framework of a multiscale investigation [274]. In this work, the effects of the Young's modulus of the interphase zone and its thickness on the fracture toughness of the PNC were included, too. According to another multiscale modeling-based research, the load transfer between CNT and polymer matrix was monitored by considering the influence of the covalent functionalization of the CNTs on the load transfer at the interface and interphase region in the CNTR PNCs [275]. In this study, the MD was chosen to simulate the chemically functionalized CNTs and the crosslinked chains of the polymer matrix were analyzed using the well-known FEM. On the other hand, a micromechanics-assisted multiscale model for the approximation of the viscoelastic properties of the CNTR PNC materials whose interfacial bonding are assumed to be from weak vdW-type was presented in Ref. [276]. The mentioned study contains two approaches for finding the creep compliance of the PNC, namely equivalent fiber and equivalent inclusion methods. It was demonstrated that the equivalent inclusion method generates more realistic data. It is noteworthy that in this study the influences of the CNT loading and length on the relaxation modulus of the PNCs were addressed, too. Based on the bottom-up method, a multiscale model was introduced to achieve a modified version of the rule of the mixture for the goal of evaluating the modulus of the CNTR PNCs without mentionable overestimation [277]. In this research, the CNT was simulated using atomistic approach and the RVE consisted of the CNT, interface, and matrix was replaced with an equivalent fiber to keep on analyzing the PNC in the context of the proposed model. Moreover, multiscale modeling of the damped dynamic behaviors of the CNTR PNCs was fulfilled in Ref. [278] by incorporating the FEM with the LJ well-known atomistic potential to account for the vdW-type interfacial bonding between CNT and polymer. Another FE-based multiscale simulation was carried out by researchers in order to analyze the compressive stress–strain curves of the CNTR polyurethane foam-matrix NCs [279]. In this investigation, the interfacial bonding was modeled based on the cohesive zone model and the distribution of the CNTs in the matrix was assumed to be random. Similar to the previously discussed study, a multiscale FE-assisted framework was utilized in [280] to probe the tensile behaviors of the polymeric foam-hosted NC materials reinforced via CNTs. In this analysis, the interfacial bonding between matrix and nanofiller was covered based on the interfacial zone model. In addition, this study revealed the optimum status of CNT reinforcement for practical applications. Indeed, it was illustrated that existence of 3.3 wt.% CNTs can be resulted in a noticeable enhancement in both Young's modulus and tensile strength of the CNTR PNCs which can be never achieved in any other case. The magnificence of the interfacial bonding between the polymer and CNT and the remarkable effect of this issue on the load transfer in the CNTR PNC materials led a group of the researchers to devote their time to the investigation of the role of both bonded and nonbonded interactions in the PNCs on the evaluation of the IFSS on the basis of the multiscale modeling incorporated with the FEM [281]. Also, this study revealed that implementation of the functionalized CNTs in the composition of the PNC can be led to a better resistance of the CNTs against pull-out stimulations. Based on a completely stochastic multiscale investigation, the same authors reported the delamination behaviors and fracture toughness values of the laminated composites consisted of CNTs as the reinforcing elements [282]. Comparison of the simulation results with the experiments accomplished by the authors certified the accuracy of the developed multiscale modeling. The reason is the fact that all of the involved variants like straight or wavy shape of the CNTs, their tendency to construct agglomerates, and the distribution of the nanofillers in the polymeric matrix are assumed to be random inputs of the problem.

1.5 CNTR PNC structures: literature review

In this section, a review on the efforts made by a large number of researchers concerning the CNTR PNCs will be presented. In fact, the main objective is to be familiar with the attempts accomplished in order to find out the mechanical behaviors of CNTR structures once they are subjected to various types of either static or dynamic excitations. In general, the historical origin of such studies is the year 2010. In this year, the nonlinear modal analysis of CNTR PNC beams was carried out on the basis of the well-known Timoshenko beam theorem [283]. In this study, the equivalent mechanical properties of the PNC were characterized in the framework of the extended form of the rule of the mixture. In another research, the same beam theory was utilized in an analytical study for the purpose of presenting reliable data addressing both vibration and static buckling characteristics of CNTR beams by considering the impact of existence of a two-parameter stiff medium under the structure [284]. In the same year, the FEM was hired in order to attain both static bending and vibration frequency of the CNTR rectangular plates with the aid of the first-order shear deformation theory (FSDT) of the plates [285]. Afterward, the dynamic stability region of the CNTR PNC beams was determined in Ref. [286]. Once again, the geometrical relations were presented based on the Timoshenko beam hypothesis and the modulus of the PNC was derived using the rule of the mixture. In another study, a numerical FE-based framework, founded on the basis of the kernel particle Ritz (kp-Ritz) approach, was implemented toward determining the thermally affected free vibration responses of CNTR PNC plates [287]. In this numerical investigation, authors utilized the Eshelby–Mori–Tanaka micromechanical algorithm to extract the material properties of the PNC. Moreover, the impact of the shear-mode deformation of the structure on the stress distribution, stability limit, and the frequency of the CNTR beam's free fluctuations was covered in Ref. [288]. Another study aimed at clarifying the nonlinear forced oscillations of CNTR PNC structures was conducted by a group of Iranian researchers based on the Timoshenko theorem [289]. In this work, the dynamic characteristics of the CNTR beam were illustrated using the generalized differential quadrature method (GDQM). In a numerical FE analysis [290], the vibrational behaviors of CNTR beams in the linear regime were monitored using both FSDT and third-order shear deformation theory (TSDT). In another research [291], a CNTR PNC was employed as the constituent layers of a laminated plate whose free vibration analysis was accomplished with the aid of the kp-Ritz method. The critical buckling temperature of the CNTR truncated conical shells was determined in Ref. [292] based on the discrete singular convolution method (DSCM).

A different investigation addressing the natural frequency behaviors of CNTR plates having elastic restrains was carried out by [293] in the context of the FSDT incorporated with the rule of the mixture. A group of numerical FE-based studies were performed by Zhang et al. [294–296] based on the improved moving least squares Ritz (IMLS-Ritz) method for the goal of presenting an approximate response for natural frequency and buckling loads of CNTR PNC plates. In another study [297], the mechanical analysis of the CNTR plates subjected to biaxial buckling excitations was accomplished by the means of the well-known Levy-type solution for plates modeled via TSDT. Furthermore, both buckling load and natural frequency of the CNTR PNC truncated conical shells were calculated with the aid of the GDQM regarding for the impact of the shear deflection up to the first order in Ref. [298]. In another numerical investigation, the FSDT was utilized in conjunction with well-known Ritz method in order to provide an accurate response for the shear buckling problem of CNTR rectangular plates [299]. In this work, the Chebyshev polynomials were employed in the solution function. The same author also implemented the Ritz-type FEM to present a reliable answer to the free vibration problem of the CNTR skew plates [300]. In this study, the motion equations of the PNC structure were derived by the means of the FSDT and the modulus of the PNC was extracted from the rule of the mixture. According to the previously mentioned kp-Ritz FE-based algorithm, the both bending and thermoelastic buckling analyses of CNTR laminated plates were accomplished in Refs. [301] and [302], respectively. In both of the abovementioned studies, the governing equations of the plate were extracted according to the FSDT. In another numerical study, the IMLS-Ritz method was implemented by the same authors to analyze the free oscillations of the quadrilateral plates consisted of CNTR PNCs whenever the continua is rested on a stiff substrate [303]. In a similar attempt, the FE-assisted thermal vibration analysis of CNTR panels in the nonlinear regime was performed in Ref. [304]. In this study, the vibrational characteristics of the continuous system were extracted using an isoparametric quadrilateral element. Besides, the influence of existence of an external pressure on the nonlinear mechanical behaviors of CNTR truncated conical shells was included in a numerical investigation, that is, accomplished based on the well-known DQM [305]. With the aid of the kp-Ritz method, the modal analysis of CNTR plates in thermal environments was performed in Ref. [306] once the extended form of the mixture's rule is utilized for the goal of finding the material properties of the PNC. In the framework of a series of consequent studies, both free and forced fluctuation analyses of CNTR shells and plates were conducted once the PNC structure's oscillations are either controlled or left by themselves [307–310]. On the other hand, a hybrid semianalytical approach was developed by researchers in order to calculate both buckling load and natural frequency of the CNTR PNC thin plates [311]. In this work, the motion equations are derived by the means of the classical theory of thin plates for the sake of simplicity. In another endeavor, the mode shapes and vibration frequency of the CNTR PNC plates of triangular shape were gathered based on the so-called IMLS-Ritz FE solution and the displacement field of the well-known FSDT [312]. Based on the well-known GDQM, the dynamic behaviors of CNTR doubly curved shells were analyzed in Ref. [313] while the Eshelby–Mori–Tanaka approach was employed to capture the influences of the CNTs' aggregation on the natural frequency of the system. The same group of researchers implemented the powerful isogeometric analysis (IGA) incorporated with the plates' FSDT in order to probe the frequency responses of CNTR PNC structures with any arbitrary shape by considering the magnificent effect of the agglomeration phenomenon on the system's dynamics [314]. In addition, the GDQM was utilized by the researchers to make it possible to provide appropriate data concerned with the thermomechanical free vibrations and buckling responses of CNTR annular sector plates [315]. The application of the DSCM in solving the aerodynamically damped instability problem of CNTR plates was shown an investigation accomplished in Ref. [316] with the aid of the plates' FSDT. Moreover, the thermal vibration analysis of laminated beams with CNTR PNC facesheets was accomplished in Refs. [317] and [318] on the basis of Timoshenko and refined beam theories, respectively. Based on the fundamentals of the FEM, the low-velocity impact behaviors of CNTR rectangular plates were evaluated with regard to the influence of temperature gradient on the dynamic response of the system [319]. To this end, a C1-continuous element with 15 degree-of-freedom at each node was employed by the authors to guarantee the accuracy of the results. At the same time, the vibrational responses of CNTR annular plates in the large deformation regime were tracked with the aid of the GDQM [320]. The annular NC plate was presumed to be surrounded by smart piezo facesheets and be rested on an elastic foundation. The dynamic behaviors and mode shapes of the CNTR PNC spherical-type panels were probed within the framework of a numerical investigation concerned with the Gram–Schmidt shape functions [321]. The same author presented an FE-based method toward analyzing the thermal postbuckling behaviors of CNTR plates by the means of the displacement field of the well-known FSDT [322]. Based on the well-known powerful layer-wise theorem, a general mechanical analysis was organized to monitor the reaction of the CNTR plates to bending, buckling, and free vibration stimulations [323]. In this study, the modified form of the Halpin–Tsai method was implemented to attain the effective properties of the CNTR PNC. In another study, a framework founded on the basis of the FEM was utilized by the authors in order to solve the nonlinear free oscillations of the CNTR plates regarding for the effects of shear deflection on the mechanical response of the continuous system [324]. According to the IGA,[2] numerical approximations of both bending and free vibration problems of CNTR skew plates were provided in Ref. [325]. In this FE investigation, the influence of the orientation angle of the reinforcing CNTs on the system's mechanical behavior is included. By the means of the FSDT, the nonlinear dynamic responses of annular sector plates consisted of a CNTR PNC core surrounded by smart facesheets were surveyed using the powerful GDQM [326]. Furthermore, researchers showed that the vibrations of the CNTR PNC plates can be controlled by adding two facesheets on their top and bottom edges [327]. In this investigation, the equivalent modulus of the PNC was derived by the means of the well-known Mori–Tanaka micromechanical scheme. According to a temperature-dependent

---

[2] It must be mentioned that IGA is different from FEM. In any IGA, both geometry and displacement field of the continuous system will be approximated with the aid of B-splines. Due to this issue, the continuity glass ceiling existing in the shape functions used in the FEM cannot be seen in IGA. In addition, the user is allowed to manipulate the control points in the B-splines of any element. So, in many cases, implementation of IG-assisted modeling can be better than solving the problem via FEM.

study, the nonlinear-to-linear frequency ratio of the CNTR beams was probed in Ref. [328] once the beam is in postbuckled state due to the thermal excitations. Besides, the IMLS-Ritz algorithm was chosen in a numerical investigation toward finding the linear natural frequency of the CNTR PNC skew plates whenever the impact of in-plane loading on the dynamic characteristics of the continua is covered [329]. The optimized orientation angle for the alignment of the CNTs in a CNTR skew plate to achieve the best buckling resistance was found by researchers according to an IGA [330]. The dynamic deflection behaviors of CNTR shells in the presence of impact pulses were monitored in Ref. [331] based on the displacement field of the shell's TSDT.

In 2018 and in continue of the studies dealing with the CNTR PNC structures, the DQM was selected by investigators for the goal of studying the free vibration responses of CNTR cylindrical-type panels [332]. In another numerical investigation, the natural frequency of sandwich beams with CNTR facesheets was computed by the means of the differential transformation method [333]. In a similar study, the differential transformation method and Timoshenko beam hypothesis were again utilized by the same authors in order to carry out the thermal buckling analysis of laminated beams possessing CNTR facesheets [334]. Also, analytical investigation of the dispersion characteristics of the elastic eaves scattered inside the CNTR beam-type elements was accomplished by researchers in the context of an exponential wave propagation method [335,336]. By the means of the Chebyshev polynomials and according to the Ritz method, the linear modal analysis of CNTR skew cylindrical panels was fulfilled in Ref. [337] by considering the effects of shear deformation on the PNC structure's dynamic response. Moreover, based on a numerical discretization-assisted methodology, the modal analysis of CNTR plates whose geometrical shape can be determined arbitrarily was accomplished in Ref. [338] using the well-known TSDT. The nonlinear analysis of both free vibration and buckling problems of CNTR laminated cylindrical panels was performed by the researchers in the context of a higher-order shear deformation theory (HSDT) incorporated with the extended rule of the mixture [339]. The responses of the CNTR laminated plates to bending, buckling, and inertial oscillation excitations were determined and discussed in Ref. [340] according to a refined-type zigzag plate hypothesis. The dynamic instability and modal analyses of CNTR cylinders conveying viscose fluid flow in thermal environments were conducted by a group of researchers with the aid of the GDQM [341]. In another study, the thermoelastic buckling responses of CNTR laminated doubly curved shells were monitored using the fundamentals of the FEM [342]. On the other hand, the influence of the CNTR shell's rotation around its revolution axis on the mode shapes of the system's free vibrations was covered based on the well-known FSDT and by the means of a numerical solution involved with the Chebyshev polynomials to be able to account for different types of BCs at the ends [343]. Moreover, the FEM was employed by researchers for the purpose of analyzing both elastic and thermal buckling behaviors of laminated porous plates possessing two CNTR PNC facesheets [344]. In the mentioned study, the material properties of the CNTR were achieved based on the Eshelby–Mori–Tanaka micromechanical procedure to capture the influences of the CNTs' agglomeration on the stability of the system. It is reported that the stability margin of the CNTR piezoelectric shells can be tailored by changing the value of the applied electric voltage [345]. In another effort, the thermal stability of the CNTR plates was explored based on the TSDT of the thick plates [346]. In this work, the plates with various geometries were analyzed via the GDQM. The aforementioned solution method was employed by another group of researchers to survey the vibrational characteristics of CNTR PNC panels via the TSDT and the extended form of the rule of the mixture [347]. Furthermore, both static buckling and free vibration problems of CNTR PNC plates were solved numerically by the researchers using the efficient DSCM [348,349]. In another study, the Ritz method was implemented to extract the natural frequency of the CNTR panels in the presence of a central cut-out in the PNC system [350]. The influence of the defects in the CNTR PNCs' microstructure on the buckling resistance of the laminated CNTR plates surrounded with piezoelectric facesheets was the main motivation of the researchers in other studies [351,352]. In these works, FE-based mesh-free method was implemented for the goal of solving the problem whose governing equations are obtained by the means of the Reddy's TSDT. Most recently and based on the fundamentals of the 3D elasticity, buckling load and natural frequency responses of CNTR cylindrical panels were extracted within the framework of a micromechanical study [353].

1.6 Introduction to multiscale hybrid NCs

FRP composites are proven to be efficient materials containing a wide range of multidisciplinary properties which match many engineering applications. However, the performance of such composites can be dramatically influenced by the interfacial bonding between the fiber and polymer. In the early 2000s, it was experimentally observed that the interfacial shear strength of these composites can be improved in a remarkable manner once nanostructured CNTs are coated on the individual fibers [354]. To this end, CNTs synthesized via CVD were grown on the surface of the CFs to achieve a novel type of composites, called multiscale hybrid (MSH) NCs [354]. The word *"multiscale"* is used for such advanced NCs due to the fact that reinforcing agents from microscale to nanoscale are gathered together in such advanced nanomaterials. To examine the mechanical behaviors of such novel composites, a single-fiber fragmentation test was accomplished by the researchers to record the mechanical reaction of the manufactured sample to tension [354]. According to the test results, a mentionable enhancement in the mechanical properties was observed. The existence of the CNTs in the hybrid nanomaterial resulted in an improvement of as great as 30% in the ILSS of the conventional FRP [355]. In addition, the tensile modulus of the MSH NC was reported to be at least as big as that of the bare FRP. On the other hand, the aforesaid features are not the only advantages of the MSH NCs. Experiments showed that these nanomaterials are able to manifest both in-plane and out-of-plane electrical conductivities better than those attributed to the FRPs, in particular whenever MWCNTs are utilized in their composition [355]. In the late 2000s, a low CNT loading was utilized in [356] to fabricate MSH NCs hosted by epoxy. In this study, CFs were employed as microscale fibers. It is worth mentioning that the sonication technique was implemented for the goal of dispersing the CNTs in the resin. According to the accomplished experimental examinations [356], addition of a very low content of CNTs to the composition of the CFRP can be resulted in a remarkable enhancement in the flexural modulus and strength of the composite. However, the magnitude of these improvements can be varied if the sonication time is tailored. It is observed that using from a rapid sonication procedure can be a better approach toward enhancing the composite's material properties [356]. Within the framework of a comparative experimental study, it was observed that using silane-functionalized MWCNTs in the composition of the MSH NC can be resulted in a much better enhancement in both flexural modulus and strength of the hybrid nanomaterial compared with the bare FRP [357]. Also, it was mentioned that silane functionalization is a better approach compared with acid functionalization for the purpose of strengthening the MSH NC [357]. Furthermore, the positive impact of adding CNT to the conventional FRP composites on the mechanical properties of the final specimen is reported in another experimental investigation, too [358]. In this study, the Raman spectra of pure and CNT-coated CFs are plotted and it was demonstrated that both D and G peaks of the CFs modified with CNTs are shifted upward in comparison with those of the bare CFs. Hence, the improvement of the tensile properties of the CFRP with addition of the CNTs seems clear [358]. Another research showed improvements of as big as 11% and 35% in the on-axis tensile strength and ductility of the MSH NCs in comparison with the typical FRPs, respectively [359]. Also, an increment of 16% in the off-axis stiffness is enriched by adding CNT to the composition of the previously known FRP [359]. The enhancement of the composite's ILSS by adding CNTs, which was previously reported in [355], was mentioned in another research concerned with the fabrication of three-phase hybrid nanomaterials consisted of CFs, CNTs, and polyhedral oligomeric silsesquioxanes [360]. In another general study, it was both experimentally and numerically illustrated that utilization of the CNTs in the composition of the MSH NCs can make it possible to improve the ILSS and fracture toughness by even up to 88% and 53% of magnitude, respectively [361]. However, it is declared that the aforementioned nominal enhancements can be achieved whenever both fiber and matrix are modified with the aid of the nanostructured CNTs [361]. It is noteworthy that addition of the CNTs to the FRP cannot guarantee the enhancement of the fracture performance of the MSH NC in any desired content of the CNTs [362]. It is reported that the interlaminar fracture toughness of the MSH NCs with CNT loading of 1 g/m^2 can be aggrandized by up to 32% of initial magnitude and decreases in higher CNT loadings [362]. Investigations revealed that MSH NCs consisted of CNT-grafted CFs, as reinforcing gadgets, can be resulted in an increase in both ILSS and mode-II fracture toughness of the laminated composites [363].

In addition to the positive effect of existence of the CNTs in the MSH NCs on the mechanical properties of such nanomaterials, such nanomaterials are able to promote eye-catching thermoelectrical properties from themselves. It is proven that utilization of CNTs in the MSH NCs sounds interesting in the engineering applications concerned with the management of thermal energy. In other words, the high thermal conductivity of the CNTs can cause remarkable improvement in the conductivity of the FRPs. Hence, MSH NCs can be efficiently employed in space shuttles, launcher systems, aircrafts, and oil-drilling applications [364]. Besides, the dynamic mechanical thermal analysis results indicate on the fact that the enhancement of the material properties of the MSH NC is a function of the time spent on coating the CNTs on the CFs. The limitless increment of the required time for coating can be resulted in a decrease in the variation of the storage modulus of the MSH NC against local temperature [365]. In an experimental study concerned with the electrical properties of the advanced composites [366], it was reported that adding a limited amount of the CNTs to the CFR composites can be resulted in a noticeable increase in the electromagnetic interference shielding effectiveness of the composite. It is proven that MSH NCs containing 5 wt.% CNTs can provide shielding effectiveness of higher than 70 dB in the frequency range of 8–12 GHz [366]. Therefore, it can be concluded that MSH NCs can be appropriate candidates in order to isolate smart devices from the environments whose existing electromagnetic fields cannot be controlled at all. Moreover, experimental examinations [367] proved that using CNTs in the GFR polymers (GFRPs) results in the appearance of MSH NCs whose lowered percolation threshold and improved dielectric constant and electrical conductivity can match the design criteria in the field of electronics as explained in [71].

According to the above merits of MSH NC materials and with regard to the magnificence of existence of reliable data addressing the structural performance of such nanomaterials in different working conditions, many of the researchers devoted their time to survey mechanical behaviors of MSH NC structures. The analysis of the nonlinear free fluctuations of the multi-layered beams manufactured from MSH NCs with smart piezoelectric feature was accomplished in the middle 2010s by the means of the well-known Timoshenko beam theorem [368]. Then, the same research group employed the conventional Euler–Bernoulli beam model toward investigating both free and forced nonlinear oscillation characteristics of MSH NCs while the constitutive equation of the nanomaterial was founded on the basis of a fractional order method to cover the impact of the nanomaterial's internal damping on the nonlinear dynamic behaviors of the system [369]. Furthermore, the nonlinear mechanical performance of MSH NC airfoils was tracked in [370] by considering the airfoil as a thin-type beam element that the influence of its rotation speed around its axis on the mechanical response of the system is regarded. A multiscale FE-assisted modeling was organized in [371] for the goal of monitoring the bending, deflection, buckling endurance, natural frequency, and mode shapes of MSH NC plate-type elements of arbitrary shapes. The effect of existence of gradient in both local temperature and moisture concentration of the environment on the nonlinear low-velocity impact behaviors of the MSH NC plates was reported in [372,373] using the well-known FEM incorporated with the HSDT. The nonlinear analyses of bending, thermal buckling, and free vibrations of the MSH NC beams reinforced with E-GFs and GPL were reported in [374] with the aid of the classical beam theory. By the means of the well-known Rayleigh-Ritz FE solution, the free vibration analysis of MSH NC beams was fulfilled in [375] by considering the negative influence of availability of clusters in the nanomaterial on the dynamic response of the system. The same modal analysis was accomplished in [376] to present an analytical estimation of the natural frequencies corresponding to the free oscillations of a MSH NC thin-type plate in the context of the Kirchhoff-Love plate theorem. In another theoretical attempt, the same plate model was employed in [377] for the goal of extracting the critical buckling load of the plates fabricated from MSH NCs. The FE investigation of the modal characteristics of MSH NCs reinforced with CF and graphene oxide was carried out in [378] within the framework of a refined-type HSDT of the beams. Moreover, with the aid of the same beam theory, the thermally influenced vibrational responses of MSH NC beams rested on stiff elastic seat was accomplished in [379] with respect to the impacts of various types of supports on the system's dynamic behavior. Another study in the area was aimed at analyzing the impact of the existence of the CNTs agglomerates on the free vibration characteristics of the MSH NC plates by the means of the refined shear deformable theory of the thick-type plates [380]. On the other hand, nonlinear free vibration problem of the doubly curved sandwich panels manufactured from a magnetorheological layer and MSH NC plies was solved analytically in

[381]. In this investigation, the well-known TSDT of the doubly curved shell-type element was implemented in order to derive the nonlinear set of the governing equations. According to the FSDT of the cylindrical shells, the influence of the aggregation of the CNTs on the dispersion of the elastic waves in the MSH NC shells was tracked in [382]. In a similar analytical study, the wave propagation responses of MSH NC beams were explored by considering the destroying impact of the CNTs agglomerates on the total stiffness of the utilized nanomaterial [383]. The impact of external magnetic field on the nonlinear dynamics of a doubly curved sandwich shell consisted of both magnetorheological and MSH nanomaterial layers was included in another study concerning the HSDT of the doubly curved shells [384]. Moreover, the thermal postbuckling analysis of the multi-layered MSH NC panels was conducted by the same researchers by considering the effect of existence of smart shape memory alloy wires in the composition of the continuous system [385]. The side influences of hygrothermal environment on the nonlinear dynamic characteristics of MSH panels containing both nanoparticle and shape memory alloy wire were covered in the framework of an analytical investigation accomplished based on the HSDT of the doubly curved shells [386]. The same authors utilized the nonlinear strain displacement relations of panels with dual curvature for the goal of probing the nonlinear frequency behaviors of MSH NC doubly curved panels possessing piezoelectric feature whenever the continuous system is placed in an environment whose temperature and moisture concentration are variable [387]. With emphasis on the important role of the CNTs' agglomeration on the mechanical performance of the MSH nanomaterials, the critical buckling problem of beam-type elements in the absence and presence of thermal loading was solved analytically in [388,389], respectively. In these studies, the influences of shear-mode deformation on the stability endurance of the continuous system was included, too. However, this issue was ignored in another study, involved in analyzing the postbuckling behaviors of MSH NC structures, organized by the same authors [390]. In the latter study, the negative impact of the construction of the CNT agglomerates on the postbuckling path of beams with/without initial midspan rise was covered in detail. In another attempt concentrating on the CNTs' aggregation phenomenon, the free vibration analysis of the cylinders manufactured from MSH NCs was performed in the context of the shell's FSDT incorporated with the Eshelby–Mori–Tanaka micromechanical method [391]. The influence of existence of a viscose fluid flow inside a MSH NC cylinder on the variation of its natural frequency was included in another study dealing with the agglomeration phenomenon [392]. By the means of the same micromechanical homogenization approach, the effect of the aggregation of the CNTs in the inclusions on the stability limit of MSH NC cylindrical shells was reported in [393]. Moreover, the negative effect of the aggregation of the

CNTs on the buckling load of the rectangular plates fabricated from MSH NC was mentioned in [394] with the aid of the refined higher-order plate hypothesis. In another endeavor, the dispersion curves of the elastic waves scattered in the MSH NC plates were discussed in [395] with respect to the unavoidable probability of the appearance of the clusters of CNT bundles in the nanomaterial. The classical theory of the panels with dual curvature was employed in [396] for the goal of tracking the nonlinear free oscillation problem of the MSH NC structures. In addition, the thermal postbuckling path of both perfect and imperfect panels consisted of smart piezoelectric MSH NCs was illustrated in [397] on the basis of the well-known TSDT of the doubly curved shells. Moreover, the powerful GDQM was employed in [398] for the purpose of finding the nonlinear frequency of the free fluctuations of MSH NC disks with central cut-out. It was recently reported that implementation of the nanostructured GPLs as the coatings of the CFs in FRPs can be resulted in a remarkable enhancement in the buckling load of the rectangular plates manufactured from such MSH NCs [399].

CHAPTER

2

Micromechanical homogenization and kinematic relations

2.1 Micromechanical homogenization

In this part of Chapter 2, the concepts and mathematical frameworks of some of the most famous homogenization schemes will be reviewed and discussed at first. In this step, it will be tried to concentrate on different types of theoretical micromechanical approaches which can help the user to extract the properties of the carbon nanotube-reinforced (CNTR) polymer nanocomposite (PNC) materials. In such theorems, different issues like random or uniform distribution of the CNTs in the matrix, alignment of the reinforcing nanofillers in the PNC, the possibility of the appearance of agglomeration phenomenon in the media, and wavy shape of the CNTs will be monitored and discussed in detail. Once this part of homogenization is completed, it will be paid attention to the derivation of the effective properties of the multiscale hybrid (MSH) PNC material. Indeed, a general framework, free from the homogenization method which was chosen in the former step, will be introduced in order to make it possible to extract the material properties of the MSH nanomaterial.

2.1.1 Homogenization of CNTR PNCs

In this section, the equivalent material properties of the CNTR PNC materials will be obtained based upon the micromechanical methods. It will be tried to present different methods to show the differences between the approximations provided by each one. To this end, the simplest method of homogenization (i.e., rule of the mixture) will be introduced at first and thereafter, models with more complexities will be discussed in detail.

2.1.1.1 Rule of the mixture

According to this method, a certain volume will be considered for and assigned to each of the constituent phases (CNTs and polymer matrix) of the PNC. Next, the effective material properties of the PNC will be gathered by calculating the summation of the amount of each phase's desired property multiplied by the volume fraction of the same phase. In this approach, the bonding between nanofiller and matrix will be assumed to be from perfect type and the nanotubes will be treated as aligned straight reinforcing elements. However, the aforesaid assumptions are valid for conventional composites instead of NCs. To compensate for this mismatch, one common way is to combine the simple form of the rule of the mixture with the atomistic molecular dynamics (MD) simulations. To this purpose, the material properties of the CNTR PNC will be calculated by the means of the MD and some efficiency coefficients will be added to the rule of the mixture. By setting the predicted MD-assisted answer with the expressions achieved from the rule of the mixture, the efficiency coefficients will be achieved. Afterward, the efficiency coefficients will be implemented for further estimations to lessen the difference between the material properties obtained from theoretical and atomistic modeling.

Based on the abovementioned instructions, the values of Young's and shear modules of the CNTR PNC material can be extracted using the below definitions [283]:

$$E_{11} = \eta_1 V_r E_{11}^r + V_m E_m \tag{2.1}$$

$$\frac{\eta_2}{E_{22}} = \frac{V_r}{E_{22}^r} + \frac{V_m}{E_m} \tag{2.2}$$

$$\frac{\eta_3}{G_{12}} = \frac{V_r}{G_{12}^r} + \frac{V_m}{G_m} \tag{2.3}$$

where E_{11} and E_{22} are the longitudinal and transverse Young's modules of the CNTR PNC and G_{12} denotes the shear modulus of the same material in the plane constructed from the intersection of longitudinal and transverse directions. In the above relations, the η_i's ($i = 1, 2, 3$) are the efficiency coefficients that must be determined from the comparison of theoretical and MD answers. Also, the terms V_r and V_m are the volume fractions of CNT and matrix, respectively. It is evident that the sum of the above volume fractions must be equal to one. Based on this approach, the shear modulus in the plane between longitudinal and transverse directions will be calculated and the shear modulus in other planes will be considered to be half of the value that will be obtained for the G_{12} [400].

In addition to the normal and shear stiffnesses obtained in Eqs. (2.1)–(2.3), the mass density, Poisson's ratio between longitudinal and transverse directions, and both longitudinal and transverse CTEs of the CNTR PNC

TABLE 2.1 The efficiency coefficients required to derive the material properties of CNTR PNCs by the means of rule of the mixture. The tabulated values are utilized in Refs. [283] and [284] based on the MD simulations carried out in Ref. [402].

	Efficiency coefficients	
Volume fraction of the CNTs, V_r	η_1	η_2
0.12	1.2833	1.0556
0.17	1.3414	1.7101
0.28	1.3238	1.7380

material can be calculated by the means of below formulas [401]:

$$\rho = \rho_r V_r + \rho_m V_m \tag{2.4}$$

$$\nu_{12} = \nu_{12}^r V_r + \nu_m V_m \tag{2.5}$$

$$\alpha_{11} = \alpha_{11}^r V_r + \alpha_m V_m \tag{2.6}$$

$$\alpha_{22} = (1 + \nu_{12}^r)\alpha_{22}^r V_r + (1 + \nu_m)\alpha_m V_m - \nu_{12}\alpha_{11} \tag{2.7}$$

in which ρ_r and ν_{12}^r are density and Poisson's ratio of the CNT, respectively. The longitudinal and transverse CTEs of the CNT are, respectively, shown with α_{11}^r and α_{22}^r. Besides, the density, Poisson's ratio, and CTE of the matrix are shown with ρ_m, ν_m, and α_m, respectively. It must be pointed out that the volume fraction of the CNTs in Eqs. (2.1)–(2.7) can be computed according to the following relation [283]:

$$V_r = \frac{W_r}{W_r + \frac{\rho_r}{\rho_m}(1 - W_r)} \tag{2.8}$$

where W_r is the mass fraction of the CNTs.

In some researches, the MD simulations were conducted for the goal of finding the exact values of the efficiency coefficients. According to some of these studies, the efficiency coefficients in some particular volume fraction of the CNTs can be found by referring to Table 2.1. It must be considered that the mentioned efficiency coefficients are gathered with the assumption of implementing poly(methyl methacrylate) (PMMA) as the matrix and single-walled CNT (SWCNT) (10,10) for reinforcement. Obviously, the efficiency coefficients can be changed if each of the above constituents are replaced with another matrix or nanofiller. Also, it is noteworthy that the efficiency coefficient η_3 is assumed to be identical with η_2 due to the lack of MD data provided in the open literature about the shear modulus of the CNTR PNCs.

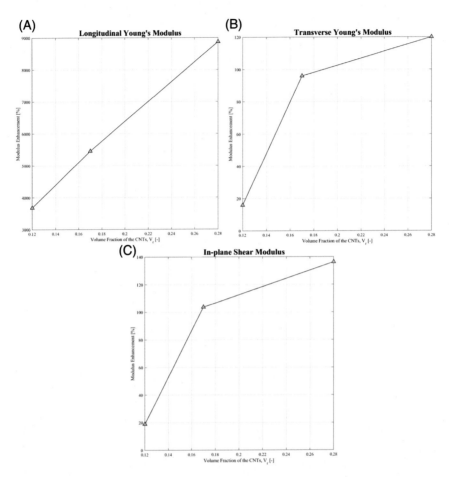

FIGURE 2.1 The influence of the content of reinforcing nanofillers in the composition of the CNTR PNC on the enhancement of the (A) longitudinal Young's modulus, (B) transverse Young's modulus, and (C) in-plane shear modulus of the nanomaterial (Color figures available in eBook only).

To show the effect of the volume fraction of the reinforcing phase on the stiffness of the polymer, the variation of the modulus against volume fraction of the CNTs can be monitored. To this end, the properties of the polymer matrix, that is, PMMA, at room temperature are: $E_m = 2.5$ GPa, $\nu_m = 0.3$, and $\rho_m = 1190$ kg/m^3 [283, 284]. Also, SWCNT (10,10), whose material properties are $E_{11}^r = 600$ GPa, $E_{22}^r = 10$ GPa, $G_{12}^r = 17.2$ GPa, $\nu_{12}^r = 0.19$, and $\rho_r = 1400$ kg/m^3 [283, 284], will be used as the reinforcement. With the aid of the abovementioned inputs, Fig. 2.1 is depicted in order to show the impact of adding the content of the CNTs in the PNC material on the improvement of the stiffness of this nanomaterial. Obviously, both Young's and shear modules of the CNTR PNC were aggrandized by increasing the volume fraction of the CNTs due to the remarkable stiffness of the nanofillers compared with that of neat polymer. According to this figure, it can be perceived that enhancement of the longitudinal Young's modulus is far different from that of transverse Young's modulus or shear modulus. The reason is maybe the assumption of considering the nanofillers to be aligned in the longitudinal direction and the fact that the longitudinal modulus of the utilized CNTs is 60 times stiffer than their transverse modulus.

2.1.1.2 Halpin–Tsai method

Herein, the material properties of the CNTR PNC will be derived according to the Halpin–Tsai method. Based on this approach, a parameter will be utilized for the purpose of capturing the influences of the inclusion's geometrical features and loading circumstances on the estimated values of the properties of any desired type of composite. This enables the user to analyze various types of materials by adopting appropriate expression for the aforesaid parameter. Similar to the rule of the mixture, the bonding condition between nanofiller and matrix will be assumed to be ideal. Based on these assumptions and with regard to the geometry of the reinforcing CNTs, the Young's modulus of a CNTR PNC can be considered to be obtained using the below definition [379]:

$$E = \left[\frac{5}{8}\left(\frac{1+2\xi V_r}{1-\xi V_r}\right) + \frac{3}{8}\left(\frac{1+2\{l_r/d_r\}\zeta V_r}{1-\zeta V_r}\right)\right]E_m \qquad (2.9)$$

in which

$$\xi = \frac{(E_r/E_m) - (d_r/4t_r)}{(E_r/E_m) + (d_r/2t_r)} \qquad (2.10)$$

$$\zeta = \frac{(E_r/E_m) - (d_r/4t_r)}{(E_r/E_m) + (l_r/2t_r)} \qquad (2.11)$$

In Eqs. (2.9)–(2.11), length, outer diameter, wall thickness, and Young's modulus of the CNTs are shown with l_r, d_r, t_r, and E_r, respectively. Obviously, E_m stands for the Young's modulus of the matrix. It is noteworthy that ξ and ζ are dimensionless parameters provided to be in charge of controlling the effect of using nanofillers for reinforcement. It is interesting to mention that the expression of these dimensionless terms can be varied if the nanofiller is changed and another element is employed to reinforce the polymer. This is the way that the impact of different types of reinforcements on the mechanical properties of the underobservation inhomogeneous material.

In Eq. (2.9), the volume fraction of the CNTs (V_r) can be calculated by using the definition previously presented in Eq. (2.8). It is worth mentioning that the equivalent mass density and Poisson's ratio of the CNTR PNC can be attained with the aid of the rule of the mixture as formerly discussed in Section 2.1.1.1 (see Eqs. (2.4) and (2.5)). However, it must be noticed that the obtained Poisson's ratio will be the general Poisson's ratio of the CNTR PNC material because of the fact that the whole NC is assumed to be an isotropic linearly elastic solid in this type of homogenization.

Furthermore, the CTE of the CNTR PNC material can be obtained by the means of the Halpin–Tsai method in the following form [379]:

$$\alpha = \frac{1}{2}\left[\left(\frac{\alpha_r E_r V_r + \alpha_m E_m V_m}{E_r V_r + E_m V_m}\right)(1-\nu) + (1+\nu_m)\alpha_m V_m + (1+\nu_r)\alpha_r V_r\right] \quad (2.12)$$

where ν is the Poisson's ratio of the CNTR PNC. Also, the CTEs corresponding with matrix and CNT are shown with α_m and α_r, respectively.

In order to track the variation of the Young's modulus of the CNTR PNC versus volume fraction of the nanofillers, properties of matrix and CNT must be known. Assume the matrix to possess properties like: $E_m = 2.1$ GPa, $\nu_m = 0.34$, and $\rho_m = 1150$ kg/m^3 [314, 375]. Also, SWCNT (10,10) with Young's modulus of $E_r = 450$ GPa, Poisson's ratio of $\nu_r = 0.2$, and mass density of $\rho_r = 2237$ kg/m^3 [375, 403] will be utilized. The thickness of the CNT's wall is assumed to be $t_r = 0.34$ nm. Based on the chirality of the employed CNT, its diameter can be computed using the below formula [62]:

$$d_r = \frac{a}{\pi}\sqrt{m^2 + mn + n^2} \quad (2.13)$$

in which a is the lattice constant of the graphene and is defined in the following form:

$$a = \sqrt{3}a_{C-C} \quad (2.14)$$

In the above definition, $a_{C-C} = 0.142$ nm is the length of a carbon–carbon bond. Now, by considering that $n = m = 10$, the diameter of the SWCNT (10,10) is $d_r = 1.356$ nm.

Using the stated properties for the matrix and CNTs, the variations of the modulus enhancement versus volume fraction of the nanofillers for different lengths of the CNTs are illustrated in Fig. 2.2. Again, it is shown that the modulus of the polymer can be increased in a remarkable manner whenever the content of the reinforcing phase is added. The reason of this issue was explained in Section 2.1.1.1. In addition to this trend, the figure reveals that implementation of longer CNTs to reinforce the polymer can be resulted in a noticeable improvement in the stiffness of the obtained nanomaterial. Although the aforesaid issue promotes its influence on the modulus of the CNTR PNC as a completely increasing one, it must be pointed out that utilization of slender nanofillers can increase the possibility of existence of nonstraight shape in the CNTs which cannot be covered in the Halpin–Tsai method.

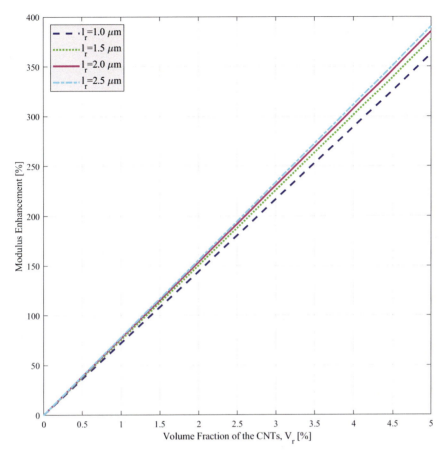

FIGURE 2.2 Effects of the volume fraction and length of the CNTs on the enhancement of the Young's modulus of the PNC material (Color figure available in eBook only).

2.1.1.3 Modified Halpin–Tsai method

In the previous section, the material properties of PNC materials reinforced with straight-shaped CNTs were derived with the aid of the Halpin–Tsai model. Although the geometrical features of the reinforcing nanofiller are paid attention in the above method, the crucial issue of existence of curve in the structure of the CNTs was dismissed. By considering the fact that the most popular synthesization method for the production of the CNTs is the CVD, the obligation of including the effect of the waviness phenomenon on the material properties of the CNTR PNCs can be better sensed. As mentioned in Section 2.3.1 of Chapter 1, possibility of the existence of defects in the CNTs is very high in this method. One of the most probable defects to appear is definitely the availability of curves in the structure of the CNT.

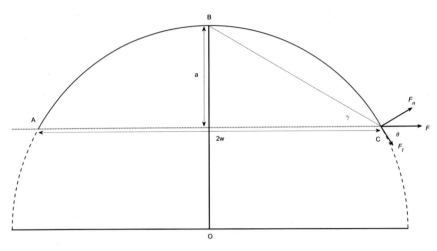

FIGURE 2.3 Schematic view of a wavy CNT subjected to load F (Color figure available in eBook only).

According to the modified form of the Halpin–Tsai method, the properties will be obtained by assuming the CNTR PNC as a linearly elastic isotropic solid. In this approach, the influence of either 2D or 3D random-type distribution of the reinforcing elements in the PNC will be covered, too. To this end, an orientation factor will be introduced to be multiplied in the modulus of the CNTs prior to insert the aforesaid modulus in the process of homogenization. It is worth mentioning that the present model is valid in the cases of analyzing PNCs with reinforcement contents of inside 2 wt.%. The reason is the nonlinear behavior of the nanomaterial in higher contents of the CNTs which is caused by the existence of the CNT agglomerates in such cases.

Assume an individual CNT to be similar with the *ABC* arc illustrated in Fig. 2.3. Based on this figure, a wave with flexural amplitude of a is considered for the CNTs. The straight distance between the ends of the CNT is $2w$. The CNT is presumed to be subjected to a load F which is applied on the CNT along with the direction that connects the ends of the nanofiller horizontally. By considering the definition of the angles θ and γ, the below geometrical relations can be expanded [404]:

$$\theta = \frac{\angle ABC}{2}, \quad \gamma = \frac{\angle AB}{2} \qquad (2.15)$$

With regard to the available symmetry in Fig. 2.3, it can be figured out that

$$\angle ABC = 2\angle AB \qquad (2.16)$$

2.1 Micromechanical homogenization

Therefore, it is clear that $\theta = 2\gamma$. Now, the waviness coefficient is defined as the ratio between the tangential component of the force applied on the CNT and the total force as [404]:

$$C_w = \frac{F_t}{F} \qquad (2.17)$$

Besides, it is known that the above parameter is equal with the cosine of the angle θ. With the aid of the trigonometric relations in the mathematics, the below identity can be written:

$$\cos\theta = \frac{1 - \tan^2\left(\frac{\theta}{2}\right)}{1 + \tan^2\left(\frac{\theta}{2}\right)} \qquad (2.18)$$

Now, by substituting the relation between angles θ and γ in Eq. (2.18), the waviness coefficient can be expressed in the following form:

$$C_w = \frac{1 - \tan^2\gamma}{1 + \tan^2\gamma} \qquad (2.19)$$

On the other hand, one can write the below identity based on Fig. 2.3:

$$\tan\gamma = \frac{a}{w} \qquad (2.20)$$

Substituting for the tangent of the angle γ from Eq. (2.20) into Eq. (2.19) gives:

$$C_w = \frac{1 - (a/w)^2}{1 + (a/w)^2} \qquad (2.21)$$

Once mathematical manipulations are carried out, the below relation can be written:

$$\frac{a}{w} = \sqrt{\frac{1 - C_w}{1 + C_w}} \qquad (2.22)$$

By substituting the right-hand side of the above relation with its Taylor expansion and choosing the first sentence of the approximation, the waviness coefficient can be obtained as below:

$$\frac{a}{w} \cong 1 - C_w \rightarrow C_w = 1 - \frac{a}{w} \qquad (2.23)$$

According to the above definition and by considering the geometrical constraint in Fig. 2.3 (i.e., $a \leq w$), it is obvious that the waviness coefficient can only possess values between zero and one ($0 \leq C_w \leq 1$).

Now, the waviness coefficient is in hand and the Young's modulus of the PNC strengthened with CNTs can be gathered using the below modified form of the Halpin–Tsai equation [404]:

$$E = \frac{1 + C\eta V_r}{1 - \eta V_r} E_m \qquad (2.24)$$

where

$$\eta = \frac{\alpha C_w (E_r/E_m) - 1}{\alpha C_w (E_r/E_m) + C} \qquad (2.25)$$

in which α stands for the orientation factor which is provided to account for the type of the random distribution of the nanofillers in the PNC. Commonly, the orientation factor of $\alpha = 1/3$ is employed for the cases that the length of the CNTs is greater than the dimensions of the media and 2D case is happened. Whereas $\alpha = 1/6$ can be used whenever the length of the CNTs is smaller than the geometrical dimensions of the underobservation material. Therefore, implementation of the latter case for the analyses about the nanocomposite structures that possess macroscopic dimensions can be a reasonable choice. In addition, C is a coefficient that is utilized in order to account for the geometrical specifics of the CNT and is equal with two times of the length-to-diameter ratio of the CNT ($C = 2l_r/d_r$).

Remark 1.
If $C_w = 0$, the amplitude of the curvy shape of the CNT will be equal with the half of the distance between two ends of the CNT. Hence, an extreme wave exists in the CNT in a way that it can be claimed that the ends of the CNT can meet each other if its length is large enough. In this case, the reinforcing efficiency of the CNTs in the PNC will be decreased utmost.

Remark 2.
If $0 < C_w < 1$, the CNT is not straight-shaped and the amplitude of its wave is smaller than the critical value which leads to have extreme wavy shape that was mentioned in Remark 1. Actually, this condition is the most logical state for the consideration of the influence of the waviness phenomenon. In this case, the realistic wavy shape of the CNTs will be captured and the possibility of tuning the intensity of the destroying influence of the curved structure of the CNTs on the material properties of the CNTR PNC will be available. It is worth mentioning that comparison of the modules obtained from this method of homogenization with those achieved from the experimental measurements indicates on the fact that assigning values between 0.3 and 0.4 to the waviness coefficient can be resulted in an acceptable prediction of the properties of the CNTR PNC materials [404].

Remark 3.
If $C_w = 1$, there will be no available curve in the structure of the CNT (i.e., corresponding with $a = 0$). In this situation, the CNTs will be considered as straight nanofillers that can help to the improvement of the properties of the PNC with remarkable efficiency. Obviously, this will be a simplifying

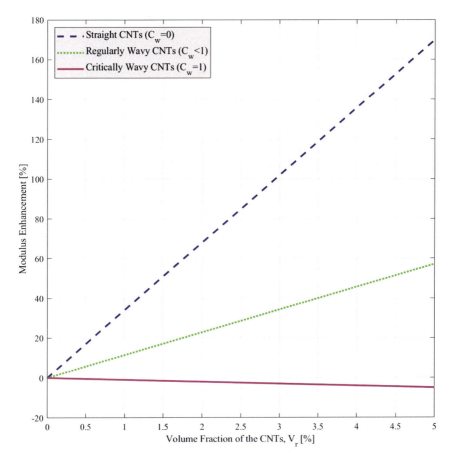

FIGURE 2.4 Effect of changing the waviness coefficient on the variation of modulus improvement of the CNTR PNC against volume fraction of the CNTs (Color figure available in eBook only).

assumption to use waviness coefficient of equal with one. Hence, this case should not be employed in the theoretical estimations due to the fact that the values obtained based on this assumption are overestimations which can never be observed in practice.

In order to numerically show the influence of the waviness phenomenon on the stiffness of the CNTR PNC, the properties introduced in previous section (see Section 2.1.1.2) are employed to draw the variation of the modulus enhancement of the CNTR nanomaterial versus volume fraction of the CNTs in Fig. 2.4. In this figure, all of three Remarks that were discussed in the above sentences can be observed. It is obvious that the waviness phenomenon reduces the reinforcing efficiency of the CNT. This phenomenon can be even resulted in negative efficiency in the most critical case. In other words, if the waviness coefficient is assumed to acquire its maximum amount, the CNT will be shaped like a circle. In this circumstance, not only addition of such a nanofiller cannot improve the stiffness of the polymer but it also leads to a reduction in the equivalent stiffness of the PNC material. Hence, it is of great magnificence to account for the waviness phenomenon whenever the mechanical behaviors of CNTR PNCs are aimed to be tracked.

2.1.1.4 Mori–Tanaka method

Herein, the equivalent properties of the CNTR PNCs will be derived with the aid of the well-known Mori–Tanaka method. In this method, it will be assumed that the CNTs used for the goal of reinforcing the polymer are ideal straight-shaped ones. In addition, the situation of having fully aligned CNTs will be regarded, too. According to the Mori–Tanaka method, any desired inclusion in the PNC will be surrounded by matrix that maybe attracted by effective stress σ_m or strain ε_m. It must be cared that the above stress and strain are the average of the stress and strain fields applied on the matrix. By considering all of the available inclusions to be the same and based on the fundamentals of the Mori–Tanaka homogenization algorithm, one can write the below expression for the elasticity tensor of PNCs reinforced with aligned, straight CNTs [200]:

$$\mathbf{C} = (V_m \mathbf{C}_m + V_r \mathbf{C}_r : \mathbf{A}) : (V_m \mathbf{I} + V_r \mathbf{A})^{-1} \tag{2.26}$$

It is worth mentioning that the boldface signs are attributed to the tensors of second/fourth order. The tensors of elasticity of the matrix and matrix and inclusion (i.e., CNTs in this case) are shown with $\mathbf{C}_m$ and $\mathbf{C}_r$, respectively. Besides, $\mathbf{I}$ stands for the fourth-order identity tensor. In Eq. (2.26), the tensor $\mathbf{A}$, also known as Dilute tensor, is provided to relate the average strains of the matrix and CNT together as below [200]:

$$\varepsilon_r = \mathbf{A} : \varepsilon_m \tag{2.27}$$

where

$$\mathbf{A} = \left[\mathbf{I} + \mathbf{S} : \mathbf{C}_m^{-1} : (\mathbf{C}_r - \mathbf{C}_m)\right]^{-1} \tag{2.28}$$

in which $\mathbf{S}$ is the Eshelby tensor. The nonzero arrays of the Eshelby tensor for the case of analyzing aligned, straight CNTs are in the following form [200, 202]:

$$S_{1111} = S_{3333} = \frac{5 - 4v_m}{8(1 - v_m)}, \quad S_{1122} = S_{3322} = \frac{v_m}{2(1 - v_m)},$$
$$S_{1133} = S_{3311} = \frac{4v_m - 1}{8(1 - v_m)}, \quad S_{1313} = \frac{3 - 4v_m}{8(1 - v_m)}, \quad S_{2323} = S_{1212} = 0.25, \tag{2.29}$$

Now, the stress–strain relationship of the CNTR PNC must be presented. If the polymeric matrix is a linearly isotropic solid whose Young's modulus and Poisson's ratio are shown with E_m and v_m, respectively, and the PNC is assumed to contain aligned, straight CNTs, the stress–strain relation can be presented in the below form [200, 202]:

$$\begin{Bmatrix} \sigma_{11} \\ \sigma_{22} \\ \sigma_{33} \\ \sigma_{23} \\ \sigma_{13} \\ \sigma_{12} \end{Bmatrix} = \begin{bmatrix} k+m & l & k-m & 0 & 0 & 0 \\ l & n & l & 0 & 0 & 0 \\ k-m & l & k+m & 0 & 0 & 0 \\ 0 & 0 & 0 & p & 0 & 0 \\ 0 & 0 & 0 & 0 & m & 0 \\ 0 & 0 & 0 & 0 & 0 & p \end{bmatrix} \begin{Bmatrix} \varepsilon_{11} \\ \varepsilon_{22} \\ \varepsilon_{33} \\ 2\varepsilon_{23} \\ 2\varepsilon_{13} \\ 2\varepsilon_{12} \end{Bmatrix} \quad (2.30)$$

In the above identity, k, n, and l are plane-strain bulk modulus perpendicular to the direction of the CNTs' alignment, uniaxial tensile modulus parallel with the direction of the CNTs' alignment, and associated cross modulus, respectively. Also, the shear modules in the planes normal and parallel to the direction of the CNTs' alignment are, respectively, shown with m and p. It is noteworthy that the above terms are called the elastic constants of Hill.

In order to derive the Hill's elastic constants, the nonzero components of the Eshelby tensor, introduced in Eq. (2.29), should be inserted into the definition of the Dilute tensor **A** to find the nonzero arrays of this tensor. Doing the aforesaid substitution results in [200, 202]:

$$A_{1111} = A_{3333} = -\frac{a_3}{a_1 a_2}, \quad A_{1133} = A_{3311} = \frac{a_4}{a_1 a_2},$$

$$A_{1122} = A_{3322} = \frac{l_r(1 - v_m - 2v_m^2) - E_m v_m}{a_1}, \quad A_{2222} = 1, \quad (2.31)$$

$$A_{2323} = A_{1212} = \frac{E_m}{E_m + 2p_r(1 + v_m)}, \quad A_{1313} = \frac{2E_m(1 - v_m)}{a_2}$$

where

$$a_1 = (2v_m - 1)[E_m + 2k_r(1 + v_m)],$$
$$a_2 = E_m + 2m_r(3 - v_m - 4v_m^2),$$
$$a_3 = E_m(1 - v_m)\{E_m(3 - 4v_m) + 2(1 + v_m)[m_r(3 - 4v_m) + k_r(2 - 4v_m)]\},$$
$$a_4 = E_m(1 - v_m)\{E_m(1 - 4v_m) + 2(1 + v_m)[m_r(3 - 4v_m) + k_r(2 - 4v_m)]\}$$
(2.32)

in which k_r, l_r, m_r, n_r, and p_r are Hill's elastic constants of the CNTs. Now, by using Eqs. (2.31) and (2.26), the Hill's elastic constants of the PNC can be extracted and written in the below form:

$$k = \frac{E_m\{E_m V_m + 2k_r(1 + v_m)[1 + V_r(1 - 2v_m)]\}}{2(1 + v_m)[E_m(1 + V_r - 2v_m) + 2V_m k_r(1 - v_m - 2v_m^2)]} \quad (2.33)$$

TABLE 2.2 The values of Hill's elastic constants of the SWCNT (10,10) [403].

k_r (GPa)	l_r (GPa)	m_r (GPa)	n_r (GPa)	p_r (GPa)
271	88	17	1089	442

$$l = \frac{E_m\{v_m V_m[E_m + 2k_r(1+v_m)] + 2l_r V_r(1-v_m^2)\}}{(1+v_m)[2k_r V_m(1-v_m-2v_m^2) + E_m(1+V_r-2v_m)]} \quad (2.34)$$

$$m = \frac{E_m[E_m V_m + 2m_r(1+v_m)(3+V_r-4v_m)]}{2(1+v_m)\{E_m[V_m + 4V_r(1-v_m)] + 2V_m m_r(3-v_m-4v_m^2)\}} \quad (2.35)$$

$$n = \frac{E_m^2 V_m(1+V_r-V_m v_m) + 2V_r V_m\left(k_r n_r - l_r^2\right)(1+v_m)^2(1-2v_m)}{(1+v_m)\{2V_m k_r(1-v_m-2v_m^2) + E_m(1+V_r-2v_m)\}}$$
$$+ \frac{E_m[2V_m^2 k_r(1-v_m) + V_r n_r(1-2v_m+V_r) - 4V_m l_r v_m]}{2V_m k_r(1-v_m-2v_m^2) + E_m(1+V_r-2v_m)} \quad (2.36)$$

$$p = \frac{E_m[E_m V_m + 2(1+V_r)p_r(1+v_m)]}{2(1+v_m)[E_m(1+V_r) + 2V_m p_r(1+v_m)]} \quad (2.37)$$

By substituting for the Hill's elastic constants of the CNTR PNC from Eqs. (2.33)–(2.37) in Eq. (2.30), the Young's modules of the PNC in both longitudinal and transverse directions can be gathered as below:

$$E_L = n - \frac{l^2}{k}, \quad E_T = \frac{4m(kn-l^2)}{kn-l^2+mn} \quad (2.38)$$

Using the material properties of the polymer matrix introduced in Section 2.1.1.2 incorporated with the Hill's elastic constants of the SWCNT (10,10) tabulated in Table 2.2, the longitudinal and transverse Young's modules of the CNTR PNC can be monitored. According to Fig. 2.5, a 5% increment in the volume fraction of the CNTs can be resulted in a huge improve in the longitudinal Young's modulus of the CNTR PNC material. However, this intensifying influence is not too extreme if the variations of the transverse modulus are tracked.

2.1.1.5 Eshelby–Mori–Tanaka method

Present section aims at capturing the magnificent influence of the aggregation of the reinforcing CNTs on the material properties of the CNTR PNC materials by the means of the Eshelby–Mori–Tanaka method. It must be considered that this model is founded based on the assumption of having straight-shaped CNTs as the reinforcing nanofillers. Hence, the waviness phenomenon which was previously discussed in Section 2.1.1.3 will be

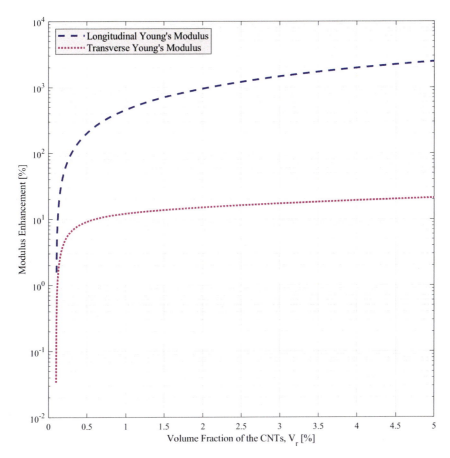

FIGURE 2.5 Variation of the modulus enhancement versus volume fraction of the reinforcing phase for both longitudinal and transverse Young's modules (Color figure available in eBook only).

dismissed in the context of Eshelby–Mori–Tanaka algorithm. In this approach, the CNTs available in the PNC will be divided into two groups. The first group is devoted to those of the CNTs that are dispersed in the polymer matrix and the latter corresponds with those that are cumulated in arbitrary inclusions inside the PNC to construct CNT agglomerates. Now, if the total volume of the CNTs is assumed to be W_r, the below mathematical decomposition can be considered [202, 313, 314]:

$$W_r = W_r^{in} + W_r^m \qquad (2.39)$$

In the above identity, W_r^{in} and W_r^m are the volume of the CNTs entangled in the inclusions and dispersed in the matrix, respectively. By considering the fact that the total volume of the structure can be obtained by addition of the volume of the matrix (W_m) to the volume of the reinforcing phase, one can write [202, 313, 314]:

$$W = W_r + W_m \qquad (2.40)$$

If both sides of the above identity are divided by the left-hand side of it, the volume fractions of the reinforcing and host phases will be obtained in the following form:

$$V_r = \frac{W_r}{W}, \\ V_m = \frac{W_m}{W} \qquad (2.41)$$

In the most of the studies, the case of having a constant value for the volume fraction of the CNTs will be monitored. However, consideration of the gradually variable volume fraction for the reinforcing phase can be led to a more realistic consideration of the agglomeration phenomenon. For instance, the volume fraction of the CNTs in a PNC can be presented in the following form [313, 314]:

$$V_r(z) = \left[\frac{\rho_r}{w_r \rho_m} - \frac{\rho_r}{\rho_m} + 1\right]^{-1} F(z) \qquad (2.42)$$

where w_r denotes the mass fraction of the CNTs and can be defined as below:

$$w_r = \frac{M_r}{M_r + M_m} \qquad (2.43)$$

In the above definition, M_r and M_m are the mass of reinforcing phase (CNTs) and matrix, respectively. In Eq. (2.42), $F(z)$ is a geometry-dependent function which is provided to govern the continuous variation of the volume fraction across the thickness direction. Generally, two types of functions will be utilized in the open literature. The difference between these models is about the location of the polymeric matrix and CNT agglomerates through the thickness direction. In one of them, the matrix will be presumed to be at the bottom surface of the material and the top surface will be dedicated to the reinforcing phase. Meanwhile, the position of the aforesaid phases will be substituted with each other in the other type of modeling. Based on the above explanations, the below expressions can be assigned to the $F(z)$ [313, 314]:

$$F(z) = \begin{cases} \left(\frac{1}{2} + \frac{z}{h}\right)^p \\ \left(\frac{1}{2} - \frac{z}{h}\right)^p \end{cases} \qquad (2.44)$$

where p is the gradient index which is in charge of controlling the distribution of the nanofillers across the thickness direction. If the polymer is at the bottom surface, the first type of abovementioned $F(z)$ must be selected. Obviously, the second case is related to the case of having polymer matrix at the top surface of the PNC. In Eq. (2.44), h is the total thickness of the material.

Now, two other variables must be introduced to make it possible to extract the stiffness of the CNTR PNCs in the existence of the aggregated CNTs. These parameters, also known as *agglomeration parameters*, can be defined with the aid of the volume of the available inclusions and volume of the CNTs inside the mentioned inclusions as follows:

$$\mu = \frac{W_{in}}{W},$$
$$\eta = \frac{W_r^{in}}{W_r} \tag{2.45}$$

In above definition, μ is the volume fraction of the inclusions (clusters) and η is the volume fraction of the CNTs inside the inclusions. According to the relative value of the agglomeration parameters, three cases can be occurred which will be discussed below.

Remark 1.
If $\mu = \eta = 1$, all of the media will be a giant inclusion which is completely filled via CNTs. Indeed, due to the fact that the volume fraction of the inclusions is equal with one, the whole material is a unique inclusion and the case of existence of small inclusions filled with bundles of CNTs does not appear. This phenomenon sounds great because it can be assumed that there is no inclusion and all of the CNTs are dispersed in one inclusion. In this case, the destroying effect of the entanglement of the CNTs on the equivalent stiffness of the PNC will be decreased critically due to the similarity of this case with the case of absence of agglomerated CNTs in the PNC.

Remark 2.
If $\mu \leq \eta < 1$, the partial agglomeration can be observed in the PNC material. In this situation, some inclusions exist in the PNC media which contain a limited content of the reinforcing phase in themselves. Hence, it can be pointed out that some of the CNTs are entangled and the rest of them are dispersed in the matrix. In comparison with the previous case, a worse condition exists in this case. So, it can be concluded that the stiffness of the PNC in this situation will be many times smaller than that of the former case.

Remark 3.
If $\mu \leq \eta$ and $\eta = 1$, all of the CNTs are entangled inside the existing inclusions. Therefore, it can be stated that the full agglomeration is appeared in this case because of the fact that no CNT can be found which

is not inside the constructed inclusions. This condition is not desired at all because of the fact that it reduces the stiffness of the PNC utmost. However, depending on the value of the μ, this condition can be discussed. If a small value is assigned to this parameter, a large number of small inclusions will appear in the media which contain limited contents of CNTs in themselves. However, if μ is increased, a movement toward Remark 1 will be observed which can help to the improvement of the stiffness of the PNC material. It can be therefore claimed that the presence of a limited number of large inclusions in the media can be led to a smaller reduction in the stiffness of the CNTR PNC rather than the case of having a large number of small inclusions in the media.

If Eqs. (2.41) and (2.45) are combined and mathematical manipulations are carried out, the below expressions can be obtained:

$$\frac{W_r^{in}}{W_{in}} = \frac{V_r \eta}{\mu} \tag{2.46}$$

$$\frac{W_r^m}{W - W_{in}} = \frac{V_r(1-\eta)}{1-\mu} \tag{2.47}$$

By the means of the well-known method of Eshelby–Mori–Tanaka [202], the below relations can be developed for the bulk and shear modules of the regions that the reinforcing CNTs are inside the inclusions:

$$K_{in}(z) = K_m + \frac{V_r \eta (\delta_r - 3K_m \alpha_r)}{3(\mu - V_r \eta + V_r \eta \alpha_r)} \tag{2.48}$$

$$G_{in}(z) = G_m + \frac{V_r \eta (\eta_r - 2G_m \beta_r)}{2(\mu - V_r \eta + V_r \eta \beta_r)} \tag{2.49}$$

where K_m and G_m stand for bulk and shear modules of the matrix, respectively. With the aid of the same algorithm, the bulk and shear modules of the rest parts of the PNC can be written in the following form [202]:

$$K_{out}(z) = K_m + \frac{V_r(1-\eta)(\delta_r - 3K_m \alpha_r)}{3[1 - \mu + V_r(1-\eta)(\alpha_r - 1)]} \tag{2.50}$$

$$G_{out}(z) = G_M + \frac{V_r(1-\eta)(\eta_r - 2G_M \beta_r)}{3[1 - \mu + V_r(1-\eta)(\beta_r - 1)]} \tag{2.51}$$

In Eqs. (2.48)–(2.51), stiff terms α_r, β_r, δ_r, and η_r are in the following form:

$$\alpha_r = \frac{3(K_m + G_m) + k_r + l_r}{3(G_m + k_r)} \tag{2.52}$$

$$\beta_r = \frac{1}{5}\left[\frac{4G_m + 2k_r + l_r}{3(G_m + k_r)} + \frac{4G_m}{G_m + p_r} + \frac{2(G_m[3K_m + G_m] + G_m[3K_m + 7G_m])}{G_m(3K_m + G_m) + m_r(3K_m + 7G_m)}\right] \tag{2.53}$$

2.1 Micromechanical homogenization

$$\delta_r = \frac{1}{3}\left[n_r + 2l_r + \frac{(2k_r + l_r)(3K_m + G_m - l_r)}{G_m + k_r}\right] \tag{2.54}$$

$$\eta_r = \frac{1}{5}\left[\begin{array}{c} \frac{2}{3}(n_r - l_r) + \frac{8G_m p_r}{G_m + p_r} + \frac{(2k_r - l_r)(2G_m + l_r)}{3(G_m + k_r)} \\ + \frac{8m_r G_m(3K_m + 4G_m)}{3K_m(m_r + G_m) + G_m(7m_r + G_m)} \end{array}\right] \tag{2.55}$$

In Eqs. (2.52)–(2.55), k_r, l_r, m_r, n_r, and p_r are the stiff Hill's constants of the reinforcing phase. A list of Hill's constants for the SWCNT (10,10) can be found in Section 2.1.1.4 (see Table 2.2).

According to [202], the effective bulk and shear modules of the PNC material can be presented in the following form:

$$K(z) = K_{out}\left[1 + \frac{\mu\left(\frac{K_{in}}{K_{out}} - 1\right)}{1 + (1-\mu)\left(\frac{K_{in}}{K_{out}} - 1\right)\frac{1+\nu_{out}}{3(1-\nu_{out})}}\right] \tag{2.56}$$

$$G(z) = G_{out}\left[1 + \frac{\mu\left(\frac{G_{in}}{G_{out}} - 1\right)}{1 + (1-\mu)\left(\frac{G_{in}}{G_{out}} - 1\right)\frac{8-10\nu_{out}}{15(1-\nu_{out})}}\right] \tag{2.57}$$

where

$$\nu_{out} = \frac{3K_{out} - 2G_{out}}{6K_{out} + 2G_{out}} \tag{2.58}$$

Herein, the fundamental relations of solid mechanics can be employed in order to express the Young's modulus and Poisson's ratio of the CNTR PNC material in the following form:

$$E(z) = \frac{9K(z)G(z)}{3K(z) + G(z)} \tag{2.59}$$

$$\nu(z) = \frac{3K(z) - 2G(z)}{6K(z) + 2G(z)} \tag{2.60}$$

One can utilize the conventional rule of the mixture for the goal of deriving the equivalent density of the CNTR PNC material:

$$\rho(z) = (\rho_r - \rho_m)V_r + \rho_m \tag{2.61}$$

Implementing the material properties of the polymer matrix and CNT as same as those utilized in Section 2.1.1.2 as well as considering the Hill's constants of the reinforcing phase to be similar with data tabulated in Table 2.2, the variation of the Young's modulus of the CNTR PNC versus dimensionless thickness of the material is plotted in Fig. 2.6. In this figure, six cases were taken into consideration. It can be figured out that free from the position of the polymer and reinforcing phase across the thickness of the material, the Young's modulus of the underobservation

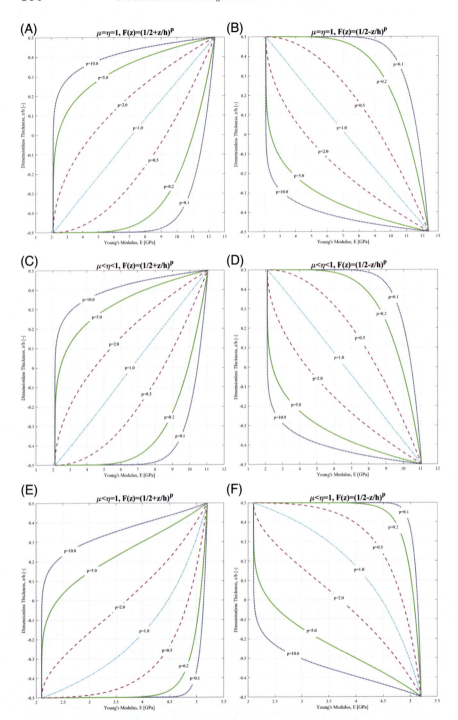

FIGURE 2.6 Variation of the equivalent Young's modulus of the CNTR PNC through the thickness of the material for various agglomeration parameters and geometry-dependent volume fractions of the reinforcing phase (Color figures available in eBook only).

nanomaterial will be lessened in any arbitrary thickness by increasing the value assigned to the gradient index. However, the difference between the cases that polymer is at the top or bottom surface of the material shows itself in the shape of the gradual change of the Young's modulus. Obviously, if polymer is at the bottom surface (i.e., $F(z) = (1/2 + z/h)^p$), the maximum amount of the Young's modulus in any desired thickness and gradient index corresponds with the top surface of the material due to the cumulative existence of the reinforcing phase in that region. In contrast, the maximum Young's modulus of the PNC will be at the bottom surface if $F(z) = (1/2 - z/h)^p$ is employed. Comparison of the subplots reveals that reduction of the volume fraction of the inclusions results in a reciprocal decrease in the Young's modulus of the PNC. This trend is in complete agreement with the explanations provided before. This decreasing effect does not depend on the spatial part of the volume fraction of the reinforcing phase at all. In other words, the concepts of partial and full agglomeration are valid in the CNTR PNCs free from the expression that might be used for the spatial distribution of the nanofillers.

2.1.1.6 3D Mori–Tanaka method

In this section, the material properties of the PNCs reinforced via unidirectionally positioned CNTs will be gathered by considering the random orientation of the CNTs in the matrix. In this type of modeling, the entire PNC material will be assumed to behave like transversely isotropic solids. It is noteworthy that all of the reinforcing CNTs will be considered to be straight, slender nanofillers with high length-to-diameter ratios. It will be shown that this ratio possesses great influence on the equivalent properties of the PNC. Due to the transversely isotropic nature of the underobservation nanomaterial, five characteristic variables must be determined to find the properties of such a material. These variables are longitudinal Young's modulus E_{11}, transverse Young's modulus E_{22}, in-plane shear modulus G_{12}, out-of-plane shear modulus G_{23}, and plane-strain bulk modulus K_{23}. Based upon the 3D Mori–Tanaka method [405], the above variants can be formulated in the following form:

$$\frac{E_{11}}{E_m} = \frac{1}{1 + V_r(A_1 + 2v_m A_2)/A} \qquad (2.62)$$

$$\frac{E_{22}}{E_m} = \frac{1}{1 + V_{CNT}[(1 - v_m)A_4 - 2v_m A_3 + (1 + v_m)A_5 A]/2A} \qquad (2.63)$$

$$\frac{G_{12}}{G_m} = 1 + \frac{V_r}{G_m/\Delta G + 2(1 - V_r)S_{1212}} \qquad (2.64)$$

$$\frac{G_{23}}{G_m} = 1 + \frac{V_r}{G_m/\Delta G + 2(1 - V_r)S_{2323}} \qquad (2.65)$$

$$\frac{K_{23}}{\bar{K}_0} = \frac{(1+\nu_m)(1-2\nu_m)}{1-\nu_m(1+2\nu_{12}) + V_r\{2(\nu_{12}-\nu_m)A_3 + [1-\nu_m(1+2\nu_{12})]A_4\}/A} \quad (2.66)$$

where V_r is the volume fraction of reinforcing phase and can be calculated using Eq. (2.8). In addition, E_m, G_m, and ν_m are, respectively, Young's modulus, shear modulus, and Poisson's ratio of the matrix. ΔG stands for the difference between the shear modules of the CNT and matrix ($\Delta G = G_r - G_m$). Also, ν_{12} is the Poisson's ratio of the matrix in the 1–2 plane which can be assumed to be identical with the matrix's Poisson's ratio due to the isotropic nature of the matrix. The term $\bar{K}_0$ is the bulk modulus of the matrix under plane strain condition and can be considered as $\bar{K}_0 = \lambda_m + G_m$.

The A_i's utilized in Eqs. (2.62)–(2.66) can be defined in the below form [405]:

$$\begin{aligned}
A_1 &= D_1(B_4 + B_5) - 2B_2, \\
A_2 &= (1+D_1)B_2 - (B_4 + B_5), \\
A_3 &= B_1 - D_1 B_3, \\
A_4 &= (1+D_1)B_1 - 2B_3, \\
A_5 &= \frac{1-D_1}{B_4-B_5}, \\
A &= 2B_2 B_3 - B_1(B_4 + B_5)
\end{aligned} \quad (2.67)$$

in which

$$\begin{aligned}
B_1 &= V_r D_1 + D_2 + (1-V_r)(D_1 S_{1111} + 2S_{2211}), \\
B_2 &= V_r + D_3 + (1-V_r)(D_1 S_{1122} + S_{2222} + S_{2233}), \\
B_3 &= V_r + D_3 + (1-V_r)[S_{1111} + (1+D_1)S_{2211}], \\
B_4 &= V_r D_1 + D_2 + (1-V_r)(S_{1122} + D_1 S_{2222} + S_{2233}), \\
B_5 &= V_r + D_3 + (1-V_r)(S_{1122} + S_{2222} + D_1 S_{2233})
\end{aligned} \quad (2.68)$$

In the above relations, S_{ijkl}'s indicate the components of the fourth-order Eshelby tensor and can be computed using the below definitions [405]:

$$S_{1111} = \frac{1}{2(1-\nu_m)}\left[1 - 2\nu_m + \frac{3\alpha^2 - 1}{\alpha^2 - 1} - g\left(1 - 2\mu_m + \frac{3\alpha^2}{\alpha^2 - 1}\right)\right] \quad (2.69)$$

$$S_{2222} = S_{3333} = \frac{3}{8(1-\nu_m)}\frac{\alpha^2}{\alpha^2-1} + \frac{1}{4(1-\nu_m)}\left(1 - 2\nu_m - \frac{9}{4(\alpha^2-1)}\right)g \quad (2.70)$$

$$S_{2233} = S_{3322} = \frac{1}{4(1-\nu_m)}\left[\frac{\alpha^2}{2(\alpha^2-1)} - \left(1 - 2\nu_m - \frac{3}{4(\alpha^2-1)}\right)g\right] \quad (2.71)$$

2.1 Micromechanical homogenization

$$S_{2211} = S_{3311} = -\frac{1}{2(1-v_m)}\frac{\alpha^2}{\alpha^2-1} + \frac{1}{4(1-v_m)}\left[\frac{3\alpha^2}{\alpha^2-1} - (1-2v_m)\right]g \tag{2.72}$$

$$S_{1122} = S_{1133} = -\frac{1}{2(1-v_m)}\left(1 - 2v_m + \frac{1}{\alpha^2-1}\right)$$
$$+ \frac{1}{2(1-v_m)}\left(1 - 2v_m + \frac{3}{2(\alpha^2-1)}\right)g \tag{2.73}$$

$$S_{2323} = S_{3232} = \frac{1}{4(1-v_m)}\left[\frac{\alpha^2}{2(\alpha^2-1)} + \left(1 - 2v_m - \frac{3}{4(\alpha^2-1)}\right)g\right] \tag{2.74}$$

$$S_{1212} = S_{1313} = \frac{1}{4(1-v_m)}\left[1 - 2v_m - \frac{\alpha^2+1}{\alpha^2-1} - 0.5\left[1 - 2v_m - \frac{3(\alpha^2+1)}{\alpha^2-1}\right]g\right] \tag{2.75}$$

where

$$g = \frac{\alpha^2\sqrt{\alpha^2-1} - \alpha\cosh^{-1}\alpha}{(\alpha^2-1)\sqrt{\alpha^2-1}} \tag{2.76}$$

In Eqs. (2.69)–(2.76), α is the length-to-diameter ratio of the CNTs. In the definitions provided in Eqs. (2.67) and (2.68), D_i's can be obtained by:

$$D_1 = 1 + \frac{2\Delta G}{\Delta\lambda},$$
$$D_2 = \frac{\lambda_m + 2G_m}{\Delta\lambda},$$
$$D_3 = \frac{\lambda_m}{\Delta\lambda} \tag{2.77}$$

in which λ_m is the Lame's constant of the matrix and $\Delta\lambda$ is the difference between the Lame's constants of the CNT and matrix ($\Delta\lambda = \lambda_r - \lambda_m$). The Lame's constants corresponding with matrix and nanotube can be defined as follows:

$$\lambda_m = \frac{v_m E_m}{(1-2v_m)(1+v_m)}, \quad \lambda_r = \frac{v_r E_r}{(1-2v_r)(1+v_r)} \tag{2.78}$$

Using the relation between Young's modulus, shear modulus, and Poisson's ratio of the linearly elastic isotropic solids, the shear modules of the matrix and CNT can be expressed in the following form:

$$G_m = \frac{E_m}{2(1+v_m)}, \quad G_r = \frac{E_r}{2(1+v_r)} \tag{2.79}$$

Now, the equivalent bulk and shear modules of the CNTR PNC material can be obtained as below [405]:

$$K = \frac{E_{11} + 4(1+v_{12})^2 K_{23}}{9} \tag{2.80}$$

$$G = \frac{E_{11} + (1 - 2\nu_{12})^2 K_{23} + 6(G_{12} + G_{23})}{15} \tag{2.81}$$

With the aid of the existing relationships between shear and bulk modules of a desired material with its Young's modulus and Poisson's ratio, one can write:

$$E = \frac{9KG}{3K + G} \tag{2.82}$$

$$\nu = \frac{E}{2G} - 1 \tag{2.83}$$

It is worth regarding that the effective mass density of the CNTR PNC can be attained by the means of the so-called rule of the mixture (see Eq. (2.61)).

For the purpose of showing how can the variation of the CNTR PNC's stiffness be captured by the 3D Mori–Tanaka method, the properties utilized in previous sections (e.g., see Section 2.1.1.5) will be re-used in this section, too. The influence of changing the volume fraction of the reinforcing phase on the enhancement of both Young's and shear modules of the PNCs can be seen in Fig. 2.7. It is clear that a linear-type profile for the modulus of the CNTR PNC is suggested by the 3D Mori–Tanaka method which seems to be a little out of practice. The reason of this claim is that by adding the content of the reinforcing elements in the PNCs, phenomena like aggregation of the CNTs will appear which lead to a decrease in the reinforcement efficiency. On the other hand, it is noteworthy that according to this method, the shear modulus will be reinforced more than the Young's one. This trend is more visible in high contents of the CNTs.

2.1.2 Homogenization of MSH PNCs

Herein, we will finalize the procedure of homogenization by introducing the equivalent properties of the PNC materials reinforced via nanosize CNTs and macrosize fibers. In Section 2.1.1, various methods of homogenization for the goal of finding the material properties of the CNTR PNCs were discussed in detail. In this section, the properties obtained in the former section will be considered as the input values. In other words, the properties of the CNTR PNC will be in the role of material properties of the matrix which is aimed to be reinforced by addition of macrosize fibers. The hierarchical steps of homogenization can be observed in Fig. 2.8.

In the what follows, the material properties of the CNTR PNC will be shown with "NCM" subscript/superscript. Also, the material properties of the MSH PNC will be presented for the most general case. In other words, an expression for the calculation of the CTE of the hybrid nanomaterial will be provided. However, if determination of the CTE is aimed, one

2.1 Micromechanical homogenization

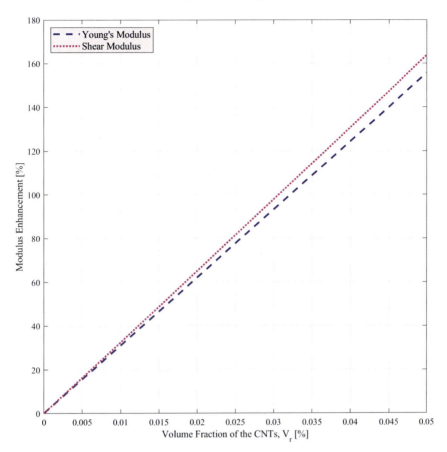

FIGURE 2.7 Illustration of the improvement made in both Young's and shear modules of the CNTR PNC by adding the content of the reinforcing phase (Color figure available in eBook only).

should notice that modified Halpin–Tsai, Mori–Tanaka, Eshelby–Mori–Tanaka, and 3D Mori–Tanaka algorithms cannot be used.

Starting with the derivation of the normal and shear stiffnesses of the MSH PNC, below relations must be utilized to derive longitudinal Young's modulus, transverse Young's modulus, and shear modulus of the MSH nanomaterial [376, 377, 379]:

$$E_{11} = V_F E_{11}^F + V_{NCM} E_{11}^{NCM} \tag{2.84}$$

$$\frac{1}{E_{22}} = \frac{1}{E_{22}^F} + \frac{V_{NCM}}{E_{22}^{NCM}} - V_F V_{NCM} - \frac{v_F^2 E_{22}^{NCM}/E_{22}^F + v_{NCM}^2 E_{22}^F/E_{22}^{NCM} - 2v_F v_{NCM}}{V_F E_{22}^F + V_{NCM} E_{22}^{NCM}} \tag{2.85}$$

2. Micromechanical homogenization and kinematic relations

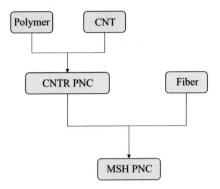

FIGURE 2.8 Schematic of the hierarchical steps required to derive the material properties of the MSH PNC material. In the first step, the material properties of CNTR PNC will be obtained. Next, the obtained values will be used as the inputs of the second step which leads to the extraction of the equivalent properties of the MSH PNC (Color figure available in eBook only).

$$\frac{1}{G_{12}} = \frac{V_F}{G_{12}^F} + \frac{V_{NCM}}{G_{12}^{NCM}} \qquad (2.86)$$

It must be declared that in the cases that the properties obtained in Section 2.8.1 are promoting isotropic behavior for the CNTR PNC, the longitudinal and transverse Young's modules of the CNTR material (E_{11}^{NCM} and E_{22}^{NCM}) must be considered to be equal with the obtained Young's modulus. Similarly, in such cases, the in-plane shear modulus of the CNTR PNC (G_{12}^{NCM}) must be considered to be equal with the extracted value of the shear modulus.

On the other hand, the equivalent mass density and Poisson's ratio of the MSH PNC can be obtained as follows [376, 377, 379]:

$$\rho = V_F \rho_F + V_{NCM} \rho_{NCM} \qquad (2.87)$$

$$\nu_{12} = V_F \nu_F + V_{NCM} \nu_{NCM} \qquad (2.88)$$

where ρ_F and ν_F are mass density and Poisson's ratio of the fiber, respectively. In addition, the CTE of the MSH PNCs can be derived using the below formulas [373, 379]:

$$\alpha_{11} = \frac{V_F E_{11}^F \alpha_{11}^F + V_{NCM} E_{11}^{NCM} \alpha_{11}^{NCM}}{V_F E_{11}^F + V_{NCM} E_{11}^{NCM}} \qquad (2.89)$$

$$\alpha_{22} = (1 + \nu_F) V_F \alpha_{22}^F + (1 + \nu_{NCM}) V_{NCM} \alpha_{22}^{NCM} - \nu_{12} \alpha_{11} \qquad (2.90)$$

In Eqs. (2.89) and (2.90), α_{11}^F and α_{22}^F denote longitudinal and transverse CTE of the reinforcing fiber, respectively. In all of the above relations, V_F is volume fraction of the fibers used as macrosize reinforcement. Clearly, summation of the volume fractions of the constituents must be equal with one (i.e., $V_F + V_{NCM} = 1$).

2.2 Kinematic relations

In this part of this text book, the main focus will be on introduction and implementation of various types of kinematic theorems to describe both static and dynamic behaviors of continuous systems. Till now, many different types of kinematic theories are introduced whose precision in estimating the accurate response of the system differs. In the classical theorems, the impact of existence of shear deformation on the mechanical behaviors of the continua will be neglected [406-536]. The reason of this issue is that it will be assumed that the shear deformation of the system is too smaller than the deformation induced by bending and therefore can be ignored. Generally, such theories can be trusted in thin-walled elements but not in the thick-walled ones. In a minorly modified theory, also known as first-order shear deformation theory (FSDT), a constant approximation for the shear deformation is recommended all over the thickness of the structure which makes it powerful to provide more accurate responses than the previously introduced classical model [537-587]. Moreover, advanced higher-order shear deformation theories (HSDTs) can enable us to take into account for the effect of shear deformation on the mechanical response of the structures by approximating the distortion of the cross-section of the continuous system by the means of a shape function [588-721].

Free from the kinematic theory which is utilized to describe the mechanical behaviors of the continuous systems, an energy-based approach will be employed all over this chapter. This method is the principle of virtual work. For static problems, it is enough to calculate the variations of the total strain energy of the system and work done on it by external loading followed by summing the aforesaid energy functionals and set the answer to be zero. The above expression can be presented in the following mathematical form [390]:

$$\delta(U + V) = 0 \qquad (2.91)$$

where U and V are strain energy and work done by external loading, respectively. Although the above identity is valid for static problems, its concept can be extended to dynamic problems, too. In fact, this principle claims that the variation of the total energy of the continuous system must be equal with zero to attain the motion equations of the system. In dynamic problems, another kind of energy will appear. In other words, the time-dependency results in appearance of kinetic energy, too. Therefore, it can be stated that in this case, the variation of the total energy of the system in any desired time interval must be zero. This physical phenomenon can be expressed in mathematical form with the aid of the dynamic form of the virtual work's principle, also known as Hamilton's principle. This principle can be written in the following form [1]:

$$\int_{t_1}^{t_2} \delta(U + V - T)dt = 0 \qquad (2.92)$$

in which T is the kinetic energy of the system. Strain energy and work done by external loading are exactly as same as those introduced after Eq. (2.91).

2.2.1 Kinematic relations of beams

2.2.1.1 Euler–Bernoulli beam theory

Herein, the Euler–Bernoulli beam theory, also known as classical beam theory, will be introduced. In this theory, the influences of the shear mode on the deformation of the structure will be considered to be small enough to be dismissed. This is due to the fact that in the slender structures (e.g., beams with slenderness ratios greater than $L/h = 50$), the deflection of the structure can be roughly attributed to the bending of the structure. According to this assumption, the cross-section of the beam will not experience distortion. Also, it will be presumed that any desired line which is normal to the cross-section of the structure, will remain normal to the cross-section after loading, too. Although it seems that this theory is an out-of-logic one, it is noteworthy that the response achieved by the means of this theory can be trusted in thin-type beams. The displacement field of the beam in the context of this theory can be expressed in the following form:

$$\begin{aligned} u_x(x, z, t) &= u(x, t) - (z - \bar{z}) \frac{\partial w(x,t)}{\partial x}, \\ u_z(x, z, t) &= w(x, t) \end{aligned} \qquad (2.93)$$

where $u(x, t)$ and $w(x, t)$ are the longitudinal displacement of the beam's neural axis and bending deflection of the beam, respectively. In the above definition, the influence of time on the displacement field is taken into consideration for the sake of generality. However, this term can be simply omitted in the static problems. In the above definition, $\bar{z}$ stands for the exact position of the neutral axis of the beam across the thickness direction. The reason of considering the influence of the precise position of the beam's neutral axis is to omit the stretching-bending coupling rigidity from the calculations to make it possible to decouple the motion equations of the problem in an easier manner. Following the below definition, $\bar{z}$ can be calculated simply [390]:

$$\bar{z} = \frac{\int_{-h/2}^{h/2} zE(z)dz}{\int_{-h/2}^{h/2} E(z)dz} \qquad (2.94)$$

2.2 Kinematic relations

In the above definition, $E(z)$ is the Young's modulus of the underobservation beam which is assumed to be a function of thickness for the sake of completeness.

According to the von Kármán nonlinear strain–displacement relations, the only nonzero component of the strain tensor of the Euler–Bernoulli beams can be expressed as below [390]:

$$\varepsilon_{xx} = \frac{\partial u}{\partial x} + \frac{1}{2}\left(\frac{\partial w}{\partial x}\right)^2 - (z - \bar{z})\frac{\partial^2 w}{\partial x^2} \qquad (2.95)$$

In the what follows, the nonlinear vibrations and postbuckling problems are aimed to be formulated. It is clear that the derivation procedure for the aforesaid problems is not the same due to the existence of time domain in the dynamic analyses. However, in both of these problems, the variation of the total strain energy of the beam must be calculated. The variation of the strain energy for the beams modeled via Euler–Bernoulli theorem can be formulated in the following form:

$$\delta U = \int_0^L \left\{ N\left[\frac{\partial \delta u}{\partial x} + \frac{\partial w}{\partial x}\frac{\partial \delta w}{\partial x}\right] - M\frac{\partial^2 \delta w}{\partial x^2} \right\} dx \qquad (2.96)$$

In above definition, N and M are normal force and bending moment, respectively. These stress-resultants can be defined via:

$$\begin{Bmatrix} N \\ M \end{Bmatrix} = \int_0^b \int_{-h/2}^{h/2} \begin{Bmatrix} 1 \\ z - \bar{z} \end{Bmatrix} \sigma_{xx} dz dy \qquad (2.97)$$

2.2.1.1.1 Postbuckling problem

In order to find the complete equation describing the postbuckling problem, the variation of the work done by external loading must be expressed in mathematical form. Now, assume a beam which is rested on a three-parameter stiff medium with linear or Winkler (k_L), shear or Pasternak (k_P), and nonlinear (k_{NL}) springs. Once this system is subjected to a buckling stimulation of P_b, the variation of work done on the beam by the external loading can be presented in the following form [390]:

$$\delta V = \int_0^L \left[k_L w - k_P \frac{\partial^2 w}{\partial x^2} + k_{NL} w^3 \right] \delta w dx - P_b \delta u |_0^L \qquad (2.98)$$

Now, all of the materials required for deriving the motion equations of the postbuckling problem are in hand. Substitution of Eqs. (2.96) and (2.98) into Eq. (2.91) and implementation of the nontrivial response of the obtained relation gives the following Euler–Lagrange equations:

$$\frac{\partial N}{\partial x} = 0 \qquad (2.99)$$

$$\frac{\partial^2 M}{\partial x^2} + \bar{N} - k_L w + k_p \frac{\partial^2 w}{\partial x^2} - k_{NL} w^3 = 0 \qquad (2.100)$$

where

$$\bar{N} = \frac{\partial}{\partial x}\left[N \frac{\partial w}{\partial x}\right] \qquad (2.101)$$

By considering the identity available in Eq. (2.99), it can be deduced that $N = $ constant. Therefore, one can write the below relation:

$$\bar{N} = N \frac{\partial^2 w}{\partial x^2} \qquad (2.102)$$

On the other hand, the normal force and bending moment of the beam can be written as below whenever Eqs. (2.95) and (2.97) are incorporated with the 1D constitutive equation in the beam-type elements whose neutral axis' exact position is known (i.e., $\sigma_{xx} = E(z)\varepsilon_{xx}$):

$$N = A_{xx}\left[\frac{\partial u}{\partial x} + \frac{1}{2}\left(\frac{\partial w}{\partial x}\right)^2\right] \qquad (2.103)$$

$$M = -D_{xx}\frac{\partial^2 w}{\partial x^2} \qquad (2.104)$$

in which extensional and bending cross-sectional rigidities of the beam can be computed with the aid of the following definitions:

$$\begin{Bmatrix} A_{xx} \\ D_{xx} \end{Bmatrix} = \int_0^b \int_{-h/2}^{h/2} \begin{Bmatrix} 1 \\ (z-\bar{z})^2 \end{Bmatrix} E(z)dzdy \qquad (2.105)$$

Now, the governing equations of the problem can be gathered by substituting for the cross-sectional rigidities of the beam from Eqs. (2.103) and (2.104) into Eqs. (2.99) and (2.100). Once the mentioned substitution is carried out, the following governing equations can be obtained:

$$A_{xx}\left[\frac{\partial^2 u}{\partial x^2} + \frac{\partial w}{\partial x}\frac{\partial^2 w}{\partial x^2}\right] = 0 \qquad (2.106)$$

$$-D_{xx}\frac{\partial^4 w}{\partial x^4} + \frac{\partial}{\partial x}\left[N \frac{\partial w}{\partial x}\right] - k_L w + k_p \frac{\partial^2 w}{\partial x^2} - k_{NL} w^3 = 0 \qquad (2.107)$$

Now, it must be tried to reduce the order of the problem to achieve a single equation describing the physics of the problem. If Eq. (2.106) is divided by the nonzero parameter A_{xx}, the below identity can be extracted:

$$\frac{\partial}{\partial x}\left[\frac{\partial u}{\partial x} + \frac{1}{2}\left(\frac{\partial w}{\partial x}\right)^2\right] = 0 \qquad (2.108)$$

2.2 Kinematic relations

Hence, the below relation can be written:

$$\frac{\partial u}{\partial x} = -\frac{1}{2}\left(\frac{\partial w}{\partial x}\right)^2 + c_0 \tag{2.109}$$

where c_0 is a constant value. Integration from the above relation over the length of the beam gives:

$$u = -\frac{1}{2}\int_0^L \left(\frac{\partial w}{\partial x}\right)^2 dx + c_0 x + c_1 \tag{2.110}$$

In Eq. (2.110), c_1 is constant. The axial displacement of the beam is now in hand. However, the constants c_0 and c_1 need to be determined. To this purpose, the below conditions at the ends of the beam must be considered [390]:

$$u(0) = 0, \quad u(L) = -\frac{P_b L}{A_{xx}} \tag{2.111}$$

Incorporation of Eqs. (2.110) and (2.111) results:

$$c_0 = \frac{1}{2L}\int_0^L \left(\frac{\partial w}{\partial x}\right)^2 dx - \frac{P_b}{A_{xx}}, \quad c_1 = 0 \tag{2.112}$$

With the aid of the above value for c_0 and by combining Eqs. (2.107) and (2.109), the following equation for the postbuckling problem of the beam can be attained:

$$A_{xx}\left[\frac{1}{2L}\int_0^L \left(\frac{\partial w}{\partial x}\right)^2 dx - \frac{P_b}{A_{xx}}\right]\frac{\partial^2 w}{\partial x^2} - D_{xx}\frac{\partial^4 w}{\partial x^4} - k_L w + k_P \frac{\partial^2 w}{\partial x^2} - k_{NL}w^3 = 0 \tag{2.113}$$

Now, the governing equation should be solved to extract the nonlinear buckling load of the beam. To this end, a powerful analytical approach will be implemented which makes it possible to capture the effects of various supports at the ends of the beam on the nonlinear stability characteristics of the beams. Based on this method [390], the bending deflection of the beam will be supposed to be in the following form:

$$w(x) = \sum_{m=1}^{\infty} W_m X_m(x) \tag{2.114}$$

in which W_m denotes the amplitude of the deflection and the function $X_m(x)$ is the eigen function which is in charge of satisfying the boundary condition (BC) at the ends of the beam. The below expressions can be utilized for the eigen functions if the BCs at both ends of the structure are either clamped or simply supported [390]:

$$X_m(x) = \begin{cases} \sin\left(\dfrac{m\pi x}{L}\right) & \text{S-S beam} \\ \sin^2\left(\dfrac{m\pi x}{L}\right) & \text{C-C beam} \end{cases} \qquad (2.115)$$

Now, if Eq. (2.114) is inserted into Eq. (2.113) and the orthogonality of the mode shapes is considered, solution of the following equation can be resulted in finding the nonlinear buckling load of the beam:

$$KW_m + \bar{K} W_m^3 = 0 \qquad (2.116)$$

where

$$K = -D_{xx} r_{40} - k_L r_{00} + (k_P - P_b) r_{20} \qquad (2.117)$$

$$\bar{K} = \frac{A_{xx}}{2L} r_{11} r_{20} - k_{NL} r_{0000} \qquad (2.118)$$

In the above relations, r_{ij}'s and r_{ijkl}'s can be calculated using the below definitions:

$$\begin{aligned} r_{ij} &= \int_0^L X_m^{(i)}(x) X_m^{(j)}(x) dx, \\ r_{ijkl} &= \int_0^L X_m^{(i)}(x) X_m^{(j)}(x) X_m^{(k)}(x) X_m^{(l)}(x) dx \end{aligned} \qquad (2.119)$$

It is worth mentioning that the thermal postbuckling problem can be solved with the aforementioned algorithm. To this end, it must be assumed that the obtained nonlinear buckling load is equal with the force applied on the beam due to the variations in the temperature profile. One can write the thermal loading applied on a beam in the following form:

$$N^T = \int_0^b \int_{-h/2}^{h/2} E(z) \alpha_{11} (T - T_0) dz dy \qquad (2.120)$$

In the above definition, the Young's modulus is assumed to be a function of the thickness for the sake of generality. At the end, by considering the initial reference temperature T_0 to be known, the postbuckling temperature of the beam can be gathered by setting the magnitude of the nonlinear buckling load of the beam to be identical with the thermal loading applied on the structure.

2.2.1.1.2 Nonlinear forced vibration problem

Herein, the influence of existence of the time domain will be covered in order to derive the governing equation of the system's nonlinear vibrations. Similar to Section 2.1.1.1, the influence of the exact position of the neutral axis of the structure will be taken into account in this section, too. In this section, the beam will be considered to be rested on a four-parameter nonlinear viscoelastic substrate which is consisted of linear, Pasternak, and nonlinear springs as well as a damper with damping coefficient of c_d. The stiffnesses of the springs will be shown with the notation formerly introduced in Section 2.1.1.1. Also, it is assumed that the beam is placed in an environment whose local temperature T maybe different from the reference temperature T_0. So, an axial thermal loading like what described in Eq. (2.120) will be considered to be applied on the beam. Plus, the structure will be considered to be subjected to a soft harmonic excitation, too. Now, the variation of work done on the system by the viscoelastic foundation and thermal loading can be formulated in the following form [390, 516]:

$$\delta V = \int_0^L \left[k_L w + (N^T - k_P) \frac{\partial^2 w}{\partial x^2} + k_{NL} w^3 + c_d \frac{\partial w}{\partial t} \right] \delta w \, dx \quad (2.121)$$

Next, the variation of the kinetic energy of the beam must be calculated. To this end, the below expression can be implemented [1]:

$$\delta T = \int_0^L \int_0^b \int_{-h/2}^{h/2} \rho(z) \left[\frac{\partial u_x}{\partial t} \frac{\partial \delta u_x}{\partial t} + \frac{\partial u_z}{\partial t} \frac{\partial \delta u_z}{\partial t} \right] dz \, dy \, dx \quad (2.122)$$

In the above formula, the mass density of the beam is considered to be a function of the spatial variable z to capture the most general condition. By substituting for the components of the beam's displacement field from Eq. (2.93) into Eq. (2.122) and regarding for the fact that the exact position of the neutral axis is captured, the below expression for the variation of the system's kinetic energy can be written:

$$\delta T = \int_{-h/2}^{h/2} \left[I_0 \left(\frac{\partial u}{\partial t} \frac{\partial \delta u}{\partial t} + \frac{\partial w}{\partial t} \frac{\partial \delta w}{\partial t} \right) + I_2 \frac{\partial^2 w}{\partial x \partial t} \frac{\partial^2 \delta w}{\partial x \partial t} \right] dx \quad (2.123)$$

In the above definition, the mass moments of inertia can be computed using the following formula:

$$\begin{Bmatrix} I_0 \\ I_2 \end{Bmatrix} = \int_0^b \int_{-h/2}^{h/2} \begin{Bmatrix} 1 \\ (z - \bar{z})^2 \end{Bmatrix} \rho(z) \, dz \, dy \quad (2.124)$$

Now, the dynamic motion equations of the problem can be obtained by substituting for the variations of strain energy, work done by external loading, and kinetic energy from Eqs. (2.96), (2.121), and (2.123), respectively, into the definition of the Hamilton's principle (see Eq. (2.92)). After doing the abovementioned substitution, the Euler–Lagrange equations of the problem can be gathered in the following form:

$$\frac{\partial N}{\partial x} = I_0 \frac{\partial^2 u}{\partial t^2} \quad (2.125)$$

$$\frac{\partial^2 M}{\partial x^2} + \bar{N} - k_L w + k_P \frac{\partial^2 w}{\partial x^2} - k_{NL} w^3 - c_d \frac{\partial w}{\partial t} = I_0 \frac{\partial^2 w}{\partial t^2} - I_2 \frac{\partial^4 w}{\partial t^2 \partial x^2} \quad (2.126)$$

in which $\bar{N}$ is similar to what introduced in Eq. (2.101). With regard to the existence of thermal loading in the problem, the definitions of normal force and bending moment (refer to Eq. (2.97)), and the modified form of the constitutive equation of 1D problems in the thermoelasticity (i.e., $\sigma_{xx} = E(z)[\varepsilon_{xx} - \alpha_{11}(T - T_0)]$), the following expressions can be written:

$$N = A_{xx}\left[\frac{\partial u}{\partial x} + \frac{1}{2}\left(\frac{\partial w}{\partial x}\right)^2\right] - N^T \qquad (2.127)$$

$$M = -D_{xx}\frac{\partial^2 w}{\partial x^2} - M^T \qquad (2.128)$$

where thermal force is previously introduced in Eq. (2.120) and the thermal moment M^T is equal with zero due to the symmetry that exists in the problem due to the consideration of the exact position of the neutral axis.

It is acceptable to assume that the in-plane inertial is small enough to be neglected in the nonlinear dynamics. Therefore, Eq. (2.125) can be rewritten in the following form:

$$\frac{\partial N}{\partial x} = \frac{\partial}{\partial x}\left\{A_{xx}\left[\frac{\partial u}{\partial x} + \frac{1}{2}\left(\frac{\partial w}{\partial x}\right)^2\right] - N^T\right\} = 0 \qquad (2.129)$$

By manipulating the left-hand side of the above identity, the below expression can be obtained:

$$\frac{\partial u}{\partial x} = -\frac{1}{2}\left(\frac{\partial w}{\partial x}\right)^2 + \frac{N^T}{A_{xx}} + \frac{c_1}{A_{xx}} \qquad (2.130)$$

where c_1 is a constant value. If the above relation is integrated over the longitudinal direction, the axial displacement can be achieved like:

$$u = -\frac{1}{2}\int_0^x \left(\frac{\partial w}{\partial x}\right)^2 dx + (N^T + c_1)\frac{x}{A_{xx}} + c_2 \qquad (2.131)$$

Similar to c_1, the term c_2 is another constant related to the integration. Now, these constant values should be calculated to derive the final form of the axial displacement. To this end, we will consider the beam to have no axial motion at both ends, that is,

$$u(0) = u(L) = 0 \qquad (2.132)$$

Once the above constraints are inserted into Eq. (2.131), constants c_1 and c_2 will be obtained as below:

$$c_1 = -\frac{A_{xx}}{2L}\int_0^L \left(\frac{\partial w}{\partial x}\right)^2 dx - N^T, \quad c_2 = 0 \qquad (2.133)$$

2.2 Kinematic relations

Substitution of Eq. (2.133) into Eq. (2.130) gives:

$$\frac{\partial u}{\partial x} = -\frac{1}{2}\left(\frac{\partial w}{\partial x}\right)^2 - \frac{1}{2L}\int_0^L \left(\frac{\partial w}{\partial x}\right)^2 dx \qquad (2.134)$$

Recalling the definition of $\bar{N}$ (refer to Eq. (2.101)) and with respect to Eq. (2.127), one can reach to the following expression:

$$\bar{N} = -\left[\frac{A_{xx}}{2L}\int_0^L \left(\frac{\partial w}{\partial x}\right)^2 dx + N^T\right]\frac{\partial^2 w}{\partial x^2} \qquad (2.135)$$

Now, incorporation of the above definition with Eqs. (2.126) and (2.128) results:

$$D_{xx}\frac{\partial^4 w}{\partial x^4} + \left[\frac{A_{xx}}{2L}\int_0^L \left(\frac{\partial w}{\partial x}\right)^2 dx + N^T\right]\frac{\partial^2 w}{\partial x^2} + k_L w - k_P \frac{\partial^2 w}{\partial x^2} + k_{NL} w^3$$

$$+ c_d \frac{\partial w}{\partial t} + I_0 \frac{\partial^2 w}{\partial t^2} - I_2 \frac{\partial^4 w}{\partial t^2 \partial x^2} = 0 \qquad (2.136)$$

The above relation can completely describe the nonlinear free vibration problem of a beam. It must be pointed out that for the goal of covering the forced oscillations of the structure, a harmonic load must be added to the right-hand side of the above formula. Once the aforesaid modification is performed, the following equation can be achieved:

$$D_{xx}\frac{\partial^4 w}{\partial x^4} + \left[\frac{A_{xx}}{2L}\int_0^L \left(\frac{\partial w}{\partial x}\right)^2 dx + N^T\right]\frac{\partial^2 w}{\partial x^2} + k_L w - k_P \frac{\partial^2 w}{\partial x^2} + k_{NL} w^3$$

$$+ c_d \frac{\partial w}{\partial t} + I_0 \frac{\partial^2 w}{\partial t^2} - I_2 \frac{\partial^4 w}{\partial t^2 \partial x^2} = f \cos \Omega t \qquad (2.137)$$

where f denotes the amplitude of the harmonic load applied on the beam. The next step is to solve this equation to find an expression for the nonlinear behaviors of the system's oscillations. To this end, first of all it is required to utilize the concept of separation of the variables to decompose the problem into two wholly different problems in spatial and temporal domains. According to this approach, the solution of the above problem must be considered in the following form:

$$w(x,t) = W(t)X(x) \tag{2.138}$$

In the above solution function, $W(t)$ is the time-dependent amplitude of the system's oscillations. Also, the geometrical constraints at the ends of the beam will be applied on the eigen function $X(x)$. In the first step, the influence of the derivatives with respect to the spatial variable x on the governing equation of the problem must be applied. Doing so, the below relation can be attained:

$$\{I_0 r_{00} - I_2 r_{20}\}\ddot{W}(t) + c_d r_{00}\dot{W}(t) + \{D_{xx} r_{40} + k_L r_{00} + (N^T - k_P) r_{20}\}W(t)$$
$$+ \left\{\frac{A_{xx}}{2L} r_{11} r_{20} + k_{NL} r_{0000}\right\} W^3(t) = \int_0^L fX(x)dx \cos \Omega t \tag{2.139}$$

where r_{ij}'s and r_{ijkl}'s can be considered similar to what previously introduced in Eq. (2.119). By dividing the left-hand side of the above relation to the coefficient behind $\ddot{W}(t)$, the following Duffing equation can be reached:

$$\ddot{W}(t) + \varsigma_1 W(t) + \mu \dot{W}(t) + \varsigma_2 W^3(t) = \hat{f} \cos \Omega t \tag{2.140}$$

in which

$$\varsigma_1 = \frac{D_{xx} r_{40} + k_L r_{00} + (N^T - k_P) r_{20}}{I_0 r_{00} - I_2 r_{20}} \tag{2.141}$$

$$\mu = \frac{c_d r_{00}}{I_0 r_{00} - I_2 r_{20}} \tag{2.142}$$

$$\varsigma_2 = \frac{(A_{xx}/2L) r_{11} r_{20} + k_{NL} r_{0000}}{I_0 r_{00} - I_2 r_{20}} \tag{2.143}$$

$$\hat{f} = \frac{\int_0^L fX(x)dx}{I_0 r_{00} - I_2 r_{20}} \tag{2.144}$$

Now, an equation with cubic-type nonlinearity in the time domain is in hand. Although different methods are introduced in the open literature for the goal of solving such a nonlinear problem, authors believe that the most efficient approach is still the method of multiple scales [722]. To solve Eq. (2.140) via method of multiple scales, it must be considered that the damping and nonlinearities are small and the excitation must be a soft one to capture the primary resonance of the system. Therefore, it is logical to multiply these terms to the small perturbation parameter ε for scaling the equation. Therefore, the final equation can be observed in the following form:

$$\ddot{W}(t) + \varsigma_1 W(t) + 2\varepsilon\bar{\mu}\dot{W}(t) + \varepsilon\varsigma_2 W^3(t) = 2\varepsilon\bar{f} \cos \Omega t \tag{2.145}$$

Obviously, one knows that in this identity $\varsigma_1 = \omega_0^2 = \omega_L^2$. It is noteworthy that in the above relation, damping and forcing terms are multiplied

2.2 Kinematic relations

by two for the sake of simplicity in the algebraic manipulations. Therefore, below quantities are introduced:

$$\bar{\mu} = \mu/2, \qquad \bar{f} = \hat{f}/2 \qquad (2.146)$$

Due to the fact that in this problem we are faced with a cubic non-linearity, we are allowed to utilize a first-order expansion to solve the above equation in the time domain. So, the following approximation will be considered:

$$W(t) = w_0(T_0, T_1) + \varepsilon w_1(T_0, T_1) + \cdots \qquad (2.147)$$

Prior to solve the problem, it is beneficial to mention that T_i's ($T_i = \varepsilon^i t$) are temporal variables which are used to take into account for the impacts of parameters of various orders. According to [722], the differentiation with respect to time can be presented in the following form:

$$\frac{d}{dt} = \frac{dT_0}{dt}\frac{\partial}{\partial T_0} + \frac{dT_1}{dt}\frac{\partial}{\partial T_1} + \cdots = D_0 + \varepsilon D_1 + \cdots$$

$$\frac{d^2}{dt^2} = D_0^2 + 2\varepsilon D_0 D_1 + \varepsilon^2 (D_1^2 + 2D_0 D_2) + \cdots \qquad (2.148)$$

By considering the above definitions for the first and second differentiations with respect to time and substituting for the response of the problem from Eq. (2.147) into Eq. (2.145) followed by setting the coefficients of any arbitrary power of the perturbation parameter to zero, the following relations can be obtained [722]:

$$\varepsilon^0 \rightarrow D_0^2 w_0 + \varsigma_1 w_0 = 0 \qquad (2.149)$$

$$\varepsilon^1 \rightarrow D_0^2 w_1 + \varsigma_1 w_1 = -2D_0 D_1 w_0 - 2\bar{\mu} D_0 w_0 - \varsigma_2 w_0^3 + \bar{f} \exp(i\Omega t) \qquad (2.150)$$

By considering that the undamped linear natural frequency of the system is $\omega_0 = \sqrt{\varsigma_1}$, the solution of Eq. (2.149) can be written in the below form [722]:

$$w_0 = A(T_1) \exp(i\omega_0 T_0) + CC \qquad (2.151)$$

where CC is the complex conjugate of the preceding terms. Once the above solution is substituted into Eq. (2.150), the following expression can be generated:

$$\begin{aligned} D_0^2 w_1 + \varsigma_1 w_1 = &-2i\omega_0 D_1 A \exp(i\omega_0 T_0) - 2i\omega_0 \bar{\mu} A \exp(i\omega_0 T_0) \\ &- \varsigma_2 \left[A^3 \exp(3i\omega_0 T_0) + 3A^2 \bar{A} \exp(i\omega_0 T_0) \right] \\ &+ \bar{f} \exp(i\Omega t) + CC \end{aligned} \qquad (2.152)$$

Elimination of the secular terms and small divisors from the above relation cannot be completed unless:

$$-\left[2i\omega_0(D_1 A + \bar{\mu} A) + 3\varsigma_2 A^2 \bar{A}\right] \exp(i\omega_0 T_0) + \bar{f} \exp(i\Omega T_0) = 0 \qquad (2.153)$$

Once the above identity is divided into the nonzero value $\exp(i\omega_0 T_0)$, the following expression can be attained:

$$-\left[2i\omega_0(D_1 A + \bar{\mu}A) + 3\varsigma_2 A^2 \bar{A}\right] + \bar{f}\exp(i\{\Omega - \omega_0\}T_0) = 0 \quad (2.154)$$

Based on the fact that we are aimed to analyze the nonlinear dynamic characteristics of the beam in excitation frequencies close to the linear natural frequency, the excitation and linear frequencies differ as large as a tiny value. Hence [722]:

$$\Omega \simeq \omega_0 \rightarrow \Omega - \omega_0 = \varepsilon\sigma \quad (2.155)$$

where σ is called detuning parameter in the literature of the nonlinear oscillations. By considering the definition of the predefined temporal variables in the multiple scales method and with the aid of the above definition, Eq. (2.154) can be re-written as follows:

$$-\left[2i\omega_0(D_1 A + \bar{\mu}A) + 3\varsigma_2 A^2 \bar{A}\right] + \bar{f}\exp(i\sigma T_1) = 0 \quad (2.156)$$

Now, A will be considered to be in the following polar form [722]:

$$A(T_1) = \frac{1}{2}a(T_1)\exp(i\beta(T_1)) \quad (2.157)$$

Substituting the above definition in Eq. (2.156) gives:

$$-\left\{i\omega_0\left[a' + ia\beta' + \bar{\mu}a\right] + \frac{3}{8}\varsigma_2 a^3\right\}\exp(i\beta) + \bar{f}\exp(i\sigma T_1) = 0 \quad (2.158)$$

in which $(.)'$ stands for differentiation with respect to T_1. By diving the above identity to $\exp(i\beta)$ and setting both real and imaginary parts of the achieved relation equal with zero, one can write the following modulation equations:

$$a' + \bar{\mu}a = \frac{\bar{f}}{\omega_0}\sin(\sigma T_1 - \beta),$$

$$a\beta' - \frac{3}{8\omega_0}\varsigma_2 a^3 = -\frac{\bar{f}}{\omega_0}\cos(\sigma T_1 - \beta) \quad (2.159)$$

The above coupled equations are nonautonomous relations and working with them is not desired. These relations can be transferred into a set of coupled autonomous equations by the means of the below variable:

$$\sigma T_1 - \beta = \gamma \quad (2.160)$$

Differentiating from the above relation with respect to T_1 gives:

$$\sigma - \beta' = \gamma' \rightarrow \beta' = \sigma - \gamma' \quad (2.161)$$

So, the autonomous form of the modulation equations will be:

$$a' = \frac{\bar{f}}{\omega_0} \sin\gamma - \bar{\mu}a,$$

$$\gamma' = \sigma - \frac{3}{8\omega_0}\varsigma_2 a^2 + \frac{\bar{f}}{a\omega_0}\cos\gamma \quad (2.162)$$

In order to probe the characteristics of the nonlinear system whenever the system's response is settled on its steady-state value, the terms a' and γ' must be set to zero. Doing so, the below identities can be obtained simply:

$$\bar{\mu}a = \frac{\bar{f}}{\omega_0}\sin\gamma,$$

$$a\sigma - \frac{3}{8\omega_0}\varsigma_2 a^3 = -\frac{\bar{f}}{\omega_0}\cos\gamma \quad (2.163)$$

Indeed, the above relations give the fixed points of the system. In the nonlinear dynamics' literature, it is common to show the coordinates of the fixed points with a "*" superscript. However, we do not use such a notation for the sake of clarity and avoid from construction of misleading in the readers' mind.

Using the well-known feature of the trigonometric functions (i.e., $\sin^2\gamma + \cos^2\gamma = 1$), one can easily derive the below relation:

$$\bar{\mu}^2 + \left(\sigma - \frac{3}{8\omega_0}\varsigma_2 a^2\right)^2 = \left(\frac{\bar{f}}{a\omega_0}\right)^2 \quad (2.164)$$

It is a little boring to try for finding an expression describing the variations of the system's oscillation amplitude versus excitation frequency. However, it is simple to find a relation to express the excitation frequency as a function of amplitude in the following form:

$$\sigma = \frac{3}{8\omega_0}\varsigma_2 a^2 \pm \sqrt{\frac{\bar{f}^2}{a^2\omega_0^2} - \bar{\mu}^2} \quad (2.165)$$

The above identity is known as frequency–response equation in the literature and gives remarkable insight about the mechanism of the system's behavior.

Now, the qualitative analysis of this problem can be fulfilled. To this end, the autonomous system introduced in Eq. (2.162) will be considered as the state space relations exhibiting the dynamics of the system. Clearly, this system is a nonlinear system and it needs to be modified in a linearization procedure. To this end, it is necessary to introduce this system in the following general form at first [723]:

$$a' = F(a, \gamma),$$
$$\gamma' = G(a, \gamma) \quad (2.166)$$

In order to linearize the above autonomous system, a small deviation from one of the fixed points of this general system will be considered. Recalling the fact that the position of the fixed points in the $a - \gamma$ plane will be shown with a^* and γ^*, the abovementioned deviations can be regarded to be in the following form:

$$A = a - a^*,$$
$$Y = \gamma - \gamma^* \quad (2.167)$$

After substituting for a and γ from the above definition into Eq. (2.166), the below new system can be achieved [723]:

$$A' = F(A + a^*, Y + \gamma^*),$$
$$Y' = G(A + a^*, Y + \gamma^*) \quad (2.168)$$

Using the Taylor expansion of the above system around (a^*, γ^*) reveals [723]:

$$A' = F(a^*, \gamma^*) + A \frac{\partial F(a, \gamma)}{\partial a}\bigg|_{(a^*, \gamma^*)} + Y \frac{\partial F(a, \gamma)}{\partial \gamma}\bigg|_{(a^*, \gamma^*)} + \text{Higher-order terms} \quad (2.169)$$

$$Y' = G(a^*, \gamma^*) + A \frac{\partial G(a, \gamma)}{\partial a}\bigg|_{(a^*, \gamma^*)} + Y \frac{\partial G(a, \gamma)}{\partial \gamma}\bigg|_{(a^*, \gamma^*)} + \text{Higher-order terms} \quad (2.170)$$

Considering the fact that functions F and G return zero at the fixed points and by summing the above relations, the below linearized form can be derived:

$$\begin{Bmatrix} A' \\ Y' \end{Bmatrix} = \begin{bmatrix} \frac{\partial F}{\partial a} & \frac{\partial F}{\partial \gamma} \\ \frac{\partial G}{\partial a} & \frac{\partial G}{\partial \gamma} \end{bmatrix}_{(a^*, \gamma^*)} \begin{Bmatrix} A \\ Y \end{Bmatrix} \quad \text{or} \quad \begin{Bmatrix} A' \\ Y' \end{Bmatrix} = [J] \begin{Bmatrix} A \\ Y \end{Bmatrix} \quad (2.171)$$

where J is the Jacobian matrix transferring the autonomous functions from $a - \gamma$ system to $A - Y$ one.

For the present case, the Jacobian matrix can be constructed in the following form:

$$J = \begin{bmatrix} -\bar{\mu} & \frac{\hat{f}}{\omega_0} \cos \gamma \\ -\frac{3\varsigma_2 a}{4\omega_0} - \frac{\hat{f}}{a^2 \omega_0} \cos \gamma & -\frac{\hat{f}}{a \omega_0} \sin \gamma \end{bmatrix}_{(a^*, \gamma^*)} \quad (2.172)$$

Using Eq. (2.163), the Jacobian can be expressed as below:

$$J = \begin{bmatrix} -\bar{\mu} & \frac{\bar{f}}{\omega_0}\cos\gamma^* \\ \frac{\sigma}{a^*} - \frac{9\varsigma_2 a^*}{8\omega_0} & -\bar{\mu} \end{bmatrix} \quad (2.173)$$

It is worth mentioning that the above linearization does not reveal accurate geometrical prediction of the system's behaviors for any case. Indeed, the response obtained from this linearization should be double-checked if the new prediction addresses a nonhyperbolic fixed point for the system. In general, the Hartman–Grobman theorem must be kept in mind while implementing this linearization technique, particularly for the nonconservative systems like what discussed in this section. Detailed data about this issue can be found in Ref. [723].

2.2.1.1.3 Wave propagation problem

The main concern of this part of the book is to provide an analytical solution for the wave propagation problem within thin-walled beams. It must be considered that this problem will be investigated in the linear domain. So, the strain–displacement relations of the beam must be re-generated. The geometrical features of the beam are similar to what we introduced in previous sections. Also, the wave media will be considered to be rested on a three-parameter visco-Pasternak medium containing Winkler (k_W), Pasternak (k_P), and damping (c_d) coefficients and subjected to a uniform transverse loading of magnitude $f(t)$.

Consider the displacement field of Euler–Bernoulli beams in the following form [532]:

$$\begin{aligned} u_x(x, z, t) &= -z\frac{\partial w(x, t)}{\partial x}, \\ u_z(x, z, t) &= w(x, t) \end{aligned} \quad (2.174)$$

where $w(x, t)$ denotes the deflection of the beam caused by bending of the structure. In the above definition, the influence of the axial displacement of the mid-axis of the beam is not captured. With the aid of the definition of the linearized strain tensor in the continuum mechanics, the only nonzero component of the beam's strain tensor can be written in the following form:

$$\varepsilon_{xx} = -z\frac{\partial^2 w}{\partial x^2} \quad (2.175)$$

So, it can be claimed that the axial strain of the beam in this condition is originated from the curvature that appears in the shape of the structure under loading. Using the definition of the variation of the strain energy of the beam, one can achieve to the following relation:

$$\delta U = -\int_0^L M \frac{\partial^2 \delta w}{\partial x^2} dx \qquad (2.176)$$

in which M is the bending moment which can be defined as below:

$$M = \int_0^b \int_{-h/2}^{h/2} z\sigma_{xx} dz dy \qquad (2.177)$$

As mentioned above, the beam will be considered to be rested on a viscoelastic foundation. So, the variation of work done on the beam by the surrounding substrate can be formulated in the following form [515, 532]:

$$\delta V = \int_0^L \left[k_W w - k_P \frac{\partial^2 w}{\partial x^2} + c_d \frac{\partial w}{\partial t} + f(t) \right] \delta w dx \qquad (2.178)$$

Also, the variation of the kinetic energy of the beam can be extracted by substituting for the displacement field of the beam from Eq. (2.174) into Eq. (2.122):

$$\delta T = \int_{-h/2}^{h/2} \left[I_0 \frac{\partial w}{\partial t} \frac{\partial \delta w}{\partial t} + I_2 \frac{\partial^2 w}{\partial x \partial t} \frac{\partial^2 \delta w}{\partial x \partial t} \right] dx \qquad (2.179)$$

where

$$\begin{Bmatrix} I_0 \\ I_2 \end{Bmatrix} = \int_0^b \int_{-h/2}^{h/2} \begin{Bmatrix} 1 \\ z^2 \end{Bmatrix} \rho(z) dz dy \qquad (2.180)$$

Now, all of the materials needed to find the motion equation of the beam are in hand. To complete the derivation procedure, it is enough to substitute Eqs. (2.176), (2.178), and (2.179) into Eq. (2.92). Then, if the nontrivial response of the obtained equation is chosen, the below Euler–Lagrange equation can be written:

$$\frac{\partial^2 M}{\partial x^2} - k_W w + k_P \frac{\partial^2 w}{\partial x^2} - c_d \frac{\partial w}{\partial t} + f(t) = I_0 \frac{\partial^2 w}{\partial t^2} - I_2 \frac{\partial^4 w}{\partial t^2 \partial x^2} \qquad (2.181)$$

Now, the above equation must be re-written in terms of the displacement field's variables to attain the governing equation of the problem. To this purpose, the 1D Hook's law ($\sigma_{xx} = E(z)\varepsilon_{xx}$) must be considered and integrated over the cross-section area of the beam with regard to the definition of the bending moment expressed in Eq. (2.177). Therefore, the bending moment can be re-written as below:

$$M = -D_{xx} \frac{\partial^2 w}{\partial x^2} \qquad (2.182)$$

2.2 Kinematic relations

In the above definition, the cross-sectional bending rigidity of the plate can be defined in the following form:

$$D_{xx} = \int_0^b \int_{-h/2}^{h/2} z^2 E(z) dz dy \tag{2.183}$$

Substitution of Eq. (2.182) into Eq. (2.181) gives:

$$D_{xx}\frac{\partial^4 w}{\partial x^4} + k_W w - k_P \frac{\partial^2 w}{\partial x^2} + c_d \frac{\partial w}{\partial t} + I_0 \frac{\partial^2 w}{\partial t^2} - I_2 \frac{\partial^4 w}{\partial t^2 \partial x^2} = f(t) \tag{2.184}$$

Once the above equation is solved, the wave response of the system will be obtained. In common, an exponential solution will be considered to derive the natural frequency, phase, and group velocities of the scattered waves. The general guidelines of implementation of this method can be found by referring to the authors' previous text book [724]. Herein, the main concept of the exponential method, that is, the expression of the spatial part of the solution will be kept and a different point of view for solving the problem in the time domain will be introduced. Indeed, the solution of the problem will be considered in the following general form:

$$w(x,t) = W(t)\exp(i\beta x) \tag{2.185}$$

In the above solution, β denotes the wave number and $W(t)$ will be considered as the amplitude of the dispersed waves. In this solution, the amplitude will be a function of time. By simple substitution of the above solution in Eq. (2.184) one can reach to the below expression:

$$M\ddot{W}(t) + C\dot{W}(t) + KW(t) = f(t) \tag{2.186}$$

in which M, C, and K are, respectively, mass, damping, and stiffness of the underobservation system and can be calculated as below:

$$M = I_0 + I_2\beta^2, \quad C = c_d, \quad K = D_{xx}\beta^4 + k_W + k_P\beta^2 \tag{2.187}$$

The above formula describes the damped fluctuation of a single degree-of-freedom whose schematic view can be observed in Fig. 2.9.

The time-dependent response of the above problem is going to be analyzed in the what follows. To this goal, the Laplace transformation technique will be employed to change the domain of the problem from time to frequency. Doing so, the transferred equation can be expressed as follows:

$$[Ms^2 + Cs + K]\bar{W} = \bar{F} \tag{2.188}$$

In the above relation, $\bar{W}$ and $\bar{F}$ stand for the transformed deflection amplitude and force in the frequency domain. Once algebraic operations are performed on the above equation, the transformed deflection can be enhanced. Afterward, the obtained expression must be transferred to the

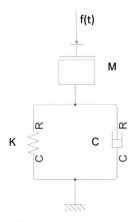

FIGURE 2.9 Schematic view of a single degree-of-freedom nonconservative oscillating system.

time domain by the means of a simple Laplace inversion in the region of convergence of the system. Once the above steps are completed, the amplitude of the wave's motion as a function of time will be in hand.

In order to clarify the aforementioned procedure, assume the deflection amplitude of the wave in the frequency domain to be known in the following form:

$$\bar{W} = \frac{1}{Ms^2 + Cs + K}\bar{F} \qquad (2.189)$$

The deflection amplitude in the time domain can be derived now. Consider the following general form for the deflection in the frequency domain:

$$\bar{W} = \frac{N(s)}{D(s)} = \frac{a_k s^k + a_{k-1} s^{k-1} + \ldots + a_1 s + a_0}{b_l s^k + b_{l-1} s^{k-1} + \ldots + b_1 s + b_0} \qquad (2.190)$$

The functions $N(s)$ and $D(s)$ are polynomials including no common zeros. The above expression has poles from simple type. In other word, the zeros of function $D(s)$ are simple roots. Therefore, these roots do not assign zero value to the derivative of the function $D(s)$. In general, these roots maybe real or complex values. If the number of real and complex poles are shown with n_r and n_c, respectively, the below expression can be hired to derive the inverse Laplace transformation of $\bar{W}$:

$$w(t) = \mathcal{L}^{-1}\bar{W} = \text{Re}\left\{\sum_{i=1}^{n_c} \frac{N(c_i)}{E(c_i)} \exp(c_i t)\right\} + \sum_{j=1}^{n_r} \frac{N(r_j)}{E(r_j)} \exp(r_j t) \qquad (2.191)$$

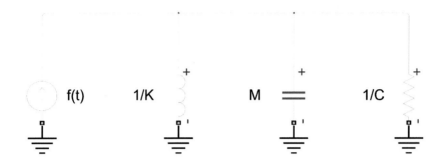

FIGURE 2.10 The schematic view of the equivalent electrical circuit corresponding with the problem whose mechanical representation was shown in Fig. 2.9.

where c_i and r_j stand for the ith complex root and jth real root, respectively. Also, the function $E(s)$ denotes the first derivative of $D(s)$ with respect to s. Following the above procedure gives an expression for the deflection amplitude of the scattered waves in the time domain.

Now, the stability of the present problem is going to be monitored by the means of a control engineering-assisted framework. To this end, first of all it is necessary to introduce the equivalent electrical circuit of the formulated problem. This circuit can be observed in Fig. 2.10. Using the Kirchhoff's circuit law for the introduced electrical system, one can write the following identity [725]:

$$\frac{s\bar{W}}{1/Ms} + \frac{s\bar{W}}{s/K} + \frac{s\bar{W}}{1/C} = \bar{F} \tag{2.192}$$

By doing simple mathematical manipulations, Eq. (2.192) can be rewritten in the following form:

$$\frac{\bar{W}}{\bar{F}} = \frac{1}{Ms^2 + Cs + K} \tag{2.193}$$

The obtained relation is indeed a new version of Eq. (2.188) and it gives the transfer function of the system, that is, identical with the ratio between the transformed input and output of the system in the frequency domain. The above transfer function corresponds with the following characteristics equation of a typical second-order system [725]:

$$s^2 + 2\xi\omega_n s + \omega_n^2 = 0 \tag{2.194}$$

where ξ is the damping ratio and ω_n is the natural frequency of the undamped (conservative) system. In order to track the stability of the system in the framework of well-known stability criteria in the control's open literature, the following variables must be known [725]:

$$\omega_d = \omega_n\sqrt{1-\xi^2} \qquad (2.195)$$

$$\omega_r = \omega_n\sqrt{1-2\xi^2} \qquad (2.196)$$

$$\alpha = \xi\omega_n \qquad (2.197)$$

in which ω_d and ω_r are the damped and resonance frequencies of the second-order equation. Also, the term α stands for the damping coefficient of the mentioned system. As the final variable important for us, the percent of overshoot (PO) of the introduced system can be calculated with the aid of the below formula [725]:

$$\text{PO}(\%) = \exp\left(-\frac{\xi\pi}{\sqrt{1-\xi^2}}\right) \qquad (2.198)$$

Using the above definitions and employing the Bode stability criterion, the stability of the system in any condition can be tracked simply.

2.2.1.2 Refined higher-order beam theory

In the Euler–Bernoulli theory, it will be assumed that there exists no distortion in the cross-section of the beam once it is subjected to loading. This simplifying assumption results in the deficiency of this theory for analysis of thick-type elements. Indeed, in the thin-walled structure, this effect shows itself in a weak manner due to the high weight of the bending mode-induced deformation in the total deformation of the structure. Such an issue is not acceptable in thick-walled elements because the ratio between length and thickness of the structure is small. To compensate for this lack in the classical theory of the beams, thickness-dependent shape functions were adopted in order to make it possible to approximate the profile of the distorted cross-section of the beam. Therefore, higher-order theories were born to include the shear-induced deformation of the beams in the calculations. In the conventional theories, the rotation of the beam's cross-section will be a new variable as well as formerly available axial displacement and bending deflection of the structure. However, in another group of higher-order theories, also known as refined HSDTs, the rotation of the cross-section is omitted and the total deflection of the continuous system is divided into two parts originated from bending and shearing modes. Similar to conventional HSDTs, the shape function exists in refined-type theories, too.

In this section of this text, refined beam theories will be introduced in their most general case for time-dependent problems. Afterward, it will be explained how to solve bending, buckling, and free vibration problems of the beams with the aid of this type of beam theories. Herein, the previous

2.2 Kinematic relations

geometrical assumptions about the geometrical variables will be kept. In other words, the beam will be assumed to be of length L, width b, and thickness h. The displacement field of a refined beam theory can be considered to be in the following form [1]:

$$u_x(x, z, t) = u(x, t) - z\frac{\partial w_b(x, t)}{\partial x} - f(z)\frac{\partial w_s(x, t)}{\partial x},$$
$$u_z(x, z, t) = w_b(x, t) + w_s(x, t)$$
(2.199)

in which $u(x, t)$ is the axial displacement of the neutral axis; $w_b(x,t)$ and $w_s(x,t)$ are, respectively, bending and shear deflections of the beam. In the above definition, $f(z)$ is the shape function which is in charge of governing the profile of both shear strain and stress through the thickness of the beam. Due to the fact that the stress-free condition on the upper and lower edges of the structure will be satisfied, no extra shear correction coefficient is hired in such theories. In this text, the below shape function will be utilized [726]:

$$f(z) = \frac{he^z}{h^2 + \pi^2}\left[\pi \sin\left(\frac{\pi z}{h}\right) + h\cos\left(\frac{\pi z}{h}\right)\right] - \frac{h^2}{h^2 + \pi^2}$$
(2.200)

Now, the nonzero components of the strain tensor of the continuous system can be presented in the following form:

$$\varepsilon_{xx} = \frac{\partial u}{\partial x} - z\frac{\partial^2 w_b}{\partial x^2} - f(z)\frac{\partial^2 w_s}{\partial x^2},$$
$$\gamma_{xz} = 2\varepsilon_{xz} = g(z)\frac{\partial w_s}{\partial x}$$
(2.201)

where

$$g(z) = 1 - \frac{df(z)}{dz}$$
(2.202)

In this section, we will be about to formulate a wide range of problems in the linear domain with the aid of refined HSDT of beams. These problems are static bending, static buckling, and free vibration of beams. Among all of these problems, the variation of the strain energy of the system is unique. Therefore, this energy functional will be expressed in mathematical language before dividing the present section to detailed subsections. The variation of the strain energy of the continuous system can be formulated as below:

$$\delta U = \int_0^L \left\{ N\frac{\partial \delta u}{\partial x} - M_b\frac{\partial^2 \delta w_b}{\partial x^2} - M_s\frac{\partial^2 \delta w_s}{\partial x^2} + Q\frac{\partial \delta w_s}{\partial x} \right\} dx$$
(2.203)

In the above identity, the stress-resultants N, M_b, M_s, and Q can be attained using the below formulas:

$$\begin{Bmatrix} N \\ M_b \\ M_s \end{Bmatrix} = \int_0^b \int_{-h/2}^{h/2} \begin{Bmatrix} 1 \\ z \\ f(z) \end{Bmatrix} \sigma_{xx} dz dy, \qquad Q = \int_0^b \int_{-h/2}^{h/2} g(z) \sigma_{xz} dz dy \qquad (2.204)$$

2.2.1.2.1 Bending problem

In this part, it is aimed to go through analyzing the static bending behaviors of beams subjected to transverse loading. To this end, consider an arbitrary bending load of magnitude $q(x)$ which is applied on the structure. The variation of the work done on the beam element by this type of loading can be formulated in the following form [727]:

$$\delta V = -\int_0^L q(x) \delta(w_b + w_s) dx \qquad (2.205)$$

Now, Eqs. (2.203) and (2.205) must be inserted into Eq. (2.91) for the goal of deriving the Euler–Lagrange equations of the beam for static bending problem. After doing the aforementioned substitution and setting the coefficients of δu, δw_b, and δw_s to zero, the following relations can be written:

$$\frac{\partial N}{\partial x} = 0 \qquad (2.206)$$

$$\frac{\partial^2 M_b}{\partial x^2} + q(x) = 0 \qquad (2.207)$$

$$\frac{\partial^2 M_s}{\partial x^2} + \frac{\partial Q}{\partial x} + q(x) = 0 \qquad (2.208)$$

Now, the above equations must be expressed based on the components of the beam's displacement field to solve the problem. To do so, each of the stress-resultants must be expressed in terms of the field variables. Recalling the constitutive equation of the linearly elastic solids (i.e., $\sigma_{xx} = E(z)\varepsilon_{xx}$, $\sigma_{xz} = G(z)\gamma_{xz}$) and integrating from this equation over the cross-section area of the beam while the definition introduced in Eq. (2.204) are kept in mind, reveals:

$$N = A_{xx} \frac{\partial u}{\partial x} - B_{xx} \frac{\partial^2 w_b}{\partial x^2} - B_{xx}^s \frac{\partial^2 w_s}{\partial x^2} \qquad (2.209)$$

$$M_b = B_{xx} \frac{\partial u}{\partial x} - D_{xx} \frac{\partial^2 w_b}{\partial x^2} - D_{xx}^s \frac{\partial^2 w_s}{\partial x^2} \qquad (2.210)$$

$$M_s = B_{xx}^s \frac{\partial u}{\partial x} - D_{xx}^s \frac{\partial^2 w_b}{\partial x^2} - H_{xx}^s \frac{\partial^2 w_s}{\partial x^2} \qquad (2.211)$$

$$Q = A_{xz} \frac{\partial w_s}{\partial x} \qquad (2.212)$$

2.2 Kinematic relations

In the above relations, the cross-sectional rigidities can be gathered using the below relation:

$$\begin{Bmatrix} A_{xx} \\ B_{xx} \\ B^s_{xx} \\ D_{xx} \\ D^s_{xx} \\ H^s_{xx} \end{Bmatrix} = \int_0^b \int_{-h/2}^{h/2} \begin{Bmatrix} 1 \\ z \\ f(z) \\ z^2 \\ zf(z) \\ f^2(z) \end{Bmatrix} E(z) dz dy, \qquad A_{xz} = \int_0^b \int_{-h/2}^{h/2} G(z) g^2(z) dz dy$$

(2.213)

By substituting Eqs. (2.209)–(2.212) into Eqs. (2.206)–(2.208), the governing equations of the problem can be achieved in the following form [1, 727]:

$$A_{xx} \frac{\partial^2 u}{\partial x^2} - B_{xx} \frac{\partial^3 w_b}{\partial x^3} - B^s_{xx} \frac{\partial^3 w_s}{\partial x^3} = 0 \qquad (2.214)$$

$$B_{xx} \frac{\partial^3 u}{\partial x^3} - D_{xx} \frac{\partial^4 w_b}{\partial x^4} - D^s_{xx} \frac{\partial^4 w_s}{\partial x^4} + q(x) = 0 \qquad (2.215)$$

$$B^s_{xx} \frac{\partial^3 u}{\partial x^3} - D^s_{xx} \frac{\partial^4 w_b}{\partial x^4} - H^s_{xx} \frac{\partial^4 w_s}{\partial x^4} + A_{xz} \frac{\partial^2 w_s}{\partial x^2} + q(x) = 0 \qquad (2.216)$$

Now, the governing equations of the problem are achieved and must be solved. To this goal, the components of the displacement field of the beam will be considered to be in the below general form [1, 727]:

$$u(x) = \sum_{m=1}^{\infty} U_m \cos(\alpha x),$$

$$w_b(x) = \sum_{m=1}^{\infty} W_{bm} \sin(\alpha x), \qquad (2.217)$$

$$w_s(x) = \sum_{m=1}^{\infty} W_{sm} \sin(\alpha x)$$

where $\alpha = m\pi/L$. In addition, U_m, W_{bm}, and W_{sm} are the unknown Fourier coefficients indicating on the amplitude of motion of each of the axial displacement, bending deflection, and shear deflection, respectively. It is worth mentioning that the above functions are presented by assuming both ends of the underobservation beam to be simply supported. Besides, the transverse bending load applied on the beam can be considered to be in the following form [727]:

$$q(x) = \sum_{m=1}^{\infty} Q_m \sin(\alpha x) \qquad (2.218)$$

in which the amplitude of loading can be computed using the below formula [727]:

$$Q_m = \frac{2}{L} \int_0^L q(x) \sin(\alpha x) dx \qquad (2.219)$$

In this text, three cases of having uniform, sinusoidal, and point load in the middle of the beam will be captured. For the mentioned cases, the loading amplitude can be obtained with the aid of the below definition:

$$\begin{aligned} Q_m &= q_0 \ (m=1) & \text{for sinusoidal load} \\ Q_m &= \frac{4q_0}{m\pi} \ (m=1,3,5,\cdots) & \text{for uniform load} \\ Q_m &= \frac{2}{L} q_0 \sin\left(\frac{m\pi}{2}\right) (m=1,2,3,\cdots) & \text{for point load in the middle} \end{aligned} \qquad (2.220)$$

In the above definition, the amplitude of the applied loading is shown with q_0. Now, substitution of Eqs. (2.217) and (2.218) into the governing Eqs. (2.214) gives:

$$\begin{bmatrix} k_{11} & k_{12} & k_{13} \\ k_{21} & k_{22} & k_{23} \\ k_{31} & k_{32} & k_{33} \end{bmatrix} \begin{Bmatrix} U_m \\ W_{bm} \\ W_{sm} \end{Bmatrix} = \begin{Bmatrix} 0 \\ Q_m \\ Q_m \end{Bmatrix} \quad \text{or} \quad \mathbf{Kd} = \mathbf{F} \qquad (2.221)$$

in which

$$\begin{aligned} k_{11} &= A_{xx}\alpha^2, & k_{12} &= -B_{xx}\alpha^3, & k_{13} &= -B_{xx}^s \alpha^3, \\ k_{22} &= D_{xx}\alpha^4, & k_{23} &= D_{xx}^s \alpha^4, & k_{33} &= H_{xx}^s \alpha^4 + A_{xz}\alpha^2 \end{aligned} \qquad (2.222)$$

In Eq. (2.221), the stiffness matrix **K** is a symmetric one. Using the fundamentals of linear algebra, the vector of the unknown amplitudes can be computed in the following form:

$$\mathbf{d} = \mathbf{K}^{-1}\mathbf{F} \qquad (2.223)$$

2.2.1.2.2 Buckling problem

In this part, the displacement field of the refined HSDT of the beam-type elements will be employed to track the linear static stability behaviors of such continuous systems. To do this analysis, the variations of strain energy and work done by external loading must be in hand. The first one is already in hand (refer to Eq. (2.203)). In order to determine the latter, it must be pointed out that the underobservation beam, that is, rested on a Winkler–Pasternak elastic seat, is presumed to be subjected to an axial compressive load P. Based on this assumption, the variation of work done on the structure by this axial stimulation can be formulated as below

[379, 388]:

$$\delta V = \int_0^L \left[k_W(w_b + w_s) - (k_P - P)\frac{\partial^2(w_b + w_s)}{\partial x^2} \right] dx \quad (2.224)$$

Now, the motion equations of the buckling problem of a beam can be simply found by substituting for the variation of strain energy and work done by external loading from Eqs. (2.203) and (2.224), respectively, into Eq. (2.91). After substitution, the following relations can be attained:

$$\frac{\partial N}{\partial x} = 0 \quad (2.225)$$

$$\frac{\partial^2 M_b}{\partial x^2} - k_W(w_b + w_s) + (k_P - P)\frac{\partial^2(w_b + w_s)}{\partial x^2} = 0 \quad (2.226)$$

$$\frac{\partial^2 M_s}{\partial x^2} + \frac{\partial Q}{\partial x} - k_W(w_b + w_s) + (k_P - P)\frac{\partial^2(w_b + w_s)}{\partial x^2} = 0 \quad (2.227)$$

The next step is to transfer the above equations to the governing equations in terms of the displacement field variables. To this end, the equivalent expressions for the stress-resultants in terms of the components of the beam's displacement field must be substituted into the above relations. In this problem, the stress-resultants are as same as those previously introduced in Eqs. (2.209)–(2.212). Once the aforesaid mathematical manipulation is completely done, the below governing equations can be achieved:

$$A_{xx}\frac{\partial^2 u}{\partial x^2} - B_{xx}\frac{\partial^3 w_b}{\partial x^3} - B^s_{xx}\frac{\partial^3 w_s}{\partial x^3} = 0 \quad (2.228)$$

$$B_{xx}\frac{\partial^3 u}{\partial x^3} - D_{xx}\frac{\partial^4 w_b}{\partial x^4} - D^s_{xx}\frac{\partial^4 w_s}{\partial x^4} - k_W(w_b + w_s) + (k_P - P)\frac{\partial^2(w_b + w_s)}{\partial x^2} = 0 \quad (2.229)$$

$$B^s_{xx}\frac{\partial^3 u}{\partial x^3} - D^s_{xx}\frac{\partial^4 w_b}{\partial x^4} - H^s_{xx}\frac{\partial^4 w_s}{\partial x^4} + A_{xz}\frac{\partial^2 w_s}{\partial x^2}$$
$$- k_W(w_b + w_s) + (k_P - P)\frac{\partial^2(w_b + w_s)}{\partial x^2} = 0 \quad (2.230)$$

In order to solve the above equation, an analytical approach will be implemented herein. Based on this method, called Galerkin's method, the components of the beam's displacement field can be considered in the following form [1, 388]:

$$u(x) = \sum_{m=1}^{\infty} U_m \frac{\partial X_m(x)}{\partial x},$$

$$w_b(x) = \sum_{m=1}^{\infty} W_{bm} X_m(x), \qquad (2.231)$$

$$w_s(x) = \sum_{m=1}^{\infty} W_{sm} X_m(x)$$

where U_m, W_{bm}, and W_{sm} are the unknown amplitudes. The eigen function $X_m(x)$ is responsible for satisfying the BCs at both ends of the beam. In this problem, two cases will be covered and discussed. In the first case, both ends of the beam will be assumed to be simply supported (i.e., S–S BC). In the second condition, clamped–clamped beams will be studied (i.e., C–C BC). According to the Galerkin's method, the proper eigen function of each of the abovementioned BCs is [1, 388]:

$$X_m(x) = \begin{cases} \sin\left(\dfrac{m\pi x}{L}\right) & S-S \text{ beam} \\ \sin^2\left(\dfrac{m\pi x}{L}\right) & C-C \text{ beam} \end{cases} \qquad (2.232)$$

Substitution of Eq. (2.231) into Eqs. (2.228)–(2.230) with regard to the orthogonality of the natural mode shapes results in the following eigenvalue equation:

$$[\mathbf{K} - \mathbf{K_b}]\mathbf{d} = 0 \qquad (2.233)$$

where $\mathbf{K}$ is the stiffness matrix and $\mathbf{d}$ stands for the vector of the unknown amplitudes. Also, $\mathbf{K_b}$ is the buckling load matrix. Solving the above eigenvalue problem requires:

$$|\mathbf{K} - \mathbf{K_b}| = 0 \qquad (2.234)$$

The nonvanishing arrays of the above symmetric matrices are [389]:

$$k_{11} = A_{xx} r_{31}, \quad k_{12} = -B_{xx} r_{31}, \quad k_{13} = -B^s_{xx} r_{31}, \quad k_{22} = -D_{xx} r_{40} - k_W r_{00} + k_p r_{20},$$
$$k_{23} = -D^s_{xx} r_{40} - k_W r_{00} + k_p r_{20}, \quad k_{33} = -H^s_{xx} r_{40} + A_{xz} r_{20} - k_W r_{00} + k_p r_{20}$$
$$(2.235)$$

and

$$k_{b,22} = k_{b,23} = k_{b,33} = P r_{20} \qquad (2.236)$$

It is worth noting that r_{ij}'s are as same as what previously introduced in Eq. (2.119).

2.2.1.2.3 Free vibration problem

This part is dedicated to the investigation of the free vibration problem in beam-type elements simulated via refined HSDT. In order to account for

2.2 Kinematic relations

the existence of the vibrating system in a thermal environment, the effect of an effective thermal loading applied on the beam will be considered. It must be mentioned that the variation of the strain energy of the system will not be derived in this section because it is exactly as same as what introduced in Eq. (2.203). If the structure is rested on a two-parameter stiff substrate containing Winkler (k_W) and Pasternak (k_P) springs, the variation of work done on the beam by external loading can be presented in the following form [379]:

$$\delta V = \int_0^L \left[k_W(w_b + w_s) - (k_P - N^T) \frac{\partial^2(w_b + w_s)}{\partial x^2} \right] \delta(w_b + w_s) dx \quad (2.237)$$

where N^T is the in-plane thermal force. To complete the derivation procedure, the variation of the kinetic energy of the system should be extracted, too. To do so, the displacement field of refined HSDT (Eq. (2.199)) must be substituted in the definition of the variation of the beam's kinetic energy (Eq. (2.122)). Once the aforesaid substitution is done, the below expression can be obtained:

$$\delta T = \int_0^L \begin{bmatrix} I_0 \left(\dfrac{\partial u}{\partial t} \dfrac{\partial \delta u}{\partial t} + \dfrac{\partial(w_b + w_s)}{\partial t} \dfrac{\partial \delta(w_b + w_s)}{\partial t} \right) \\ -I_1 \left(\dfrac{\partial u}{\partial t} \dfrac{\partial^2 \delta w_b}{\partial x \partial t} + \dfrac{\partial \delta u}{\partial t} \dfrac{\partial^2 w_b}{\partial x \partial t} \right) \\ +I_2 \dfrac{\partial^2 w_b}{\partial x \partial t} \dfrac{\partial^2 \delta w_b}{\partial x \partial t} - J_1 \left(\dfrac{\partial u}{\partial t} \dfrac{\partial^2 \delta w_s}{\partial x \partial t} + \dfrac{\partial \delta u}{\partial t} \dfrac{\partial^2 w_s}{\partial x \partial t} \right) \\ +J_2 \left(\dfrac{\partial^2 w_b}{\partial x \partial t} \dfrac{\partial^2 \delta w_s}{\partial x \partial t} + \dfrac{\partial^2 \delta w_b}{\partial x \partial t} \dfrac{\partial^2 w_s}{\partial x \partial t} \right) + K_2 \dfrac{\partial^2 w_s}{\partial x \partial t} \dfrac{\partial^2 \delta w_s}{\partial x \partial t} \end{bmatrix} dx \quad (2.238)$$

In the above definition, the mass moments of inertia can be calculated using the below formula:

$$\begin{Bmatrix} I_0 \\ I_1 \\ J_1 \\ I_2 \\ J_2 \\ K_2 \end{Bmatrix} = \int_0^b \int_{-h/2}^{h/2} \begin{Bmatrix} 1 \\ z \\ f(z) \\ z^2 \\ zf(z) \\ f^2(z) \end{Bmatrix} \rho(z) dz dy \quad (2.239)$$

At this stage, all of the variations needed to derive the motion equations of the beam are in hand. To find the motion equations, it is enough to substitute Eqs. (2.203), (2.237), and (2.238) into the definition of the dynamic form of the virtual work's principle (refer to Eq. (2.92)). Hence, the following Euler–Lagrange equations can be extracted by doing the above substitution and setting the coefficients of δu, δw_b, and δw_s to zero:

$$\frac{\partial N}{\partial x} - I_0 \frac{\partial^2 u}{\partial t^2} + I_1 \frac{\partial^3 w_b}{\partial x \partial t^2} + J_1 \frac{\partial^3 w_s}{\partial x \partial t^2} = 0 \quad (2.240)$$

$$\frac{\partial^2 M_b}{\partial x^2} - k_W(w_b + w_s) + (k_P - N^T)\frac{\partial^2(w_b + w_s)}{\partial x^2} - I_0\frac{\partial^2(w_b + w_s)}{\partial t^2}$$
$$-I_1\frac{\partial^3 u}{\partial x \partial t^2} + I_2\frac{\partial^4 w_b}{\partial x^2 \partial t^2} + J_2\frac{\partial^4 w_s}{\partial x^2 \partial t^2} = 0 \qquad (2.241)$$

$$\frac{\partial^2 M_s}{\partial x^2} + \frac{\partial Q}{\partial x} - k_W(w_b + w_s) + (k_P - N^T)\frac{\partial^2(w_b + w_s)}{\partial x^2} - I_0\frac{\partial^2(w_b + w_s)}{\partial t^2}$$
$$-J_1\frac{\partial^3 u}{\partial x \partial t^2} + J_2\frac{\partial^4 w_b}{\partial x^2 \partial t^2} + K_2\frac{\partial^4 w_s}{\partial x^2 \partial t^2} = 0 \qquad (2.242)$$

Now, the above relation must be re-written in terms of the components of the displacement field of the beam. To this end, the stress-resultants previously introduced in Eq. (2.204) must be written in terms of the displacement field of the beam. This procedure is previously passed and the results can be found in Eqs. (2.209)–(2.212). However, in the mentioned equations, the simple form of the constitutive equation is utilized and the thermal effects are ignored. To cover this issue, the constitutive equation must be considered to be $\sigma_{xx} = E(z)[\varepsilon_{xx} - \alpha_{11}(T - T_0)]$, $\sigma_{xz} = G(z)\gamma_{xz}$. In the mentioned constitutive equations, the reference and local temperatures are shown with T_0 and T, respectively. By integrating from the abovementioned constitutive equations over the cross-section of the beam the below relations can be extracted:

$$N = A_{xx}\frac{\partial u}{\partial x} - B_{xx}\frac{\partial^2 w_b}{\partial x^2} - B^s_{xx}\frac{\partial^2 w_s}{\partial x^2} - N^T \qquad (2.243)$$

$$M_b = B_{xx}\frac{\partial u}{\partial x} - D_{xx}\frac{\partial^2 w_b}{\partial x^2} - D^s_{xx}\frac{\partial^2 w_s}{\partial x^2} - M^T_b \qquad (2.244)$$

$$M_s = B^s_{xx}\frac{\partial u}{\partial x} - D^s_{xx}\frac{\partial^2 w_b}{\partial x^2} - H^s_{xx}\frac{\partial^2 w_s}{\partial x^2} - M^T_s \qquad (2.245)$$

$$Q = A_{xz}\frac{\partial w_s}{\partial x} \qquad (2.246)$$

In the above relations, the thermal-resultants N^T, M^T_b, and M^T_s can be defined as below:

$$\begin{Bmatrix} N^T \\ M^T_b \\ M^T_s \end{Bmatrix} = \int_0^b \int_{-h/2}^{h/2} \begin{Bmatrix} 1 \\ z \\ f(z) \end{Bmatrix} E(z)\alpha_{11}(T - T_0) dz dy \qquad (2.247)$$

It is noteworthy that the cross-sectional rigidities are similar to those introduced in Eq. (2.213). The governing equations of the thermally influenced free vibration problem can be derived by substituting Eqs. (2.243)–(2.246) in Eqs. (2.240)–(2.242). So, the following governing equations can be written for the present problem:

$$A_{xx}\frac{\partial^2 u}{\partial x^2} - B_{xx}\frac{\partial^3 w_b}{\partial x^3} - B^s_{xx}\frac{\partial^3 w_s}{\partial x^3} - I_0\frac{\partial^2 u}{\partial t^2} + I_1\frac{\partial^3 w_b}{\partial x \partial t^2} + J_1\frac{\partial^3 w_s}{\partial x \partial t^2} = 0 \qquad (2.248)$$

$$B_{xx}\frac{\partial^3 u}{\partial x^3} - D_{xx}\frac{\partial^4 w_b}{\partial x^4} - D^s_{xx}\frac{\partial^4 w_s}{\partial x^4} - k_W(w_b + w_s) + (k_P - N^T)\frac{\partial^2(w_b + w_s)}{\partial x^2}$$
$$-I_0\frac{\partial^2(w_b + w_s)}{\partial t^2} - I_1\frac{\partial^3 u}{\partial x \partial t^2} + I_2\frac{\partial^4 w_b}{\partial x^2 \partial t^2} + J_2\frac{\partial^4 w_s}{\partial x^2 \partial t^2} = 0 \quad (2.249)$$
$$B^s_{xx}\frac{\partial^3 u}{\partial x^3} - D^s_{xx}\frac{\partial^4 w_b}{\partial x^4} - H^s_{xx}\frac{\partial^4 w_s}{\partial x^4} + A_{xz}\frac{\partial^2 w_s}{\partial x^2} - k_W(w_b + w_s)$$
$$+ (k_P - N^T)\frac{\partial^2(w_b + w_s)}{\partial x^2} - I_0\frac{\partial^2(w_b + w_s)}{\partial t^2} - J_1\frac{\partial^3 u}{\partial x \partial t^2} + J_2\frac{\partial^4 w_b}{\partial x^2 \partial t^2}$$
$$+ K_2\frac{\partial^4 w_s}{\partial x^2 \partial t^2} = 0 \quad (2.250)$$

Now, the above set of coupled differential equations must be solved to find the natural frequency of the free vibrations of the beam. To solve this problem, the well-known Galerkin's analytical method in dynamic form will be hired. Based on this method, the solution of the displacement field's variables can be expressed as follows [1, 379]:

$$u(x,t) = \sum_{m=1}^{\infty} U_m \frac{\partial X_m(x)}{\partial x} \exp(i\omega_m t),$$
$$w_b(x,t) = \sum_{m=1}^{\infty} W_{bm} X_m(x) \exp(i\omega_m t), \quad (2.251)$$
$$w_s(x,t) = \sum_{m=1}^{\infty} W_{sm} X_m(x) \exp(i\omega_m t)$$

in which $X_m(x)$ denotes the eigen function that satisfies the essential BCs at both ends of the beam. The eigen function corresponding with S–S and C–C BCs will like what introduced in Eq. (2.232) will be implemented in this problem, too. Like previous sections, U_m, W_{bm}, and W_{sm} are the unknown amplitudes of vibration. The natural frequency of the system's free vibrations is shown with ω_m. Once the solution functions of Eq. (2.251) are inserted into the governing equations of the problem (Eqs. (2.248)–(2.250)) and the orthogonality of the natural modes is taken into consideration, the following eigenvalue equation can be achieved:

$$[\mathbf{K} - \omega_m^2 \mathbf{M}]\mathbf{d} = 0 \quad (2.252)$$

in which $\mathbf{K}$ and $\mathbf{M}$ are, respectively, stiffness and mass symmetric matrices. Also, $\mathbf{d}$ is the amplitude column vector. In the above eigenvalue problem, the nonzero components of stiffness and mass matrices can be written in the following form [379]:

$$\begin{aligned}
k_{11} &= A_{xx} r_{31}, \quad k_{12} = -B_{xx} r_{31}, \quad k_{13} = -B^s_{xx} r_{31},\\
k_{22} &= -D_{xx} r_{40} - k_W r_{00} + (k_P - N^T) r_{20},\\
k_{23} &= -D^s_{xx} r_{40} - k_W r_{00} + (k_P - N^T) r_{20},\\
k_{33} &= -H^s_{xx} r_{40} + A_{xz} r_{20} - k_W r_{00} + (k_P - N^T) r_{20}
\end{aligned} \quad (2.253)$$

and

$$m_{11} = -I_0 r_{11}, \quad m_{12} = I_1 r_{11}, \quad m_{13} = J_1 r_{11}, \quad m_{22} = -I_0 r_{00} + I_2 r_{20},$$
$$m_{23} = -I_0 r_{00} + J_2 r_{20}, \quad m_{33} = -I_0 r_{00} + K_2 r_{20}$$
(2.254)

Now, solving the eigenvalue problem introduced in Eq. (2.252) gives the natural frequency of the beam. To this end, the following mathematical identity must be valid:

$$\left| \mathbf{K} - \omega_m^2 \mathbf{M} \right| = 0 \tag{2.255}$$

2.2.2 Kinematic relations of plates

2.2.2.1 Kirchhoff–Love plate theory

In this part, the kinematic relations of the Kirchhoff–Love plates will be discussed. Like Euler–Bernoulli beams, in the classical theory of the plates the influences of shear deformation will be dismissed because of the assumption of its small being in comparison with the deformation caused by bending. Indeed, the distortions will be neglected and therefore this theory cannot be implemented to analyze thick-walled plates. The results achieved from this theory can be trusted if the underobservation structure is a thin-walled one whose length-to-thickness ratio is considered to be larger than $a/h = 100$, for example. According to this theory, the displacement field of a plate-type element can be presented in the following form [533]:

$$u_x(x, y, z, t) = u(x, y, t) - (z - \bar{z}) \frac{\partial w(x, y, t)}{\partial x},$$
$$u_y(x, y, z, t) = v(x, y, t) - (z - \bar{z}) \frac{\partial w(x, y, t)}{\partial y}, \tag{2.256}$$
$$u_z(x, y, z, t) = w(x, y, t)$$

where $u(x, y, t)$ and $v(x, y, t)$ denote displacement components in longitudinal and transverse directions, respectively. Also, $w(x, y, t)$ is the bending deflection of the plate. In the above definition, the term $\bar{z}$ stands for the exact position of the plate's neutral surface which can be determined using Eq. (2.94). The magnificence of calculating the precise position of the plate's neutral surface will be better sensed once the derivation of the governing equation of the plate's nonlinear forced vibration problem is started. Indeed, this will allow us to decouple the forces and moments and this will be desirable for us. In addition, the formulated problem will be a more practical one due to the consideration of the exact location of the neutral surface of the plate.

Following nonlinear strain–displacement relations of von Kármán, the nonzero components of the strain tensor of the rectangular plates can be expressed as below [506, 520]:

$$\begin{aligned}\varepsilon_{xx} &= \frac{\partial u}{\partial x} + \frac{1}{2}\left(\frac{\partial w}{\partial x}\right)^2 - (z-\bar{z})\frac{\partial^2 w}{\partial x^2},\\ \varepsilon_{yy} &= \frac{\partial v}{\partial y} + \frac{1}{2}\left(\frac{\partial w}{\partial y}\right)^2 - (z-\bar{z})\frac{\partial^2 w}{\partial y^2},\\ \gamma_{xx} &= \frac{\partial u}{\partial y} + \frac{\partial v}{\partial x} + \frac{\partial w}{\partial x}\frac{\partial w}{\partial y} - 2(z-\bar{z})\frac{\partial^2 w}{\partial x \partial y}\end{aligned} \quad (2.257)$$

2.2.2.1.1 Nonlinear forced vibration problem

Herein, the nonlinear forced vibration problem of a rectangular plate will be formulated. To this end, the well-known Hamilton's principle will be employed. So, the variations of the strain energy, kinetic energy, and work done by external loadings must be derived based on the Kirchhoff–Love theorem. The plate will be considered to be rested on a four-variable nonlinear viscoelastic medium containing linear or Winkler (k_L), shear or Pasternak (k_P), and nonlinear (k_{NL}) stiff coefficients and a damper of damping coefficient c_d.

According to the strains introduced in Eq. (2.257) and with regard to the fact that the considered plate is consisted of a linearly elastic material, the variation of the strain energy of the plate can be formulated in the following form:

$$\delta U = \int_0^b \int_0^a \left\{ \begin{array}{l} N_{xx}\left[\dfrac{\partial \delta u}{\partial x} + \dfrac{\partial w}{\partial x}\dfrac{\partial \delta w}{\partial x}\right] + N_{yy}\left[\dfrac{\partial \delta v}{\partial y} + \dfrac{\partial w}{\partial y}\dfrac{\partial \delta w}{\partial y}\right] \\ + N_{xy}\left[\dfrac{\partial \delta u}{\partial y} + \dfrac{\partial \delta v}{\partial x} + \dfrac{\partial \delta w}{\partial x}\dfrac{\partial w}{\partial y} + \dfrac{\partial w}{\partial x}\dfrac{\partial \delta w}{\partial y}\right] \\ - M_{xx}\dfrac{\partial^2 \delta w}{\partial x^2} - M_{yy}\dfrac{\partial^2 \delta w}{\partial y^2} - 2M_{xy}\dfrac{\partial^2 \delta w}{\partial x \partial y} \end{array} \right\} dxdy$$

(2.258)

In the above definition, the forces N_{ij} and moments M_{ij} can be expressed in the following form:

$$\left\{\begin{array}{cc} N_{xx} & M_{xx} \\ N_{yy} & M_{yy} \\ N_{xy} & M_{xy} \end{array}\right\} = \int_{-h/2}^{h/2} \left\{\begin{array}{cc} \sigma_{xx} & (z-\bar{z})\sigma_{xx} \\ \sigma_{yy} & (z-\bar{z})\sigma_{yy} \\ \sigma_{xy} & (z-\bar{z})\sigma_{xy} \end{array}\right\} dz \quad (2.259)$$

The variation of the work done on the plate by the nonlinear foundation can be written like [506, 520, 728]:

$$\delta V = \int_0^b \int_0^a \left[k_L w - k_P\left(\frac{\partial^2 w}{\partial x^2} + \frac{\partial^2 w}{\partial y^2}\right) + k_{NL} w^3 + c_d \frac{\partial w}{\partial t}\right]\delta w \, dxdy \quad (2.260)$$

Plate elements possess three displacements and therefore the variation of the kinetic energy of such elements can be expressed as below [1]:

$$\delta T = \int_{-h/2}^{h/2} \int_0^b \int_0^a \rho(z) \left[\frac{\partial u_x}{\partial t} \frac{\partial \delta u_x}{\partial t} + \frac{\partial u_y}{\partial t} \frac{\partial \delta u_y}{\partial t} + \frac{\partial u_z}{\partial t} \frac{\partial \delta u_z}{\partial t} \right] dx dy dz \quad (2.261)$$

By substituting the displacement field of the Kirchhoff–Love plates (Eq. (2.256)) in the above definition, the variation of the kinetic energy of the plate will be in the following form:

$$\delta T = \int_0^b \int_0^a \left[I_0 \left(\frac{\partial u}{\partial t} \frac{\partial \delta u}{\partial t} + \frac{\partial v}{\partial t} \frac{\partial \delta v}{\partial t} + \frac{\partial w}{\partial t} \frac{\partial \delta w}{\partial t} \right) + I_2 \left(\frac{\partial^2 w}{\partial x \partial t} \frac{\partial^2 \delta w}{\partial x \partial t} + \frac{\partial^2 w}{\partial y \partial t} \frac{\partial^2 \delta w}{\partial y \partial t} \right) \right] dx dy \quad (2.262)$$

In the above relation, the mass moments of inertia can be calculated via:

$$\left\{ \begin{array}{c} I_0 \\ I_2 \end{array} \right\} = \int_{-h/2}^{h/2} \left\{ \begin{array}{c} 1 \\ (z - \bar{z})^2 \end{array} \right\} \rho(z) dz \quad (2.263)$$

Now, the Euler–Lagrange equations of the plate's motion can be derived by substituting for the variations of strain energy, work done by external loading, and kinetic energy of the plate from Eqs. (2.258), (2.260), and (2.262), respectively, into Eq. (2.92). So, one can write the following equations of motion:

$$\frac{\partial N_{xx}}{\partial x} + \frac{\partial N_{xy}}{\partial y} = I_0 \frac{\partial^2 u}{\partial t^2} \quad (2.264)$$

$$\frac{\partial N_{xy}}{\partial x} + \frac{\partial N_{yy}}{\partial y} = I_0 \frac{\partial^2 v}{\partial t^2} \quad (2.265)$$

$$\frac{\partial^2 M_{xx}}{\partial x^2} + 2 \frac{\partial^2 M_{xy}}{\partial x \partial y} + \frac{\partial^2 M_{yy}}{\partial y^2} + \frac{\partial}{\partial x} \left(N_{xx} \frac{\partial w}{\partial x} + N_{xy} \frac{\partial w}{\partial y} \right)$$
$$+ \frac{\partial}{\partial y} \left(N_{xy} \frac{\partial w}{\partial x} + N_{yy} \frac{\partial w}{\partial y} \right) - k_L w + k_P \left(\frac{\partial^2 w}{\partial x^2} + \frac{\partial^2 w}{\partial y^2} \right) \quad (2.266)$$
$$- k_{NL} w^3 - c_d \frac{\partial w}{\partial t} = I_0 \frac{\partial^2 w}{\partial t^2} - I_2 \left(\frac{\partial^4 w}{\partial x^2 \partial t^2} + \frac{\partial^4 w}{\partial y^2 \partial t^2} \right)$$

Now, the governing equations must be derived by expressing the stress-resultants in terms of the displacement field of the plate. In 2D plane stress problems like analysis of plates, the following constitutive equation can be considered [376]:

$$\left\{ \begin{array}{c} \sigma_{xx} \\ \sigma_{yy} \\ \sigma_{xy} \end{array} \right\} = \left[\begin{array}{ccc} Q_{11} & Q_{12} & 0 \\ Q_{21} & Q_{22} & 0 \\ 0 & 0 & Q_{66} \end{array} \right] \left\{ \begin{array}{c} \varepsilon_{xx} \\ \varepsilon_{yy} \\ \gamma_{xy} \end{array} \right\} \quad (2.267)$$

2.2 Kinematic relations

in which

$$Q_{11} = \frac{E_{11}}{1 - \nu_{12}\nu_{21}}, \quad Q_{12} = Q_{21} = \frac{\nu_{12}E_{22}}{1 - \nu_{12}\nu_{21}}, \quad Q_{22} = \frac{E_{22}}{1 - \nu_{12}\nu_{21}}, \quad Q_{66} = G_{12}$$
(2.268)

By considering the definitions presented in Eq. (2.259) and the constitutive equation of the linearly elastic solids in plane stress condition (see Eq. (2.267)), forces and moments can be related to the components of the plate's displacement field by:

$$\begin{Bmatrix} N_{xx} \\ N_{yy} \\ N_{xy} \end{Bmatrix} = \begin{bmatrix} A_{11} & A_{12} & 0 \\ A_{12} & A_{22} & 0 \\ 0 & 0 & A_{66} \end{bmatrix} \begin{Bmatrix} \frac{\partial u}{\partial x} + \frac{1}{2}\left(\frac{\partial w}{\partial x}\right)^2 \\ \frac{\partial v}{\partial y} + \frac{1}{2}\left(\frac{\partial w}{\partial y}\right)^2 \\ \frac{\partial u}{\partial y} + \frac{\partial v}{\partial x} + \frac{\partial w}{\partial x}\frac{\partial w}{\partial y} \end{Bmatrix}$$
(2.269)

$$\begin{Bmatrix} M_{xx} \\ M_{yy} \\ M_{xy} \end{Bmatrix} = -\begin{bmatrix} D_{11} & D_{12} & 0 \\ D_{12} & D_{22} & 0 \\ 0 & 0 & D_{66} \end{bmatrix} \begin{Bmatrix} \frac{\partial^2 w}{\partial x^2} \\ \frac{\partial^2 w}{\partial y^2} \\ 2\frac{\partial^2 w}{\partial x \partial y} \end{Bmatrix}$$
(2.270)

In the above definitions, the through-the-thickness extensional and bending rigidities of the plate can be expressed in the following form:

$$\begin{Bmatrix} A_{11} & D_{11} \\ A_{12} & D_{12} \\ A_{22} & D_{22} \\ A_{66} & D_{66} \end{Bmatrix} = \int_{-h/2}^{h/2} \begin{Bmatrix} Q_{11} & Q_{11}(z-\bar{z})^2 \\ Q_{12} & Q_{12}(z-\bar{z})^2 \\ Q_{22} & Q_{22}(z-\bar{z})^2 \\ Q_{66} & Q_{66}(z-\bar{z})^2 \end{Bmatrix} dz$$
(2.271)

In the procedure of solving a set of nonlinear equations describing the nonlinear vibrations of a system, the in-plane inertia of the plate will be assumed to be small enough to be ignored in comparison to the inertia existing in the third equation of motion. So, Eqs. (2.264) and (2.265) can be re-written in the following form:

$$\frac{\partial N_{xx}}{\partial x} + \frac{\partial N_{xy}}{\partial y} = 0$$
(2.272)

$$\frac{\partial N_{xy}}{\partial x} + \frac{\partial N_{yy}}{\partial y} = 0$$
(2.273)

Using the above equations incorporated with expanding Eq. (2.266) results in the following new form of the third Euler–Lagrange equation:

$$\frac{\partial^2 M_{xx}}{\partial x^2} + 2\frac{\partial^2 M_{xy}}{\partial x \partial y} + \frac{\partial^2 M_{yy}}{\partial y^2} + N_{xx}\frac{\partial^2 w}{\partial x^2} + 2N_{xy}\frac{\partial^2 w}{\partial x \partial y} + N_{yy}\frac{\partial^2 w}{\partial y^2} - k_L w$$
$$+ k_P\left(\frac{\partial^2 w}{\partial x^2} + \frac{\partial^2 w}{\partial y^2}\right) - k_{NL}w^3 - c_d\frac{\partial w}{\partial t} = I_0\frac{\partial^2 w}{\partial t^2} - I_2\left(\frac{\partial^4 w}{\partial x^2 \partial t^2} + \frac{\partial^4 w}{\partial y^2 \partial t^2}\right)$$
(2.274)

Now, the governing equations of the problem can be obtained by inserting Eqs. (2.269) and (2.270) into Eqs. (2.272)–(2.274). Doing the aforementioned substitution gives:

$$A_{11}\frac{\partial^2 u}{\partial x^2} + A_{66}\frac{\partial^2 u}{\partial y^2} + (A_{12} + A_{66})\frac{\partial^2 v}{\partial x \partial y} + A_{11}\frac{\partial w}{\partial x}\frac{\partial^2 w}{\partial x^2} +$$
$$(A_{12} + A_{66})\frac{\partial w}{\partial y}\frac{\partial^2 w}{\partial x \partial y} + A_{66}\frac{\partial w}{\partial x}\frac{\partial^2 w}{\partial y^2} = 0$$
(2.275)

$$(A_{12} + A_{66})\frac{\partial^2 u}{\partial x \partial y} + A_{66}\frac{\partial^2 v}{\partial x^2} + A_{22}\frac{\partial^2 v}{\partial y^2} + A_{66}\frac{\partial^2 w}{\partial x^2}\frac{\partial w}{\partial y} +$$
$$(A_{12} + A_{66})\frac{\partial w}{\partial x}\frac{\partial^2 w}{\partial x \partial y} + A_{22}\frac{\partial w}{\partial y}\frac{\partial^2 w}{\partial y^2} = 0$$
(2.276)

$$-D_{11}\frac{\partial^4 w}{\partial x^4} - 2(D_{12} + 2D_{66})\frac{\partial^4 w}{\partial x^2 \partial y^2} - D_{22}\frac{\partial^4 w}{\partial y^4} + A_{11}\frac{\partial u}{\partial x}\frac{\partial^2 w}{\partial x^2}$$
$$+ \frac{A_{11}}{2}\left(\frac{\partial w}{\partial x}\right)^2\frac{\partial^2 w}{\partial x^2} + A_{12}\frac{\partial v}{\partial y}\frac{\partial^2 w}{\partial x^2} + \frac{A_{12}}{2}\left(\frac{\partial w}{\partial y}\right)^2\frac{\partial^2 w}{\partial x^2} + 2A_{66}\frac{\partial u}{\partial y}\frac{\partial^2 w}{\partial x \partial y}$$
$$+ 2A_{66}\frac{\partial v}{\partial x}\frac{\partial^2 w}{\partial x \partial y} + 2A_{66}\frac{\partial w}{\partial x}\frac{\partial w}{\partial y}\frac{\partial^2 w}{\partial x \partial y} + A_{12}\frac{\partial u}{\partial x}\frac{\partial^2 w}{\partial y^2} + \frac{A_{12}}{2}\left(\frac{\partial w}{\partial x}\right)^2\frac{\partial^2 w}{\partial y^2}$$
$$+ A_{22}\frac{\partial v}{\partial y}\frac{\partial^2 w}{\partial y^2} + \frac{A_{22}}{2}\left(\frac{\partial w}{\partial y}\right)^2\frac{\partial^2 w}{\partial y^2} - k_L w + k_P\left(\frac{\partial^2 w}{\partial x^2} + \frac{\partial^2 w}{\partial y^2}\right)$$
$$- k_{NL}w^3 - c_d\frac{\partial w}{\partial t} - I_0\frac{\partial^2 w}{\partial t^2} + I_2\left(\frac{\partial^4 w}{\partial x^2 \partial t^2} + \frac{\partial^4 w}{\partial y^2 \partial t^2}\right) = 0$$
(2.277)

Now, the above governing equations must be solved to find the solution of the problem. To this end, the following solution functions for the components of the plate's displacement field should be considered [376]:

$$u(x, y, t) = \sum_{m=1}^{\infty}\sum_{n=1}^{\infty} U_{mn}(t)\frac{\partial X_m(x)}{\partial x}Y_n(y),$$
$$v(x, y, t) = \sum_{m=1}^{\infty}\sum_{n=1}^{\infty} V_{mn}(t)X_m(x)\frac{\partial Y_n(y)}{\partial y},$$
$$W(x, y, t) = \sum_{m=1}^{\infty}\sum_{n=1}^{\infty} W_{mn}(t)X_m(x)Y_n(y)$$
(2.278)

2.2 Kinematic relations

In the above solution, the time-dependent amplitudes are shown with $U_{mn}(t)$, $V_{mn}(t)$, and $W_{mn}(t)$. Also, the geometry-dependent functions $X_m(x)$ and $Y_n(y)$ are the eigen functions in longitudinal and transverse directions, respectively. The satisfaction of the essential BCs at the edges of the plate will be the main responsibility of the mentioned functions.

If Eq. (2.278) is substituted into Eqs. (2.275)–(2.277) with regard to the orthogonality of the natural modes, the following relations can be attained:

$$\alpha_1 U_{mn}(t) + \alpha_2 V_{mn}(t) = \alpha_3 W_{mn}^2(t) \tag{2.279}$$

$$\beta_1 U_{mn}(t) + \beta_2 V_{mn}(t) = \beta_3 W_{mn}^2(t) \tag{2.280}$$

$$M\ddot{W}_{mn}(t) + C\dot{W}_{mn}(t) + \gamma_1 W_{mn}(t) + \gamma_2 W_{mn}^3(t) + \gamma_3 U_{mn}(t)W_{mn}(t)$$
$$+ \gamma_4 V_{mn}(t)W_{mn}(t) = 0 \tag{2.281}$$

where

$$\begin{aligned}
&\alpha_1 = A_{11}r_{3100} + A_{66}r_{1120}, \quad \alpha_2 = (A_{12} + A_{66})r_{1120}, \\
&\alpha_3 = -[A_{11}r_{121000} + A_{12}r_{011110} + A_{66}(r_{101110} + r_{101020})]
\end{aligned} \tag{2.282}$$

$$\begin{aligned}
&\beta_1 = (A_{12} + A_{66})r_{2011}, \quad \beta_2 = A_{66}r_{2011} + A_{22}r_{0031}, \\
&\beta_3 = -[A_{66}(r_{200011} + r_{110011}) + A_{12}r_{110011} + A_{22}r_{000121}]
\end{aligned} \tag{2.283}$$

$$\begin{aligned}
M &= -I_0 r_{0000} + I_2(r_{2000} + r_{0020}), \quad C = -c_d r_{0000}, \\
\gamma_1 &= -D_{11}r_{4000} - 2(D_{12} + 2D_{66})r_{2020} - D_{22}r_{0040} - k_L r_{0000} + k_P(r_{2000} + r_{0020}), \\
\gamma_2 &= [A_{11}r_{11200000} + A_{12}(r_{00201100} + r_{11000020}) + A_{22}r_{00001120}]/2 \\
&\quad + 2A_{66}r_{10100110} - k_{NL}r_{00000000}, \\
\gamma_3 &= A_{11}r_{202000} + 2A_{66}r_{111100} + A_{12}r_{200200}, \quad \gamma_4 = A_{12}r_{022000} \\
&\quad + 2A_{66}r_{111100} + A_{22}r_{020200}
\end{aligned} \tag{2.284}$$

In the above definitions, we have [376]:

$$\begin{aligned}
r_{ijkl} &= \int_0^b \int_0^a \frac{\partial^i X_m(x)}{\partial x^i} \frac{\partial^j X_m(x)}{\partial x^j} \frac{\partial^k Y_n(x)}{\partial y^k} \frac{\partial^l Y_n(x)}{\partial y^l} dx dy \\
r_{ijklmn} &= \int_0^b \int_0^a \frac{\partial^i X_m(x)}{\partial x^i} \frac{\partial^j X_m(x)}{\partial x^j} \frac{\partial^k X_m(x)}{\partial x^k} \frac{\partial^l Y_n(x)}{\partial y^l} \frac{\partial^m Y_n(x)}{\partial y^m} \frac{\partial^n Y_n(x)}{\partial y^n} dx dy \\
r_{ijklmnpq} &= \int_0^b \int_0^a \frac{\partial^i X_m(x)}{\partial x^i} \frac{\partial^j X_m(x)}{\partial x^j} \frac{\partial^k X_m(x)}{\partial x^k} \frac{\partial^l X_m(x)}{\partial x^l} \\
&\quad \frac{\partial^m Y_n(x)}{\partial y^m} \frac{\partial^n Y_n(x)}{\partial y^n} \frac{\partial^p Y_n(x)}{\partial y^p} \frac{\partial^q Y_n(x)}{\partial y^q} dx dy
\end{aligned} \tag{2.285}$$

Solving Eqs. (2.279) and (2.280) for $U_{mn}(t)$ and $V_{mn}(t)$ in terms of $W_{mn}^2(t)$, we have:

$$U_{mn}(t) = C_1 W_{mn}^2(t), \quad V_{mn}(t) = C_2 W_{mn}^2(t) \tag{2.286}$$

in which

$$C_1 = \frac{\alpha_3(\alpha_2\beta_1 - \alpha_1\beta_2) - \alpha_2(\alpha_3\beta_1 - \alpha_1\beta_3)}{\alpha_1(\alpha_2\beta_1 - \alpha_1\beta_2)},$$
$$C_2 = \frac{\alpha_3\beta_1 - \alpha_1\beta_3}{\alpha_2\beta_1 - \alpha_1\beta_2} \tag{2.287}$$

By inserting Eq. (2.286) into Eq. (2.281) and adding a harmonic external excitation ($f\cos\Omega t$) to the right-hand side of the obtained relation, the following equation for the nonlinear forced vibrations of the rectangular plate will be gathered:

$$\ddot{W}_{mn}(t) + \varsigma_1 W_{mn}(t) + \mu \dot{W}_{mn}(t) + \varsigma_2 W_{mn}^3(t) = \hat{f}\cos\Omega t \tag{2.288}$$

where

$$\varsigma_1 = \frac{\gamma_1}{M}, \quad \varsigma_2 = \frac{\gamma_2 + \gamma_3 C_1 + \gamma_4 C_2}{M}, \quad \mu = \frac{C}{M}, \quad \hat{f} = \frac{\int_0^b \int_0^a f X(x) Y(y) dx dy}{M} \tag{2.289}$$

By considering the fact that both damping and nonlinearity are small terms and using a coefficient of two behind damping and excitation amplitude for simplifying the algebraic manipulations, the below Duffing equation can be achieved:

$$\ddot{W}(t) + \varsigma_1 W(t) + 2\varepsilon\bar{\mu}\dot{W}(t) + \varepsilon\varsigma_2 W^3(t) = 2\varepsilon\bar{f}\cos\Omega t \tag{2.290}$$

where

$$\bar{\mu} = \mu/2, \quad \bar{f} = \hat{f}/2 \tag{2.291}$$

It must be considered that in this problem we are aimed at monitoring the nonlinear forced vibrations of a plate stimulated with soft-type excitation to track the primary resonance of the system. Now, the obtained Duffing equation must be solved in the time domain to determine the nonlinear dynamics characteristics of the resonating plate. To this end, the method of multiple scales will be employed. Due to the fact that the present problem is completely identical with the problem solved in Section 2.1.1.2, the solution procedure will not be expressed again and the readers are referred to take a look at Section 2.1.1.2 of the present text for detailed explanations.

2.2.2.1.2 Wave propagation problem

In this subsection, the problem of transversely loaded rectangular plates attacked by the dispersion of elastic waves will be expressed in mathematical language. The geometrical configurations of the plate will be considered similar to what expressed in previous subsection. The rectangular plate will be considered to be embedded on a medium with both stiff and viscose parameters. Two Winkler and Pasternak coefficients (k_W and k_P,

2.2 Kinematic relations

respectively) will be in charge of simulating the linear and shear springs of the substrate. Also, the damper will be considered with the aid of the damping coefficient c_d. As stated above, the plate will be presumed to be subjected to a transverse load of amplitude $f(t)$. In this case, the below form of the displacement field will be considered [515]:

$$
\begin{aligned}
u_x(x,y,z,t) &= -z \frac{\partial w(x,y,t)}{\partial x}, \\
u_y(x,y,z,t) &= -z \frac{\partial w(x,y,t)}{\partial y}, \\
u_z(x,y,z,t) &= w(x,y,t)
\end{aligned}
\qquad (2.292)
$$

in which $w(x, y, t)$ stands for the bending deflection of the plate. Based on the above displacement field and with regard to the definition of the linear strain tensor in the continuum mechanics, the below expressions can be written for the nonzero components of the strain tensor:

$$
\varepsilon_{xx} = -z \frac{\partial^2 w}{\partial x^2}, \quad \varepsilon_{yy} = -z \frac{\partial^2 w}{\partial y^2}, \quad \gamma_{xy} = -2z \frac{\partial^2 w}{\partial x \partial y} \qquad (2.293)
$$

So, the variation of the strain energy of the plate can be now formulated using the above strains. For a linearly elastic plate, the below form can be considered:

$$
\delta U = -\int_0^b \int_0^a \left[M_{xx} \frac{\partial^2 \delta w}{\partial x^2} + 2 M_{xy} \frac{\partial^2 \delta w}{\partial x \partial y} + M_{yy} \frac{\partial^2 \delta w}{\partial y^2} \right] dx dy \qquad (2.294)
$$

In the above definition, the bending induced moments can be calculated via:

$$
\begin{Bmatrix} M_{xx} \\ M_{yy} \\ M_{xy} \end{Bmatrix} = \int_{-h/2}^{h/2} z \begin{Bmatrix} \sigma_{xx} \\ \sigma_{yy} \\ \sigma_{xy} \end{Bmatrix} dz \qquad (2.295)
$$

The variation of the work done by the viscoelastic foundation and transverse loading on the plate-type element can be expressed in the following form [515]:

$$
\delta V = \int_0^b \int_0^a \left[k_W w - k_P \left(\frac{\partial^2 w}{\partial x^2} + \frac{\partial^2 w}{\partial y^2} \right) + c_d \frac{\partial w}{\partial t} + f(t) \right] \delta w \, dx dy \qquad (2.296)
$$

Regarding the definition of the variation of the kinetic energy (refer to Eq. (2.261)) and substituting the displacement field introduced in Eq. (2.292) in it, the below expression for the variation of the plate's kinetic energy can be gathered:

$$
\delta T = \int_0^b \int_0^a \left[I_0 \frac{\partial w}{\partial t} \frac{\partial \delta w}{\partial t} + I_2 \left(\frac{\partial^2 w}{\partial x \partial t} \frac{\partial^2 \delta w}{\partial x \partial t} + \frac{\partial^2 w}{\partial y \partial t} \frac{\partial^2 \delta w}{\partial y \partial t} \right) \right] dx dy \qquad (2.297)
$$

in which the mass moments of inertia used in the above relation can be defined via:

$$\begin{Bmatrix} I_0 \\ I_2 \end{Bmatrix} = \int_{-h/2}^{h/2} \begin{Bmatrix} 1 \\ z^2 \end{Bmatrix} \rho(z) dz \qquad (2.298)$$

Now, the equation of the plate's motion can be attained simply. To this goal, it is enough to substitute for the variations of strain energy, work done on the plate, and kinetic energy from Eqs. (2.294), (2.296), and (2.297) into the Hamilton's principle (Eq. (2.92)) and choose the nontrivial response of the problem, that is, corresponding with setting the coefficient of δw to zero. Thus, we have:

$$\frac{\partial^2 M_{xx}}{\partial x^2} + 2\frac{\partial^2 M_{xy}}{\partial x \partial y} + \frac{\partial^2 M_{yy}}{\partial y^2} - k_W w + k_P\left(\frac{\partial^2 w}{\partial x^2} + \frac{\partial^2 w}{\partial y^2}\right) - c_d \frac{\partial w}{\partial t} =$$
$$I_0 \frac{\partial^2 w}{\partial t^2} - I_2\left(\frac{\partial^4 w}{\partial x^2 \partial t^2} + \frac{\partial^4 w}{\partial y^2 \partial t^2}\right)$$

(2.299)

In order to find the final governing equation of the problem in terms of the bending deflection of the plate, it is required to express the bending moments in terms of the plate's deflection. To do so, Eq. (2.267) must be integrated with respect to the thickness direction by considering the definitions introduced in Eq. (2.295). Once this operation is completed, a relation similar to what we previously introduced in Eq. (2.270) will be achieved. However, in this case, the through-the-thickness bending rigidities of the plate must be defined in the following form:

$$\begin{Bmatrix} D_{11} \\ D_{12} \\ D_{22} \\ D_{66} \end{Bmatrix} = \int_{-h/2}^{h/2} z^2 \begin{Bmatrix} Q_{11} \\ Q_{12} \\ Q_{22} \\ Q_{66} \end{Bmatrix} dz \qquad (2.300)$$

where Q_{ij}'s are similar to what introduced in Eq. (2.268). Now, the governing equation of the problem can be easily derived by substituting for the bending moments from Eq. (2.270) into Eq. (2.299). Therefore, the governing equation of the problem is:

$$D_{11}\frac{\partial^4 w}{\partial x^4} + 2(D_{12} + 2D_{66})\frac{\partial^4 w}{\partial x^2 \partial y^2} + D_{22}\frac{\partial^4 w}{\partial y^4} + k_W w - k_P\left(\frac{\partial^2 w}{\partial x^2} + \frac{\partial^2 w}{\partial y^2}\right)$$
$$+ c_d \frac{\partial w}{\partial t} + I_0 \frac{\partial^2 w}{\partial t^2} - I_2\left(\frac{\partial^4 w}{\partial x^2 \partial t^2} + \frac{\partial^4 w}{\partial y^2 \partial t^2}\right) = f(t)$$

(2.301)

Now, the response of the problem can be obtained by solving the above equation. To solve this equation, the well-known wave propagation method introduced in Section 2.1.1.3 will be employed. However, it must be noticed that the present problem's spatial solution differs from that of the former problem we faced with in Section 2.1.1.3. In the present problem, the problem must be solved in the $x-y$ plane and therefore, two wave numbers will be required. To clarify the explained sentences, it is better to start the solution procedure. The solution to the bending deflection of the plate can be expressed as follows [724]:

$$w(x,y,t) = W(t)\exp\{i(\beta_1 x + \beta_2 y)\} \qquad (2.302)$$

in which β_1 and β_2 are the longitudinal and transverse wave numbers, respectively. Again, the amplitude of the system's deflection $W(t)$ is considered to be time-varying. By inserting the introduced solution function in the governing equation of the problem (Eq. (2.301)), an equation similar to Eq. (2.186) will be achieved whose mass, damping, and stiffness can be written as below [515, 724]:

$$M = I_0 + I_2(\beta_1^2 + \beta_2^2), \quad C = c_d,$$
$$K = D_{11}\beta_1^4 + 2(D_{12} + 2D_{66})\beta_1^2\beta_2^2 + D_{22}\beta_2^4 + k_W + k_P(\beta_1^2 + \beta_2^2) \qquad (2.303)$$

The rest of the solution is exactly as same as what instructed in Section 2.1.1.3 and it will be avoided to repeat all of those details herein. So, the readers are referred to Section 2.1.1.3 to keep on solving the problem for a plate.

2.2.2.2 Refined higher-order plate theory

In previous section, the fundamental concept and kinematic relations of the classical theory of the rectangular plates were discussed in both linear and nonlinear problems. As stated before, classical theories dismiss the distortion of the continuous system's cross-section. According to this issue, only thin-walled plates can be analyzed via Kirchhoff–Love theorem and it cannot be utilized to investigate either a moderately thick or a thick plate. To this end, a plate theory similar to the Timoshenko beam theory was introduced which was able to estimate the shear deformation as a constant value in all of the thickness domain. However, this cannot guarantee the complete inclusion of the shear-induced deformation of the plates in the mechanical analyses. Thus, HSDTs were developed to approximate the shear strain's profile across the thickness of the plate via a thickness-dependent shape function. However, implementation of HSDT of plates instead of Kirchhoff–Love hypothesis results in enlargement of the final eigenvalue problem in a way that three governing equations will be replaced with five ones. This issue however can be softer treated by employing refined HSDTs instead of conventional ones. In such theories, no variable will be dedicated to the rotations of the cross-section area around longitudinal or transverse axes. In such theories, the entire deflection of the plate will be divided into two major parts. The larger part will be responsible for the deflection caused by bending and the smaller one is the shear deflection of the plate. According to refined plate theories, the displacement field of the plate will be in the following form [1]:

$$u_x(x, y, z, t) = u(x, y, t) - z\frac{\partial w_b(x, y, t)}{\partial x} - f(z)\frac{\partial w_s(x, y, t)}{\partial x},$$

$$u_y(x, y, z, t) = v(x, y, t) - z\frac{\partial w_b(x, y, t)}{\partial y} - f(z)\frac{\partial w_s(x, y, t)}{\partial y}, \quad (2.304)$$

$$u_z(x, y, z, t) = w_b(x, y, t) + w_s(x, y, t)$$

in which $u(x, y, t)$ and $v(x, y, t)$ are, respectively, the displacements of the mid-surface of the plate in longitudinal and transverse directions. Also, $w_b(x,y, t)$ and $w_s(x,y, t)$ stand for the bending and shear deflections, respectively. Similar to refined higher-order theorems describing the kinematic behaviors of beams, in the present problem the function $f(z)$ is inserted in the displacement field of the plate for the goal of simulating the distorted shape of the cross-section of the plate using a thickness-dependent expression. Similar to Section 2.1.2, the shape function will be considered like what mentioned in Eq. (2.200). The linear strains of the plate can be expressed in the following form [1]:

$$\begin{aligned}
\varepsilon_{xx} &= \frac{\partial u}{\partial x} - z\frac{\partial^2 w_b}{\partial x^2} - f(z)\frac{\partial^2 w_s}{\partial x^2}, \\
\varepsilon_{yy} &= \frac{\partial v}{\partial y} - z\frac{\partial^2 w_b}{\partial y^2} - f(z)\frac{\partial^2 w_s}{\partial y^2}, \\
\gamma_{xy} &= \frac{\partial u}{\partial y} + \frac{\partial v}{\partial x} - 2z\frac{\partial^2 w_b}{\partial x \partial y} - 2f(z)\frac{\partial^2 w_s}{\partial x \partial y}, \\
\gamma_{xz} &= g(z)\frac{\partial w_s}{\partial x}, \quad \gamma_{yz} = 2\varepsilon_{yz} = g(z)\frac{\partial w_s}{\partial y}
\end{aligned} \quad (2.305)$$

in which $g(z) = 1 - df(z)/dz$. In the future steps, the problems of static bending, buckling, and free vibrations of the rectangular plates are going to be analyzed. The interface between all of the mentioned problems is the calculation of the variation of the plate's strain energy functional. By the means of the assumption of probing a linearly elastic solid plate, the variation of the plate's strain energy can be formulated as below:

$$\delta U = \int_0^b \int_0^a \left[\begin{array}{c} N_{xx}\dfrac{\partial \delta u}{\partial x} - M_{xx}^b\dfrac{\partial^2 \delta w_b}{\partial x^2} - M_{xx}^s\dfrac{\partial^2 \delta w_s}{\partial x^2} + N_{yy}\dfrac{\partial \delta v}{\partial y} \\ -M_{yy}^b\dfrac{\partial^2 \delta w_b}{\partial y^2} - M_{yy}^s\dfrac{\partial^2 \delta w_s}{\partial y^2} + N_{xy}\left(\dfrac{\partial \delta u}{\partial y} + \dfrac{\partial \delta v}{\partial x}\right) - \\ 2M_{xy}^b\dfrac{\partial^2 \delta w_b}{\partial x \partial y} - 2M_{xy}^s\dfrac{\partial^2 \delta w_s}{\partial x \partial y} + Q_{xz}\dfrac{\partial \delta w_s}{\partial x} + Q_{yz}\dfrac{\partial \delta w_s}{\partial y} \end{array} \right] dxdy$$

$$(2.306)$$

2.2 Kinematic relations

In the above definition, the stress-resultants can be achieved using the below formula:

$$\left\{\begin{array}{ccc} N_{xx} & M_{xx}^b & M_{xx}^s \\ N_{yy} & M_{yy}^b & M_{yy}^s \\ N_{xy} & M_{xy}^b & M_{xy}^s \end{array}\right\} = \int_{-h/2}^{h/2} \left\{\begin{array}{ccc} \sigma_{xx} & z\sigma_{xx} & f(z)\sigma_{xx} \\ \sigma_{yy} & z\sigma_{yy} & f(z)\sigma_{yy} \\ \sigma_{xy} & z\sigma_{xy} & f(z)\sigma_{xy} \end{array}\right\} dz, \tag{2.307}$$

$$\left\{\begin{array}{c} Q_{xz} \\ Q_{yz} \end{array}\right\} = \int_{-h/2}^{h/2} g(z) \left\{\begin{array}{c} \sigma_{xz} \\ \sigma_{yz} \end{array}\right\} dz$$

2.2.2.2.1 Bending problem

The major concern of this subsection is to monitor the bending characteristics of the plates modeled with the aid of the refined theory. Consider a plate subjected to a transverse load of magnitude $q(x, y)$. Thus, the variation of work done on the continuous system by this load can be expressed in the following form [729]:

$$\delta V = -\int_0^b \int_0^a q(x, y) \delta(w_b + w_s) dx dy \tag{2.308}$$

Now, the Euler–Lagrange equations of the plate under such a static loading can be obtained by simple substitution of Eqs. (2.306) and (2.308) in the definition of the principle of virtual work (see Eq. (2.91)). By doing the aforementioned substitution, the below relations can be achieved:

$$\frac{\partial N_{xx}}{\partial x} + \frac{\partial N_{xy}}{\partial y} = 0 \tag{2.309}$$

$$\frac{\partial N_{xy}}{\partial x} + \frac{\partial N_{yy}}{\partial y} = 0 \tag{2.310}$$

$$\frac{\partial^2 M_{xx}^b}{\partial x^2} + 2\frac{\partial^2 M_{xy}^b}{\partial x \partial y} + \frac{\partial^2 M_{yy}^b}{\partial y^2} + q(x, y) = 0 \tag{2.311}$$

$$\frac{\partial^2 M_{xx}^s}{\partial x^2} + 2\frac{\partial^2 M_{xy}^s}{\partial x \partial y} + \frac{\partial^2 M_{yy}^s}{\partial y^2} + \frac{\partial Q_{xz}}{\partial x} + \frac{\partial Q_{yz}}{\partial y} + q(x, y) = 0 \tag{2.312}$$

Now, the equations of motion are available and we must extract the governing equations of the problem from them. In order to do this, it is needed to express the stress-resultants in terms of the displacement field variables. The constitutive equation in a plane stress problem for Kirchhoff–Love plates was introduced in Eq. (2.267). This relation must be modified to cover the shear strain and stress, too. Therefore, the below constitutive equation is recommended to be utilized [1,394]:

$$\begin{Bmatrix} \sigma_{xx} \\ \sigma_{yy} \\ \sigma_{xy} \\ \sigma_{xz} \\ \sigma_{yz} \end{Bmatrix} = \begin{bmatrix} Q_{11} & Q_{12} & 0 & 0 & 0 \\ Q_{21} & Q_{22} & 0 & 0 & 0 \\ 0 & 0 & Q_{66} & 0 & 0 \\ 0 & 0 & 0 & Q_{55} & 0 \\ 0 & 0 & 0 & 0 & Q_{44} \end{bmatrix} \begin{Bmatrix} \varepsilon_{xx} \\ \varepsilon_{yy} \\ \gamma_{xy} \\ \gamma_{xz} \\ \gamma_{yz} \end{Bmatrix} \qquad (2.313)$$

where

$$Q_{11} = \frac{E_{11}}{1 - \nu_{12}\nu_{21}}, \quad Q_{12} = \frac{\nu_{12}E_{22}}{1 - \nu_{12}\nu_{21}}, \quad Q_{22} = \frac{E_{22}}{1 - \nu_{12}\nu_{21}}, \qquad (2.314)$$

$$Q_{44} = G_{23}, \quad Q_{55} = G_{13}, \quad Q_{66} = G_{12}$$

By integrating from Eq. (2.313) over the thickness of the plate with regard to the definitions presented in Eq. (2.307), the below expressions for the stress-resultants can be gathered:

$$\begin{Bmatrix} N_{xx} \\ N_{yy} \\ N_{xy} \end{Bmatrix} = \begin{bmatrix} A_{11} & A_{12} & 0 \\ A_{21} & A_{22} & 0 \\ 0 & 0 & A_{66} \end{bmatrix} \begin{Bmatrix} \frac{\partial u}{\partial x} \\ \frac{\partial v}{\partial y} \\ \frac{\partial u}{\partial y} + \frac{\partial v}{\partial x} \end{Bmatrix}$$

$$- \begin{bmatrix} B_{11} & B_{12} & 0 \\ B_{21} & B_{22} & 0 \\ 0 & 0 & B_{66} \end{bmatrix} \begin{Bmatrix} \frac{\partial^2 w_b}{\partial x^2} \\ \frac{\partial^2 w_b}{\partial y^2} \\ 2\frac{\partial^2 w_b}{\partial x \partial y} \end{Bmatrix} - \begin{bmatrix} B^s_{11} & B^s_{12} & 0 \\ B^s_{21} & B^s_{22} & 0 \\ 0 & 0 & B^s_{66} \end{bmatrix} \begin{Bmatrix} \frac{\partial^2 w_s}{\partial x^2} \\ \frac{\partial^2 w_s}{\partial y^2} \\ 2\frac{\partial^2 w_s}{\partial x \partial y} \end{Bmatrix}$$

$$(2.315)$$

$$\begin{Bmatrix} M^b_{xx} \\ M^b_{yy} \\ M^b_{xy} \end{Bmatrix} = \begin{bmatrix} B_{11} & B_{12} & 0 \\ B_{21} & B_{22} & 0 \\ 0 & 0 & B_{66} \end{bmatrix} \begin{Bmatrix} \frac{\partial u}{\partial x} \\ \frac{\partial v}{\partial y} \\ \frac{\partial u}{\partial y} + \frac{\partial v}{\partial x} \end{Bmatrix}$$

$$- \begin{bmatrix} D_{11} & D_{12} & 0 \\ D_{21} & D_{22} & 0 \\ 0 & 0 & D_{66} \end{bmatrix} \begin{Bmatrix} \frac{\partial^2 w_b}{\partial x^2} \\ \frac{\partial^2 w_b}{\partial y^2} \\ 2\frac{\partial^2 w_b}{\partial x \partial y} \end{Bmatrix} - \begin{bmatrix} D^s_{11} & D^s_{12} & 0 \\ D^s_{21} & D^s_{22} & 0 \\ 0 & 0 & D^s_{66} \end{bmatrix} \begin{Bmatrix} \frac{\partial^2 w_s}{\partial x^2} \\ \frac{\partial^2 w_s}{\partial y^2} \\ 2\frac{\partial^2 w_s}{\partial x \partial y} \end{Bmatrix}$$

$$(2.316)$$

$$\begin{Bmatrix} M_{xx}^s \\ M_{yy}^s \\ M_{xy}^s \end{Bmatrix} = \begin{bmatrix} B_{11}^s & B_{12}^s & 0 \\ B_{21}^s & B_{22}^s & 0 \\ 0 & 0 & B_{66}^s \end{bmatrix} \begin{Bmatrix} \frac{\partial u}{\partial x} \\ \frac{\partial v}{\partial y} \\ \frac{\partial u}{\partial y} + \frac{\partial v}{\partial x} \end{Bmatrix}$$

$$- \begin{bmatrix} D_{11}^s & D_{12}^s & 0 \\ D_{21}^s & D_{22}^s & 0 \\ 0 & 0 & D_{66}^s \end{bmatrix} \begin{Bmatrix} \frac{\partial^2 w_b}{\partial x^2} \\ \frac{\partial^2 w_b}{\partial y^2} \\ 2\frac{\partial^2 w_b}{\partial x \partial y} \end{Bmatrix} \qquad (2.317)$$

$$- \begin{bmatrix} H_{11}^s & H_{12}^s & 0 \\ H_{21}^s & H_{22}^s & 0 \\ 0 & 0 & H_{66}^s \end{bmatrix} \begin{Bmatrix} \frac{\partial^2 w_s}{\partial x^2} \\ \frac{\partial^2 w_s}{\partial y^2} \\ 2\frac{\partial^2 w_s}{\partial x \partial y} \end{Bmatrix}$$

$$\begin{Bmatrix} Q_{xz} \\ Q_{yz} \end{Bmatrix} = \begin{bmatrix} A_{55}^s & 0 \\ 0 & A_{44}^s \end{bmatrix} \begin{Bmatrix} \frac{\partial w_s}{\partial x} \\ \frac{\partial w_s}{\partial y} \end{Bmatrix} \qquad (2.318)$$

In the above relations, the through-the-thickness rigidities of the plate can be written as below:

$$\begin{Bmatrix} A_{11} & B_{11} & B_{11}^s & D_{11} & D_{11}^s & H_{11}^s \\ A_{12} & B_{12} & B_{12}^s & D_{12} & D_{12}^s & H_{12}^s \\ A_{22} & B_{22} & B_{22}^s & D_{22} & D_{22}^s & H_{22}^s \\ A_{66} & B_{66} & B_{66}^s & D_{66} & D_{66}^s & H_{66}^s \end{Bmatrix}$$
$$= \int_{-h/2}^{h/2} \begin{Bmatrix} Q_{11} & zQ_{11} & f(z)Q_{11} & z^2Q_{11} & zf(z)Q_{11} & f^2(z)Q_{11} \\ Q_{12} & zQ_{12} & f(z)Q_{12} & z^2Q_{12} & zf(z)Q_{12} & f^2(z)Q_{12} \\ Q_{22} & zQ_{22} & f(z)Q_{22} & z^2Q_{22} & zf(z)Q_{22} & f^2(z)Q_{22} \\ Q_{66} & zQ_{66} & f(z)Q_{66} & z^2Q_{66} & zf(z)Q_{66} & f^2(z)Q_{66} \end{Bmatrix} dz$$

$$(2.319)$$

$$\begin{Bmatrix} A_{55}^s \\ A_{44}^s \end{Bmatrix} = \int_{-h/2}^{h/2} g^2(z) \begin{Bmatrix} Q_{55} \\ Q_{44} \end{Bmatrix} dz \qquad (2.320)$$

Now, the governing equations can be derived by inserting Eqs. (2.315)–(2.318) into Eqs. (2.309)–(2.312). Once the above substitution is performed, the below governing equations will be derived:

$$A_{11}\frac{\partial^2 u}{\partial x^2} + A_{66}\frac{\partial^2 u}{\partial y^2} + (A_{12}+A_{66})\frac{\partial^2 v}{\partial x \partial y} - B_{11}\frac{\partial^3 w_b}{\partial x^3}$$
$$-(B_{12}+2B_{66})\frac{\partial^3 w_b}{\partial x \partial y^2} - B^s_{11}\frac{\partial^3 w_s}{\partial x^3} - (B^s_{12}+2B^s_{66})\frac{\partial^3 w_s}{\partial x \partial y^2} = 0 \quad (2.321)$$

$$(A_{12}+A_{66})\frac{\partial^2 u}{\partial x \partial y} + A_{66}\frac{\partial^2 v}{\partial x^2} + A_{22}\frac{\partial^2 v}{\partial y^2} - (B_{12}+2B_{66})\frac{\partial^3 w_b}{\partial x^2 \partial y}$$
$$-B_{22}\frac{\partial^3 w_b}{\partial y^3} - (B^s_{12}+2B^s_{66})\frac{\partial^3 w_s}{\partial x^2 \partial y} - B^s_{22}\frac{\partial^3 w_s}{\partial y^3} = 0 \quad (2.322)$$

$$B_{11}\frac{\partial^3 u}{\partial x^3} + (B_{12}+2B_{66})\left(\frac{\partial^3 u}{\partial x \partial y^2} + \frac{\partial^3 v}{\partial x^2 \partial y}\right) + B_{22}\frac{\partial^3 v}{\partial y^3} - D_{11}\frac{\partial^4 w_b}{\partial x^4}$$
$$-2(D_{12}+2D_{66})\frac{\partial^4 w_b}{\partial x^2 \partial y^2} - D_{22}\frac{\partial^4 w_b}{\partial y^4} - D^s_{11}\frac{\partial^4 w_s}{\partial x^4} \quad (2.323)$$
$$-2(D^s_{12}+2D^s_{66})\frac{\partial^4 w_s}{\partial x^2 \partial y^2} - D^s_{22}\frac{\partial^4 w_s}{\partial y^4} + q(x,y) = 0$$

$$B^s_{11}\frac{\partial^3 u}{\partial x^3} + (B^s_{12}+2B^s_{66})\left(\frac{\partial^3 u}{\partial x \partial y^2} + \frac{\partial^3 v}{\partial x^2 \partial y}\right) + B^s_{22}\frac{\partial^3 v}{\partial y^3} - D^s_{11}\frac{\partial^4 w_b}{\partial x^4}$$
$$-2(D^s_{12}+2D^s_{66})\frac{\partial^4 w_b}{\partial x^2 \partial y^2} - D^s_{22}\frac{\partial^4 w_b}{\partial y^4} - H^s_{11}\frac{\partial^4 w_s}{\partial x^4}$$
$$-2(H^s_{12}+2H^s_{66})\frac{\partial^4 w_s}{\partial x^2 \partial y^2} - H^s_{22}\frac{\partial^4 w_s}{\partial y^4} + A^s_{55}\frac{\partial^2 w_s}{\partial x^2} \quad (2.324)$$
$$+A^s_{44}\frac{\partial^2 w_s}{\partial y^2} + q(x,y) = 0$$

Now, the above governing equations must be solved. To do so, the well-known Navier's method for fully simply supported plates will be implemented. Based on this method, the solution of the components of the plate's displacement field will be considered to be in the following form [729]:

$$u(x,y) = \sum_{m=1}^{\infty}\sum_{n=1}^{\infty} U_{mn} \cos(\alpha x) \sin(\beta y),$$
$$v(x,y) = \sum_{m=1}^{\infty}\sum_{n=1}^{\infty} V_{mn} \sin(\alpha x) \cos(\beta y),$$
$$w_b(x,y) = \sum_{m=1}^{\infty}\sum_{n=1}^{\infty} W_{bmn} \sin(\alpha x) \sin(\beta y),$$
$$w_s(x,y) = \sum_{m=1}^{\infty}\sum_{n=1}^{\infty} W_{smn} \sin(\alpha x) \sin(\beta y)$$
(2.325)

2.2 Kinematic relations

where $\alpha = m\pi/a$ and $\beta = n\pi/b$ while m and n are mode numbers in longitudinal and transverse directions, respectively. Also, the displacement amplitudes are shown with U_{mn}, V_{mn}, W_{bmn}, and W_{smn}, respectively. Besides, the transverse loading applied on the plate can be assumed to possess the below form [729]:

$$q(x,y) = \sum_{m=1}^{\infty}\sum_{n=1}^{\infty} Q_{mn} \sin(\alpha x) \sin(\beta y) \quad (2.326)$$

in which [729]

$$Q_{mn} = \frac{4}{ab} \int_0^b \int_0^a q(x,y) \sin(\alpha x) \sin(\beta y) dx dy \quad (2.327)$$

Based on the type of loading applied on the plate, the loading amplitude Q_{mn} can be expressed as follows:

$Q_{mn} = q_0 \; (m = n = 1)$ for sinusoidal load

$Q_{mn} = \dfrac{16q_0}{mn\pi^2} \; (m, n = 1, 3, 5, \cdots)$ for uniform load

$Q_{mn} = \dfrac{4q_0}{ab} \sin\left(\dfrac{m\pi}{2}\right) \sin\left(\dfrac{n\pi}{2}\right)$

$(m, n = 1, 2, 3, \cdots)$ for point load at the center

(2.328)

Once Eq. (2.325) is inserted into Eqs. (2.321)–(2.324), the below relation can be achieved:

$$\begin{bmatrix} k_{11} & k_{12} & k_{13} & k_{14} \\ k_{21} & k_{22} & k_{23} & k_{24} \\ k_{31} & k_{32} & k_{33} & k_{34} \\ k_{41} & k_{42} & k_{43} & k_{44} \end{bmatrix} \begin{Bmatrix} U_{mn} \\ V_{mn} \\ W_{bmn} \\ W_{smn} \end{Bmatrix} = \begin{Bmatrix} 0 \\ 0 \\ Q_{mn} \\ Q_{mn} \end{Bmatrix} \quad \text{or} \quad \mathbf{Kd} = \mathbf{F} \quad (2.329)$$

in which $\mathbf{d}$ and $\mathbf{F}$ are column vectors denoting deformation amplitudes and forces, respectively. In the above identity, the nonvanishing components of the symmetric stiffness matrix are [729]:

$k_{11} = (A_{11}\alpha^2 + A_{66}\beta^2), \quad k_{12} = (A_{12} + A_{66})\alpha\beta,$
$k_{13} = [B_{11}\alpha^3 + (B_{12} + 2B_{66})\alpha\beta^2], \quad k_{14} = [B^s_{11}\alpha^3 + (B^s_{12} + 2B^s_{66})\alpha\beta^2],$
$k_{22} = (A_{66}\alpha^2 + A_{22}\beta^2), \quad k_{23} = [(B_{12} + 2B_{66})\alpha^2\beta + B_{22}\beta^3],$
$k_{24} = [(B^s_{12} + 2B^s_{66})\alpha^2\beta + B^s_{22}\beta^3],$
$k_{33} = [D_{11}\alpha^4 + 2(D_{12} + 2D_{66})\alpha^2\beta^2 + D_{22}\beta^4],$
$k_{34} = [D^s_{11}\alpha^4 + 2(D^s_{12} + 2D^s_{66})\alpha^2\beta^2 + D^s_{22}\beta^4],$
$k_{44} = [H^s_{11}\alpha^4 + 2(H^s_{12} + 2H^s_{66})\alpha^2\beta^2 + H^s_{22}\beta^4 + A^s_{55}\alpha^2 + A^s_{44}\beta^2]$

(2.330)

2.2.2.2.2 Buckling problem

Once the inverse of the stiffness matrix is multiplied to Eq. (2.329) from right, the vector of amplitudes can be achieved:

$$\mathbf{d} = \mathbf{K}^{-1}\mathbf{F} \quad (2.331)$$

2.2.2.2.2 Buckling problem

The static stability problem of a rectangular plate simulated via a refined-type hypothesis will be solved in this subsection. To this end, the virtual work's principle will be employed like previous cases concerned with the static analyses. The variation of the strain energy of the plate is already in hand (refer to Eq. (2.306)). In general, various types of buckling stimulations can be considered by changing the ratio between the in-plane buckling loads applied on the plate. However, in the present text, the biaxial buckling problem will be surveyed and the buckling load ratio will be considered to be one. Hence, the buckling load in both longitudinal and transverse directions will be shown with P. So, the variation of work done on the plate by the axial compression and stiff springs of the Winkler–Pasternak substrate can be written in the following form [377, 394]:

$$\delta V = \int_0^b \int_0^a \left[k_W(w_b + w_s) - (k_P - P)\left(\frac{\partial^2(w_b + w_s)}{\partial x^2} + \frac{\partial^2(w_b + w_s)}{\partial y^2}\right) \right] \delta(w_b + w_s) dx dy \quad (2.332)$$

Now, the equations of the motion can be extracted by substituting for the variation of strain energy and work done by axial compression from Eqs. (2.306) and (2.332) in Eq. (2.91). So, we have:

$$\frac{\partial N_{xx}}{\partial x} + \frac{\partial N_{xy}}{\partial y} = 0 \quad (2.333)$$

$$\frac{\partial N_{xy}}{\partial x} + \frac{\partial N_{yy}}{\partial y} = 0 \quad (2.334)$$

$$\frac{\partial^2 M^b_{xx}}{\partial x^2} + 2\frac{\partial^2 M^b_{xy}}{\partial x \partial y} + \frac{\partial^2 M^b_{yy}}{\partial y^2} - k_W(w_b + w_s) + (k_P - P)\left(\frac{\partial^2(w_b + w_s)}{\partial x^2} + \frac{\partial^2(w_b + w_s)}{\partial y^2}\right) = 0 \quad (2.335)$$

$$\frac{\partial^2 M^s_{xx}}{\partial x^2} + 2\frac{\partial^2 M^s_{xy}}{\partial x \partial y} + \frac{\partial^2 M^s_{yy}}{\partial y^2} + \frac{\partial Q_{xz}}{\partial x} + \frac{\partial Q_{yz}}{\partial y} - k_W(w_b + w_s) + (k_P - P)\left(\frac{\partial^2(w_b + w_s)}{\partial x^2} + \frac{\partial^2(w_b + w_s)}{\partial y^2}\right) = 0 \quad (2.336)$$

In order to solve the buckling problem, the above Euler–Lagrange equations must be expressed in terms of the components of the plate's displacement field. In order to extract the governing equations, it is

2.2 Kinematic relations

enough to substitute for the stress-resultants from Eqs. (2.315) to (2.318) in Eqs. (2.333)–(2.336). After substitution, the following governing equations can be achieved:

$$A_{11}\frac{\partial^2 u}{\partial x^2} + A_{66}\frac{\partial^2 u}{\partial y^2} + (A_{12} + A_{66})\frac{\partial^2 v}{\partial x \partial y} - B_{11}\frac{\partial^3 w_b}{\partial x^3}$$
$$-(B_{12} + 2B_{66})\frac{\partial^3 w_b}{\partial x \partial y^2} - B_{11}^s\frac{\partial^3 w_s}{\partial x^3} - (B_{12}^s + 2B_{66}^s)\frac{\partial^3 w_s}{\partial x \partial y^2} = 0$$
(2.337)

$$(A_{12} + A_{66})\frac{\partial^2 u}{\partial x \partial y} + A_{66}\frac{\partial^2 v}{\partial x^2} + A_{22}\frac{\partial^2 v}{\partial y^2} - (B_{12} + 2B_{66})\frac{\partial^3 w_b}{\partial x^2 \partial y}$$
$$-B_{22}\frac{\partial^3 w_b}{\partial y^3} - (B_{12}^s + 2B_{66}^s)\frac{\partial^3 w_s}{\partial x^2 \partial y} - B_{22}^s\frac{\partial^3 w_s}{\partial y^3} = 0$$
(2.338)

$$B_{11}\frac{\partial^3 u}{\partial x^3} + (B_{12} + 2B_{66})\left(\frac{\partial^3 u}{\partial x \partial y^2} + \frac{\partial^3 v}{\partial x^2 \partial y}\right) + B_{22}\frac{\partial^3 v}{\partial y^3} - D_{11}\frac{\partial^4 w_b}{\partial x^4} -$$
$$2(D_{12} + 2D_{66})\frac{\partial^4 w_b}{\partial x^2 \partial y^2} - D_{22}\frac{\partial^4 w_b}{\partial y^4} - D_{11}^s\frac{\partial^4 w_s}{\partial x^4} - 2(D_{12}^s + 2D_{66}^s)\frac{\partial^4 w_s}{\partial x^2 \partial y^2}$$
$$-D_{22}^s\frac{\partial^4 w_s}{\partial y^4} - k_W(w_b + w_s) + (k_P - P)\left(\frac{\partial^2 (w_b + w_s)}{\partial x^2} + \frac{\partial^2 (w_b + w_s)}{\partial y^2}\right) = 0$$
(2.339)

$$B_{11}^s\frac{\partial^3 u}{\partial x^3} + (B_{12}^s + 2B_{66}^s)\left(\frac{\partial^3 u}{\partial x \partial y^2} + \frac{\partial^3 v}{\partial x^2 \partial y}\right) + B_{22}^s\frac{\partial^3 v}{\partial y^3} - D_{11}^s\frac{\partial^4 w_b}{\partial x^4}$$
$$-2(D_{12}^s + 2D_{66}^s)\frac{\partial^4 w_b}{\partial x^2 \partial y^2} - D_{22}^s\frac{\partial^4 w_b}{\partial y^4} - H_{11}^s\frac{\partial^4 w_s}{\partial x^4}$$
$$-2(H_{12}^s + 2H_{66}^s)\frac{\partial^4 w_s}{\partial x^2 \partial y^2} - H_{22}^s\frac{\partial^4 w_s}{\partial y^4} + A_{55}^s\frac{\partial^2 w_s}{\partial x^2} + A_{44}^s\frac{\partial^2 w_s}{\partial y^2}$$
$$-k_W(w_b + w_s) + (k_P - P)\left(\frac{\partial^2 (w_b + w_s)}{\partial x^2} + \frac{\partial^2 (w_b + w_s)}{\partial y^2}\right) = 0$$
(2.340)

The buckling load of the plate can be now extracted from the above set of couple equations. To this purpose, the Galerkin's method will be employed. Based on this method in static condition, the solution functions corresponding with the components of the displacement field of the plate can be stated in the below form [1, 394]:

$$u(x,y) = \sum_{m=1}^{\infty}\sum_{n=1}^{\infty} U_{mn} \frac{\partial X_m(x)}{\partial x} Y_n(y),$$

$$v(x,y) = \sum_{m=1}^{\infty}\sum_{n=1}^{\infty} V_{mn} X_m(x) \frac{\partial Y_n(y)}{\partial y},$$

$$w_b(x,y) = \sum_{m=1}^{\infty}\sum_{n=1}^{\infty} W_{bmn} X_m(x) Y_n(y),$$

$$w_s(x,y) = \sum_{m=1}^{\infty}\sum_{n=1}^{\infty} W_{smn} X_m(x) Y_n(y)$$

(2.341)

where U_{mn}, V_{mn}, W_{bmn}, and W_{smn} are the amplitudes of the plate's displacements. The eigen functions $X_m(x)$ and $Y_n(y)$ are provided to satisfy the essential BCs on the longitudinal and transverse edges of the plate, respectively. For plates with all edges either simply supported or clamped, the eigen functions in longitudinal and transverse directions can be considered in the following form [1, 394]:

$$\text{SSSS Plate:} \quad X_m(x) = \sin\left(\frac{m\pi x}{a}\right), \quad Y_n(y) = \sin\left(\frac{n\pi y}{b}\right)$$

$$\text{CCCC Plate:} \quad X_m(x) = \sin^2\left(\frac{m\pi x}{a}\right), \quad Y_n(y) = \sin^2\left(\frac{n\pi y}{b}\right)$$

(2.342)

Once the solution functions presented in Eq. (2.341) are inserted into Eqs. (2.337)–(2.340) and the orthogonality of the mode shapes is taken into account, the following eigenvalue buckling problem will be obtained:

$$[\mathbf{K} - \mathbf{K_b}]\mathbf{d} = \mathbf{0} \qquad (2.343)$$

in which $\mathbf{K}$ is the stiffness matrix and $\mathbf{K_b}$ stands for the buckling load matrix. Also, the column vector $\mathbf{d}$ contains the unknown amplitudes. In order to find the buckling load of the plate, the following identity must be valid:

$$|\mathbf{K} - \mathbf{K_b}| = 0 \qquad (2.344)$$

2.2 Kinematic relations

It is worth regarding that both of the stiffness and buckling load matrices are symmetric matrices whose corresponding arrays are [394]:

$$k_{11} = A_{11}r_{3100} + A_{66}r_{1120}, \quad k_{12} = (A_{12} + A_{66})r_{1120},$$
$$k_{13} = -[B_{11}r_{3100} + (B_{12} + 2B_{66})r_{1120}], \quad k_{14} = -[B_{11}^s r_{3100} + (B_{12}^s + 2B_{66}^s)r_{1120}],$$
$$k_{22} = A_{66}r_{2011} + A_{22}r_{0031}, \quad k_{23} = -[(B_{12} + 2B_{66})r_{2011} + B_{22}r_{0031}],$$
$$k_{24} = -[(B_{12}^s + 2B_{66}^s)r_{2011} + B_{22}^s r_{0031}],$$
$$k_{33} = -[D_{11}r_{4000} + 2(D_{12} + 2D_{66})r_{2020} + D_{22}r_{0040} + k_W r_{0000} - k_P(r_{2000} + r_{0020})],$$
$$k_{34} = -[D_{11}^s r_{4000} + 2(D_{12}^s + 2D_{66}^s)r_{2020} + D_{22}^s r_{0040} + k_W r_{0000} - k_P(r_{2000} + r_{0020})],$$
$$k_{44} = -\begin{bmatrix} H_{11}^s r_{4000} + 2(H_{12}^s + 2H_{66}^s)r_{2020} + H_{22}^s r_{0040} - \\ A_{55}^s r_{2000} - A_{44}^s r_{0020} + k_W r_{0000} - k_P(r_{2000} + r_{0020}) \end{bmatrix}$$

(2.345)

and

$$k_{b,33} = k_{b,34} = k_{b,44} = P(r_{2000} + r_{0020}) \qquad (2.346)$$

In the above definitions, the r_{ijkl}'s are exactly as same as what introduced in Eq. (2.285).

2.2.2.2.3 Free vibration problem

In this part of this text, the vibrational behaviors of the plates rested on an elastic medium and placed in a thermally affected environment will be probed. Most of the algebraic relations are available from the derivations in previous sections. Similar to previous subsection, the linear and shear stiff coefficients of the spring are shown via k_W and k_P, respectively. In addition, the local temperature will be considered to be shown with T, while T_0 stands for the reference temperature. In Eq. (2.306), the variation of the strain energy is introduced. Also, the variation of the work done by external loading is very close to what expressed in Eq. (2.332) if the buckling stimulation P is replaced with thermal loading of magnitudes N_{xx}^T and N_{yy}^T in longitudinal and transverse directions, respectively. Doing so, the variation of the work done by external loading can be formulated as below [1]:

$$\delta V = \int_0^b \int_0^a \left[k_W(w_b + w_s) - (k_P - N_{xx}^T)\frac{\partial^2(w_b + w_s)}{\partial x^2} - (k_P - N_{yy}^T)\frac{\partial^2(w_b + w_s)}{\partial y^2} \right]$$
$$\delta(w_b + w_s) dx dy \qquad (2.347)$$

So, it is enough to derive the variation of the kinetic energy of the plate to find the Euler–Lagrange equations. By substituting the displacement field of the refined HSDT from Eq. (2.304) to Eq. (2.261), the variation of the kinetic energy of the plate can be written as follows:

$$\delta T = \int_0^b \int_0^a \left[\begin{array}{l} I_0 \left(\dfrac{\partial u}{\partial t}\dfrac{\partial \delta u}{\partial t} + \dfrac{\partial v}{\partial t}\dfrac{\partial \delta v}{\partial t} + \dfrac{\partial (w_b+w_s)}{\partial t}\dfrac{\partial \delta (w_b+w_s)}{\partial t} \right) - \\[6pt] I_1 \left(\dfrac{\partial u}{\partial t}\dfrac{\partial^2 \delta w_b}{\partial x \partial t} + \dfrac{\partial^2 w_b}{\partial x \partial t}\dfrac{\partial \delta u}{\partial t} + \dfrac{\partial v}{\partial t}\dfrac{\partial^2 \delta w_b}{\partial y \partial t} + \dfrac{\partial^2 w_b}{\partial y \partial t}\dfrac{\partial \delta v}{\partial t} \right) \\[6pt] -J_1 \left(\dfrac{\partial u}{\partial t}\dfrac{\partial^2 \delta w_s}{\partial x \partial t} + \dfrac{\partial^2 w_s}{\partial x \partial t}\dfrac{\partial \delta u}{\partial t} + \dfrac{\partial v}{\partial t}\dfrac{\partial^2 \delta w_s}{\partial y \partial t} + \dfrac{\partial^2 w_s}{\partial y \partial t}\dfrac{\partial \delta v}{\partial t} \right) + \\[6pt] I_2 \left(\dfrac{\partial^2 w_b}{\partial x \partial t}\dfrac{\partial^2 \delta w_b}{\partial x \partial t} + \dfrac{\partial^2 w_b}{\partial y \partial t}\dfrac{\partial^2 \delta w_b}{\partial y \partial t} \right) + K_2 \left(\dfrac{\partial^2 w_s}{\partial x \partial t}\dfrac{\partial^2 \delta w_s}{\partial x \partial t} + \dfrac{\partial^2 w_s}{\partial y \partial t}\dfrac{\partial^2 \delta w_s}{\partial y \partial t} \right) \\[6pt] +J_2 \left(\dfrac{\partial^2 w_b}{\partial x \partial t}\dfrac{\partial^2 \delta w_s}{\partial x \partial t} + \dfrac{\partial^2 w_s}{\partial x \partial t}\dfrac{\partial^2 \delta w_b}{\partial x \partial t} + \dfrac{\partial^2 w_b}{\partial y \partial t}\dfrac{\partial^2 \delta w_s}{\partial y \partial t} + \dfrac{\partial^2 w_s}{\partial y \partial t}\dfrac{\partial^2 \delta w_b}{\partial y \partial t} \right) \end{array} \right] dx dy$$
(2.348)

In the above identity, the mass moments of inertia can be computed via:

$$\begin{Bmatrix} I_0 \\ I_1 \\ J_1 \\ I_2 \\ J_2 \\ K_2 \end{Bmatrix} = \int_{-h/2}^{h/2} \begin{Bmatrix} 1 \\ z \\ f(z) \\ z^2 \\ zf(z) \\ f^2(z) \end{Bmatrix} \rho(z) dz \qquad (2.349)$$

By substituting for the variations of the strain energy, work done by external loading, and kinetic energy from Eqs. (2.306), (2.347), and (2.348), respectively, in Eq. (2.92) reveals:

$$\frac{\partial N_{xx}}{\partial x} + \frac{\partial N_{xy}}{\partial y} = I_0 \frac{\partial^2 u}{\partial t^2} - I_1 \frac{\partial^3 w_b}{\partial x \partial t^2} - J_1 \frac{\partial^3 w_s}{\partial x \partial t^2} \qquad (2.350)$$

$$\frac{\partial N_{xy}}{\partial x} + \frac{\partial N_{yy}}{\partial y} = I_0 \frac{\partial^2 v}{\partial t^2} - I_1 \frac{\partial^3 w_b}{\partial y \partial t^2} - J_1 \frac{\partial^3 w_s}{\partial y \partial t^2} \qquad (2.351)$$

$$\frac{\partial^2 M_{xx}^b}{\partial x^2} + 2\frac{\partial^2 M_{xy}^b}{\partial x \partial y} + \frac{\partial^2 M_{yy}^b}{\partial y^2} - k_W(w_b+w_s) + \left(k_P - N_{xx}^T\right)\frac{\partial^2(w_b+w_s)}{\partial x^2}$$
$$+ \left(k_P - N_{yy}^T\right)\frac{\partial^2(w_b+w_s)}{\partial y^2} = I_0 \frac{\partial^2(w_b+w_s)}{\partial t^2} + I_1\left(\frac{\partial^3 u}{\partial x \partial t^2} + \frac{\partial^3 v}{\partial y \partial t^2}\right) \qquad (2.352)$$
$$- I_2\left(\frac{\partial^4 w_b}{\partial x^2 \partial t^2} + \frac{\partial^4 w_b}{\partial y^2 \partial t^2}\right) - J_2\left(\frac{\partial^4 w_s}{\partial x^2 \partial t^2} + \frac{\partial^4 w_s}{\partial y^2 \partial t^2}\right)$$

2.2 Kinematic relations

$$\frac{\partial^2 M_{xx}^s}{\partial x^2} + 2\frac{\partial^2 M_{xy}^s}{\partial x \partial y} + \frac{\partial^2 M_{yy}^s}{\partial y^2} + \frac{\partial Q_{xz}}{\partial x} + \frac{\partial Q_{yz}}{\partial y} - k_W(w_b + w_s) +$$
$$\left(k_P - N_{xx}^T\right)\frac{\partial^2(w_b + w_s)}{\partial x^2} + \left(k_P - N_{yy}^T\right)\frac{\partial^2(w_b + w_s)}{\partial y^2} = I_0 \frac{\partial^2(w_b + w_s)}{\partial t^2}$$
$$+ J_1\left(\frac{\partial^3 u}{\partial x \partial t^2} + \frac{\partial^3 v}{\partial y \partial t^2}\right) - J_2\left(\frac{\partial^4 w_b}{\partial x^2 \partial t^2} + \frac{\partial^4 w_b}{\partial y^2 \partial t^2}\right) - K_2\left(\frac{\partial^4 w_s}{\partial x^2 \partial t^2} + \frac{\partial^4 w_s}{\partial y^2 \partial t^2}\right)$$
(2.353)

Now, the above equations must be solved to find the natural frequency of the free oscillations of the plate. To this purpose, the stress-resultants of the plate must be expressed in terms of the components of the displacement field of the plate followed by substitution of the obtained relation in the Euler–Lagrange equations. Due to the fact that the thermal effects are captured in this problem, the plane stress constitutive equation introduced in Eq. (2.313) cannot be implemented. Considering the thermal strain caused by temperature change, the constitutive equation of the plate can be rewritten in the following form [1, 380]:

$$\begin{Bmatrix} \sigma_{xx} \\ \sigma_{yy} \\ \sigma_{xy} \\ \sigma_{xz} \\ \sigma_{yz} \end{Bmatrix} = \begin{bmatrix} Q_{11} & Q_{12} & 0 & 0 & 0 \\ Q_{21} & Q_{22} & 0 & 0 & 0 \\ 0 & 0 & Q_{66} & 0 & 0 \\ 0 & 0 & 0 & Q_{55} & 0 \\ 0 & 0 & 0 & 0 & Q_{44} \end{bmatrix} \begin{Bmatrix} \varepsilon_{xx} - \alpha_{11}(T - T_0) \\ \varepsilon_{yy} - \alpha_{22}(T - T_0) \\ \gamma_{xy} \\ \gamma_{xz} \\ \gamma_{yz} \end{Bmatrix} \quad (2.354)$$

By integrating from the above definition over the thickness of the plate with regard to the definitions of the stress-resultants (Eq. (2.307)), the following relations can be achieved:

$$\begin{Bmatrix} N_{xx} \\ N_{yy} \\ N_{xy} \end{Bmatrix} = \begin{bmatrix} A_{11} & A_{12} & 0 \\ A_{21} & A_{22} & 0 \\ 0 & 0 & A_{66} \end{bmatrix} \begin{Bmatrix} \frac{\partial u}{\partial x} \\ \frac{\partial v}{\partial y} \\ \frac{\partial u}{\partial y} + \frac{\partial v}{\partial x} \end{Bmatrix}$$
$$- \begin{bmatrix} B_{11} & B_{12} & 0 \\ B_{21} & B_{22} & 0 \\ 0 & 0 & B_{66} \end{bmatrix} \begin{Bmatrix} \frac{\partial^2 w_b}{\partial x^2} \\ \frac{\partial^2 w_b}{\partial y^2} \\ 2\frac{\partial^2 w_b}{\partial x \partial y} \end{Bmatrix} - \begin{bmatrix} B_{11}^s & B_{12}^s & 0 \\ B_{21}^s & B_{22}^s & 0 \\ 0 & 0 & B_{66}^s \end{bmatrix} \begin{Bmatrix} \frac{\partial^2 w_s}{\partial x^2} \\ \frac{\partial^2 w_s}{\partial y^2} \\ 2\frac{\partial^2 w_s}{\partial x \partial y} \end{Bmatrix}$$
$$- \begin{Bmatrix} N_{xx}^T \\ N_{yy}^T \\ 0 \end{Bmatrix}$$

(2.355)

2. Micromechanical homogenization and kinematic relations

$$\begin{Bmatrix} M^b_{xx} \\ M^b_{yy} \\ M^b_{xy} \end{Bmatrix} = \begin{bmatrix} B_{11} & B_{12} & 0 \\ B_{21} & B_{22} & 0 \\ 0 & 0 & B_{66} \end{bmatrix} \begin{Bmatrix} \dfrac{\partial u}{\partial x} \\ \dfrac{\partial v}{\partial y} \\ \dfrac{\partial u}{\partial y} + \dfrac{\partial v}{\partial x} \end{Bmatrix}$$

$$- \begin{bmatrix} D_{11} & D_{12} & 0 \\ D_{21} & D_{22} & 0 \\ 0 & 0 & D_{66} \end{bmatrix} \begin{Bmatrix} \dfrac{\partial^2 w_b}{\partial x^2} \\ \dfrac{\partial^2 w_b}{\partial y^2} \\ 2\dfrac{\partial^2 w_b}{\partial x \partial y} \end{Bmatrix} \qquad (2.356)$$

$$- \begin{bmatrix} D^s_{11} & D^s_{12} & 0 \\ D^s_{21} & D^s_{22} & 0 \\ 0 & 0 & D^s_{66} \end{bmatrix} \begin{Bmatrix} \dfrac{\partial^2 w_s}{\partial x^2} \\ \dfrac{\partial^2 w_s}{\partial y^2} \\ 2\dfrac{\partial^2 w_s}{\partial x \partial y} \end{Bmatrix} - \begin{Bmatrix} M^{b,T}_{xx} \\ M^{b,T}_{yy} \\ 0 \end{Bmatrix}$$

$$\begin{Bmatrix} M^s_{xx} \\ M^s_{yy} \\ M^s_{xy} \end{Bmatrix} = \begin{bmatrix} B^s_{11} & B^s_{12} & 0 \\ B^s_{21} & B^s_{22} & 0 \\ 0 & 0 & B^s_{66} \end{bmatrix} \begin{Bmatrix} \dfrac{\partial u}{\partial x} \\ \dfrac{\partial v}{\partial y} \\ \dfrac{\partial u}{\partial y} + \dfrac{\partial v}{\partial x} \end{Bmatrix}$$

$$- \begin{bmatrix} D^s_{11} & D^s_{12} & 0 \\ D^s_{21} & D^s_{22} & 0 \\ 0 & 0 & D^s_{66} \end{bmatrix} \begin{Bmatrix} \dfrac{\partial^2 w_b}{\partial x^2} \\ \dfrac{\partial^2 w_b}{\partial y^2} \\ 2\dfrac{\partial^2 w_b}{\partial x \partial y} \end{Bmatrix} \qquad (2.357)$$

$$- \begin{bmatrix} H^s_{11} & H^s_{12} & 0 \\ H^s_{21} & H^s_{22} & 0 \\ 0 & 0 & H^s_{66} \end{bmatrix} \begin{Bmatrix} \dfrac{\partial^2 w_s}{\partial x^2} \\ \dfrac{\partial^2 w_s}{\partial y^2} \\ 2\dfrac{\partial^2 w_s}{\partial x \partial y} \end{Bmatrix} - \begin{Bmatrix} M^{s,T}_{xx} \\ M^{s,T}_{yy} \\ 0 \end{Bmatrix}$$

$$\begin{Bmatrix} Q_{xz} \\ Q_{yz} \end{Bmatrix} = \begin{bmatrix} A^s_{55} & 0 \\ 0 & A^s_{44} \end{bmatrix} \begin{Bmatrix} \dfrac{\partial w_s}{\partial x} \\ \dfrac{\partial w_s}{\partial y} \end{Bmatrix} \qquad (2.358)$$

In Eqs. (2.355)–(2.358), through-the-thickness rigidities of the plate are exactly as same as those formerly introduced in Eqs. (2.319) and (2.320). The thermal-resultants in Eqs. (2.355)–(2.357) can be defined in the following

2.2 Kinematic relations

form:

$$\begin{Bmatrix} N_{xx}^T & M_{xx}^{b,T} & M_{xx}^{s,T} \\ N_{yy}^T & M_{yy}^{b,T} & M_{yy}^{s,T} \end{Bmatrix} = \int_{-h/2}^{h/2} \begin{Bmatrix} \alpha_{11} & z\alpha_{11} & f(z)\alpha_{11} \\ \alpha_{22} & z\alpha_{22} & f(z)\alpha_{22} \end{Bmatrix}(T - T_0)dz \quad (2.359)$$

Inserting Eqs. (2.355)–(2.358) into Eqs. (2.350)–(2.353) gives the below governing equations:

$$\begin{aligned} & A_{11}\frac{\partial^2 u}{\partial x^2} + A_{66}\frac{\partial^2 u}{\partial y^2} + (A_{12}+A_{66})\frac{\partial^2 v}{\partial x \partial y} - B_{11}\frac{\partial^3 w_b}{\partial x^3} - (B_{12}+2B_{66})\frac{\partial^3 w_b}{\partial x \partial y^2} \\ & - B_{11}^s\frac{\partial^3 w_s}{\partial x^3} - (B_{12}^s + 2B_{66}^s)\frac{\partial^3 w_s}{\partial x \partial y^2} = I_0\frac{\partial^2 u}{\partial t^2} - I_1\frac{\partial^3 w_b}{\partial x \partial t^2} - J_1\frac{\partial^3 w_s}{\partial x \partial t^2} \end{aligned} \quad (2.360)$$

$$\begin{aligned} & (A_{12}+A_{66})\frac{\partial^2 u}{\partial x \partial y} + A_{66}\frac{\partial^2 v}{\partial x^2} + A_{22}\frac{\partial^2 v}{\partial y^2} - (B_{12}+2B_{66})\frac{\partial^3 w_b}{\partial x^2 \partial y} - B_{22}\frac{\partial^3 w_b}{\partial y^3} \\ & - (B_{12}^s + 2B_{66}^s)\frac{\partial^3 w_s}{\partial x^2 \partial y} - B_{22}^s\frac{\partial^3 w_s}{\partial y^3} = I_0\frac{\partial^2 v}{\partial t^2} - I_1\frac{\partial^3 w_b}{\partial y \partial t^2} - J_1\frac{\partial^3 w_s}{\partial y \partial t^2} \end{aligned} \quad (2.361)$$

$$\begin{aligned} & B_{11}\frac{\partial^3 u}{\partial x^3} + (B_{12}+2B_{66})\left(\frac{\partial^3 u}{\partial x \partial y^2} + \frac{\partial^3 v}{\partial x^2 \partial y}\right) + B_{22}\frac{\partial^3 v}{\partial y^3} - D_{11}\frac{\partial^4 w_b}{\partial x^4} \\ & - 2(D_{12}+2D_{66})\frac{\partial^4 w_b}{\partial x^2 \partial y^2} - D_{22}\frac{\partial^4 w_b}{\partial y^4} - D_{11}^s\frac{\partial^4 w_s}{\partial x^4} - 2(D_{12}^s + 2D_{66}^s)\frac{\partial^4 w_s}{\partial x^2 \partial y^2} \\ & - D_{22}^s\frac{\partial^4 w_s}{\partial y^4} - k_W(w_b + w_s) + (k_P - N_{xx}^T)\frac{\partial^2(w_b + w_s)}{\partial x^2} \\ & + (k_P - N_{yy}^T)\frac{\partial^2(w_b + w_s)}{\partial y^2} = I_0\frac{\partial^2(w_b + w_s)}{\partial t^2} + I_1\left(\frac{\partial^3 u}{\partial x \partial t^2} + \frac{\partial^3 v}{\partial y \partial t^2}\right) \\ & - I_2\left(\frac{\partial^4 w_b}{\partial x^2 \partial t^2} + \frac{\partial^4 w_b}{\partial y^2 \partial t^2}\right) - J_2\left(\frac{\partial^4 w_s}{\partial x^2 \partial t^2} + \frac{\partial^4 w_s}{\partial y^2 \partial t^2}\right) \end{aligned} \quad (2.362)$$

$$\begin{aligned} & B_{11}^s\frac{\partial^3 u}{\partial x^3} + (B_{12}^s + 2B_{66}^s)\left(\frac{\partial^3 u}{\partial x \partial y^2} + \frac{\partial^3 v}{\partial x^2 \partial y}\right) + B_{22}^s\frac{\partial^3 v}{\partial y^3} - D_{11}^s\frac{\partial^4 w_b}{\partial x^4} \\ & - 2(D_{12}^s + 2D_{66}^s)\frac{\partial^4 w_b}{\partial x^2 \partial y^2} - D_{22}^s\frac{\partial^4 w_b}{\partial y^4} - H_{11}^s\frac{\partial^4 w_s}{\partial x^4} - 2(H_{12}^s + 2H_{66}^s)\frac{\partial^4 w_s}{\partial x^2 \partial y^2} \\ & - H_{22}^s\frac{\partial^4 w_s}{\partial y^4} + A_{55}^s\frac{\partial^2 w_s}{\partial x^2} + A_{44}^s\frac{\partial^2 w_s}{\partial y^2} - k_W(w_b + w_s) + (k_P - N_{xx}^T)\frac{\partial^2(w_b + w_s)}{\partial x^2} \\ & + (k_P - N_{yy}^T)\frac{\partial^2(w_b + w_s)}{\partial y^2} = I_0\frac{\partial^2(w_b + w_s)}{\partial t^2} + J_1\left(\frac{\partial^3 u}{\partial x \partial t^2} + \frac{\partial^3 v}{\partial y \partial t^2}\right) \\ & - J_2\left(\frac{\partial^4 w_b}{\partial x^2 \partial t^2} + \frac{\partial^4 w_b}{\partial y^2 \partial t^2}\right) - K_2\left(\frac{\partial^4 w_s}{\partial x^2 \partial t^2} + \frac{\partial^4 w_s}{\partial y^2 \partial t^2}\right) \end{aligned}$$

$$(2.363)$$

Now, the above set of coupled governing equations must be solved to determine the linear natural frequency of the plate's free vibrations. To this end, the dynamic form of the Galerkin's analytical method will be utilized. According to this method, the solution functions matching the components of the displacement field of the plate can be presented in the following form [1, 380]:

$$u(x,y,t) = \sum_{m=1}^{\infty}\sum_{n=1}^{\infty} U_{mn} \frac{\partial X_m(x)}{\partial x} Y_n(y) \exp(i\omega_{mn}t),$$

$$v(x,y,t) = \sum_{m=1}^{\infty}\sum_{n=1}^{\infty} V_{mn} X_m(x) \frac{\partial Y_n(y)}{\partial y} \exp(i\omega_{mn}t),$$

$$w_b(x,y,t) = \sum_{m=1}^{\infty}\sum_{n=1}^{\infty} W_{bmn} X_m(x) Y_n(y) \exp(i\omega_{mn}t),$$

$$w_s(x,y,t) = \sum_{m=1}^{\infty}\sum_{n=1}^{\infty} W_{smn} X_m(x) Y_n(y) \exp(i\omega_{mn}t)$$

(2.364)

where U_{mn}, V_{mn}, W_{bmn}, and W_{smn} are the unknown deformation amplitudes and ω_{mn} is the natural frequency of the system. Also, $X_m(x)$ and $Y_n(y)$ are the eigen functions responsible for satisfying the BCs in the longitudinal and transverse directions, respectively. The eigen functions corresponding with the SSSS and CCCC BCs can be found by referring to Eq. (2.342). Substitution of the above solution in Eqs. (2.360)–(2.363) with regard to the orthogonality of the natural mode shapes of the plate reveals the below eigenvalue equation:

$$[\mathbf{K} - \omega_{mn}^2 \mathbf{M}]\mathbf{d} = 0 \qquad (2.365)$$

where **K** and **M** are the symmetric stiffness and mass matrices, respectively. In addition, the amplitude vector is shown via **d**. In order to find the natural frequency of the resonating plate, the below identity must be valid:

$$|\mathbf{K} - \omega_{mn}^2 \mathbf{M}| = 0 \qquad (2.366)$$

The corresponding arrays of the stiffness and mass matrices can be expressed in the following form [380]:

$$k_{11} = A_{11}r_{3100} + A_{66}r_{1120}, \quad k_{12} = (A_{12} + A_{66})r_{1120},$$
$$k_{13} = -[B_{11}r_{3100} + (B_{12} + 2B_{66})r_{1120}], \quad k_{14} = -[B_{11}^s r_{3100} + (B_{12}^s + 2B_{66}^s)r_{1120}],$$
$$k_{22} = A_{66}r_{2011} + A_{22}r_{0031}, \quad k_{23} = -[(B_{12} + 2B_{66})r_{2011} + B_{22}r_{0031}],$$
$$k_{24} = -[(B_{12}^s + 2B_{66}^s)r_{2011} + B_{22}^s r_{0031}],$$
$$k_{33} = -\begin{bmatrix} D_{11}r_{4000} + 2(D_{12} + 2D_{66})r_{2020} + D_{22}r_{0040} + \\ k_W r_{0000} - (k_P - N_{xx}^T)r_{2000} - (k_P - N_{xx}^T)r_{0020} \end{bmatrix},$$
$$k_{34} = -\begin{bmatrix} D_{11}^s r_{4000} + 2(D_{12}^s + 2D_{66}^s)r_{2020} + D_{22}^s r_{0040} + \\ k_W r_{0000} - (k_P - N_{xx}^T)r_{2000} - (k_P - N_{xx}^T)r_{0020} \end{bmatrix},$$
$$k_{44} = -\begin{bmatrix} H_{11}^s r_{4000} + 2(H_{12}^s + 2H_{66}^s)r_{2020} + H_{22}^s r_{0040} - A_{55}^s r_{2000} - \\ A_{44}^s r_{0020} + k_W r_{0000} - (k_P - N_{xx}^T)r_{2000} - (k_P - N_{xx}^T)r_{0020} \end{bmatrix}$$

(2.367)

and

$$m_{11} = -I_0 r_{1100}, \quad m_{12} = 0, \quad m_{13} = I_1 r_{1100}, \quad m_{14} = J_1 r_{1100},$$
$$m_{22} = -I_0 r_{0011}, \quad m_{23} = I_1 r_{0011}, \quad m_{24} = J_1 r_{0011},$$
$$m_{33} = -I_0 r_{0000} + I_2(r_{2000} + r_{0020}), \quad m_{34} = -I_0 r_{0000} + J_2(r_{2000} + r_{0020}),$$
$$m_{44} = -I_0 r_{0000} + K_2(r_{2000} + r_{0020})$$

(2.368)

where r_{ijkl}'s are similar to what previously introduced in Eq. (2.285).

2.2.3 Kinematic relations of shells

2.2.3.1 Classical shell theory

Herein, we introduce the classical theory of truncated conical shells to describe their motion. The underobservation structure can be imagined by revolving an inclined line of finite length around an axis parallel to the longitudinal direction of the inclined line. Doing so and due to the incline of the line, a shell-type element will be attained whose cross-sections are shaped like circles whose diameter differs from each other. In order to provide reliable results, the conical shell will be considered to a thin-walled structure whose thickness is many times smaller than its length and diameter. Also, it will be assumed that the shell is surrounded by the means of linear and shear springs of stiffnesses k_W and k_P, respectively. Based on the above assumptions, the displacement field of this type of shells can be considered to be in the following form [730]:

$$u_x(x, \theta, z, t) = -z \frac{\partial w(x, \theta, t)}{\partial x},$$

$$u_\theta(x, \theta, z, t) = -\frac{z}{x \sin \alpha} \frac{\partial w(x, \theta, t)}{\partial \theta}, \qquad (2.369)$$

$$u_z(x, \theta, z, t) = w(x, \theta, t)$$

where $w(x, \theta, t)$ denotes the bending deflection of the shell and α is the semivertex cone angle which is the angle between each edge of the shell and its longitudinal axis. Using the definition of the linear strain tensor of a continuous system in the polar coordinate system, the nonzero strains of the conical shell can be derived using the below definitions [730]:

$$\varepsilon_{xx} = \frac{\partial u_x}{\partial x},$$

$$\varepsilon_{\theta\theta} = \frac{1}{x \sin \alpha \left(1 + \frac{z}{x} \tan \alpha\right)} \left(\frac{\partial u_\theta}{\partial \theta} + u_x \sin \alpha + u_z \cos \alpha\right),$$

$$\gamma_{x\theta} = \frac{1}{x \sin \alpha \left(1 + \frac{z}{x} \tan \alpha\right)} \left(\frac{\partial u_x}{\partial \theta} + x \sin \alpha (1 + z/x \tan \alpha) \frac{\partial u_\theta}{\partial x} - u_\theta \sin \alpha\right)$$

$$(2.370)$$

Regarding the fact that we are about to analyze thin-walled shells, that is, corresponding with $1 + \frac{z}{x} \tan \alpha \simeq 1$, and by substituting Eq. (2.369) in the above definition, the nonzero linear strains of a conical shell can be written in the following form:

$$\varepsilon_{xx} = -z \frac{\partial^2 w}{\partial x^2},$$

$$\varepsilon_{\theta\theta} = \frac{1}{x \sin \alpha} \left(-\frac{z}{x \sin \alpha} \frac{\partial^2 w}{\partial \theta^2} - z \sin \alpha \frac{\partial w}{\partial x} + w \cos \alpha\right), \qquad (2.371)$$

$$\gamma_{x\theta} = \frac{1}{x \sin \alpha} \left(-2z \frac{\partial^2 w}{\partial x \partial \theta} + \frac{z}{x} \frac{\partial w}{\partial \theta}\right)$$

Now, it is turn to find an expression for the variation of the strain energy of the conical shell. To this purpose, it is required to introduce the definition of this energy functional at first. This functional can be derived by:

$$\delta U = \int_0^{2\pi} \int_{x_0}^{x_0+L} \int_{-h/2}^{h/2} [\sigma_{xx} \delta \varepsilon_{xx} + \sigma_{\theta\theta} \delta \varepsilon_{\theta\theta} + \sigma_{x\theta} \delta \gamma_{x\theta}] x \sin \alpha \, dz \, dx \, d\theta \qquad (2.372)$$

2.2 Kinematic relations

Substitution of Eq. (2.371) in Eq. (2.372) gives:

$$\delta U = \int_0^{2\pi} \int_{x_0}^{x_0+L} \left[\begin{array}{c} -M_{xx} x \sin\alpha \dfrac{\partial^2 \delta w}{\partial x^2} - \dfrac{M_{\theta\theta}}{x \sin\alpha} \dfrac{\partial^2 \delta w}{\partial \theta^2} - M_{\theta\theta} \sin\alpha \dfrac{\partial \delta w}{\partial x} \\ + N_{\theta\theta} \cos\alpha \, \delta w - 2 M_{x\theta} \dfrac{\partial^2 \delta w}{\partial x \partial \theta} + \dfrac{M_{x\theta}}{x} \dfrac{\partial \delta w}{\partial \theta} \end{array} \right] dx d\theta$$
(2.373)

in which the forces and moments utilized in above relation can be introduced in the following form:

$$N_{\theta\theta} = \int_{-h/2}^{h/2} \sigma_{\theta\theta} dz,$$

$$\begin{Bmatrix} M_{xx} \\ M_{\theta\theta} \\ M_{x\theta} \end{Bmatrix} = \int_{-h/2}^{h/2} z \begin{Bmatrix} \sigma_{xx} \\ \sigma_{\theta\theta} \\ \sigma_{x\theta} \end{Bmatrix} dz$$
(2.374)

The conical shell is assumed to be rested on a two-parameter stiff foundation. So, the work done by the elastic medium on the shell can be expressed in as follows:

$$\delta V = \int_0^{2\pi} \int_{x_0}^{x_0+L} \int_{-h/2}^{h/2} \left[k_W x \sin\alpha + k_P \left(\dfrac{\partial^2 w}{\partial x^2} + \dfrac{1}{x \sin\alpha} \dfrac{\partial^2 w}{\partial \theta^2} \right) \right] \delta w \, dz \, dx \, d\theta$$
(2.375)

Consider the variation of the kinetic energy of the conical shells to be as [730]:

$$\delta T = \int_0^{2\pi} \int_{x_0}^{x_0+L} \int_{-h/2}^{h/2} \rho(z) \left[\dfrac{\partial u_x}{\partial t} \dfrac{\partial \delta u_x}{\partial t} + \dfrac{\partial u_\theta}{\partial t} \dfrac{\partial \delta u_\theta}{\partial t} + \dfrac{\partial u_z}{\partial t} \dfrac{\partial \delta u_z}{\partial t} \right] x \sin\alpha \, dz \, dx \, d\theta$$
(2.376)

Now, the variation of the kinetic energy of the shell can be re-written in the following form if Eq. (2.369) is inserted into Eq. (2.376):

$$\delta T = \int_0^{2\pi} \int_{x_0}^{x_0+L} \left[I_0 x \sin\alpha \dfrac{\partial w}{\partial t} \dfrac{\partial \delta w}{\partial t} + I_2 x \sin\alpha \dfrac{\partial^2 w}{\partial x \partial t} \dfrac{\partial^2 \delta w}{\partial x \partial t} + \dfrac{I_2}{x \sin\alpha} \dfrac{\partial^2 w}{\partial \theta \partial t} \dfrac{\partial^2 \delta w}{\partial \theta \partial t} \right] dx d\theta$$
(2.377)

in which the mass moments of inertia can be defined as:

$$\begin{Bmatrix} I_0 \\ I_2 \end{Bmatrix} = \int_{-h/2}^{h/2} \rho(z) \begin{Bmatrix} 1 \\ z^2 \end{Bmatrix} dz$$
(2.378)

Now, it is enough to substitute the energy functionals of the system in the definition of the Hamilton's principle to gather the motion equation of the conical shell. So, by inserting Eqs. (2.373), (2.375), and (2.377) into

Eq. (2.92), we have:

$$\frac{\partial^2 (xM_{xx})}{\partial x^2}\sin\alpha + \frac{1}{x\sin\alpha}\frac{\partial^2 M_{\theta\theta}}{\partial\theta^2} + \frac{\partial M_{\theta\theta}}{\partial x}\sin\alpha - N_{\theta\theta}\cos\alpha$$
$$+2\frac{\partial^2 M_{x\theta}}{\partial x\partial\theta} - \frac{1}{x}\frac{\partial M_{x\theta}}{\partial\theta} - k_W wx\sin\alpha - k_P\left(\frac{\partial^2 w}{\partial x^2} + \frac{1}{x\sin\alpha}\frac{\partial^2 w}{\partial\theta^2}\right) \quad (2.379)$$
$$+I_2 x\sin\alpha\frac{\partial^4 w}{\partial x^2\partial t^2} + \frac{I_2}{x\sin\alpha}\frac{\partial^4 w}{\partial\theta^2\partial t^2} + I_0 x\sin\alpha\frac{\partial^2 w}{\partial t^2} = 0$$

In order to find the governing equation of the problem, the forces and moments implemented in the above equation must be presented in terms of the bending deflection of the conical shell. To do so, it is needed to introduce the constitutive equation of the conical shell at first. Consider the following form of the Hook's law [730]:

$$\begin{Bmatrix}\sigma_{xx}\\ \sigma_{\theta\theta}\\ \sigma_{x\theta}\end{Bmatrix} = \begin{bmatrix}Q_{11} & Q_{12} & 0\\ Q_{12} & Q_{22} & 0\\ 0 & 0 & Q_{66}\end{bmatrix}\begin{Bmatrix}\varepsilon_{xx}\\ \varepsilon_{\theta\theta}\\ \gamma_{x\theta}\end{Bmatrix} \quad (2.380)$$

where Q_{ij}'s are similar to those we introduced in Eq. (2.268) before. Integrating from the above relation over the thickness of the conical shell and using the definitions introduced in Eq. (2.374), the below expressions can be extracted:

$$N_{\theta\theta} = \frac{A_{22}}{x\tan\alpha}w - B_{12}\frac{\partial^2 w}{\partial x^2} - \frac{B_{22}}{x^2\sin^2\alpha}\frac{\partial^2 w}{\partial\theta^2} - \frac{B_{22}}{x}\frac{\partial w}{\partial x} \quad (2.381)$$

$$M_{xx} = \frac{D_{12}}{x\tan\alpha}w - D_{11}\frac{\partial^2 w}{\partial x^2} - \frac{D_{12}}{x^2\sin^2\alpha}\frac{\partial^2 w}{\partial\theta^2} - \frac{D_{12}}{x}\frac{\partial w}{\partial x} \quad (2.382)$$

$$M_{\theta\theta} = \frac{B_{22}}{x\tan\alpha}w - D_{12}\frac{\partial^2 w}{\partial x^2} - \frac{D_{22}}{x^2\sin^2\alpha}\frac{\partial^2 w}{\partial\theta^2} - \frac{D_{22}}{x}\frac{\partial w}{\partial x} \quad (2.383)$$

$$M_{x\theta} = \frac{D_{66}}{x^2\sin\alpha}\frac{\partial w}{\partial\theta} - \frac{2D_{66}}{x\sin\alpha}\frac{\partial^2 w}{\partial x\partial\theta} \quad (2.384)$$

In the above relations, the through-the-thickness rigidities of the conical shell can be calculated using the following formulas:

$$\begin{Bmatrix}A_{22}\\ B_{22}\\ D_{22}\end{Bmatrix} = \int_{-h/2}^{h/2} Q_{22}\begin{Bmatrix}1\\ z\\ z^2\end{Bmatrix}dz,\quad \begin{Bmatrix}B_{12}\\ D_{12}\end{Bmatrix} = \int_{-h/2}^{h/2} Q_{12}\begin{Bmatrix}z\\ z^2\end{Bmatrix}dz,\quad \begin{Bmatrix}D_{11}\\ D_{66}\end{Bmatrix}$$
$$= \int_{-h/2}^{h/2} z^2\begin{Bmatrix}Q_{11}\\ Q_{66}\end{Bmatrix}dz \quad (2.385)$$

Now, the governing equation can be achieved by substituting for the forces and moments from Eqs. (2.381) to (2.384) in Eq. (2.379). Doing so,

2.2 Kinematic relations

the below governing equation will be enhanced:

$$-\left[\frac{B_{22}\sin\alpha}{x^2\tan\alpha}+\frac{A_{22}\cos\alpha}{x\tan\alpha}\right]w+\left[\frac{D_{22}\sin\alpha}{x^2}+\frac{2B_{22}\sin\alpha}{x\tan\alpha}\right]\frac{\partial w}{\partial x}$$
$$+\left[2B_{12}\cos\alpha-\frac{D_{22}\sin\alpha}{x}+k_p\right]\frac{\partial^2 w}{\partial x^2}-\left[2D_{11}\sin\alpha+\frac{D_{12}}{x\sin\alpha}+D_{12}\sin\alpha\right]\frac{\partial^3 w}{\partial x^3}$$
$$-D_{11}x\sin\alpha\frac{\partial^4 w}{\partial x^4}+\left[\frac{2(D_{12}-D_{22}+2D_{66})}{x^2\sin\alpha}\right]\frac{\partial^3 w}{\partial x\partial\theta^2}-\left[\frac{2(D_{12}+2D_{66})}{x\sin\alpha}\right]\frac{\partial^4 w}{\partial x^2\partial\theta^2} \quad (2.386)$$
$$+\left[\frac{2B_{22}}{x^2\sin\alpha\tan\alpha}+\frac{2(D_{22}-D_{12}-2D_{66})}{x^3\sin\alpha}+\frac{k_p}{x\sin\alpha}\right]\frac{\partial^2 w}{\partial\theta^2}-\left[\frac{D_{22}}{x^3\sin^3\alpha}\right]\frac{\partial^4 w}{\partial\theta^4}$$
$$+k_W wx\sin\alpha+I_0 x\sin\alpha\frac{\partial^2 w}{\partial t^2}+\frac{I_2}{x\sin\alpha}\frac{\partial^4 w}{\partial\theta^2\partial t^2}+I_2 x\sin\alpha\frac{\partial^4 w}{\partial x^2\partial t^2}=0$$

Now, the above relation must be solved to derive the natural frequency of the free vibrations of the system. To this goal, the problem will be solved in temporal and spatial domains separately. First of all, assume the solution of the problem to be in the below general form:

$$w(x,\theta,t)=W(x)\sin(m\theta)\exp(i\omega t) \quad (2.387)$$

where m and ω stand for circumferential wave number and natural frequency, respectively. Also, the influence of the variations over the longitudinal direction will be captured in the geometry-dependent function $W(x)$. If the above solution function is inserted into Eq. (2.386), the following expression can be achieved:

$$-\left[\frac{B_{22}\sin\alpha}{x^2\tan\alpha}+\frac{A_{22}\cos\alpha}{x\tan\alpha}\right]W(x)+\left[\frac{D_{22}\sin\alpha}{x^2}+\frac{2B_{22}\sin\alpha}{x\tan\alpha}\right]\frac{\partial W(x)}{\partial x}+$$
$$\left[2B_{12}\cos\alpha-\frac{D_{22}\sin\alpha}{x}+k_p\right]\frac{\partial^2 W(x)}{\partial x^2}-\left[2D_{11}\sin\alpha+\frac{D_{12}}{x\sin\alpha}+D_{12}\sin\alpha\right]\frac{\partial^3 W(x)}{\partial x^3}-$$
$$D_{11}x\sin\alpha\frac{\partial^4 W(x)}{\partial x^4}-\frac{2m^2(D_{12}-D_{22}+2D_{66})}{x^2\sin\alpha}\frac{\partial W(x)}{\partial x}+\frac{2m^2(D_{12}+2D_{66})}{x\sin\alpha}\frac{\partial^2 W(x)}{\partial x^2}$$
$$-\left[\frac{2m^2 B_{22}}{x^2\sin\alpha\tan\alpha}+\frac{2m^2(D_{22}-D_{12}-2D_{66})}{x^3\sin\alpha}+\frac{m^2 k_p}{x\sin\alpha}\right]W(x)-\frac{m^4 D_{22}}{x^3\sin^3\alpha}W(x)+$$
$$k_W x\sin\alpha-I_0 x\omega^2\sin\alpha W(x)+\frac{I_2 m^2\omega^2}{x\sin\alpha}W(x)-I_2 x\omega^2\sin\alpha\frac{\partial^2 W(x)}{\partial x^2}=0$$
$$(2.388)$$

The above relation is a 1D equation which can be solved with the aid of numerical methods. In this section, the well-known generalized differential quadrature method is selected to solve the above problem in the x domain. However, the solution procedure cannot be completed unless a brief review on the implemented solution is presented. In generalized differential quadrature method, a finite number of grid points will be chosen to discretize the problem and solve it numerically. One of the most popular generalized differential quadrature-assisted algorithms is implementation of the roots of the Chebyshev polynomials as the grid points. Following this assumption, the grid points can be obtained via:

$$x_i = \frac{L}{2}\left[1 - \cos\left(\frac{i-1}{N-1}\pi\right)\right], \quad i = 1, 2, \cdots, N-1 \tag{2.389}$$

By the means of the sample points gathered from the above definition, the derivatives of any arbitrary function $f(x)$ with respect to x can be calculated via:

$$F^{(n)}(x_i) = \sum_{j=1}^{N} C_{ij}^{(n)} f(x_j), \quad n = 1, 2, \ldots, N-1 \tag{2.390}$$

where the weighting coefficients corresponding with the first-order derivative of the original function can be attained by:

$$C_{ij}^{(1)} = \frac{M(x_i)}{(x_i - x_j)M(x_j)}, \quad i, j = 1, 2, \ldots, N \text{ and } j \neq i \tag{2.391}$$

in which

$$M(x_i) = \prod_{\substack{j=1 \\ j \neq i}}^{N} (x_i - x_j) \tag{2.392}$$

In addition, the higher-order weighting coefficients can be achieved using the below formulas:

$$C_{ij}^{(n)} = n\left(C_{ii}^{(n-1)}C_{ij}^{(1)} - \frac{C_{ij}^{(n-1)}}{x_i - x_j}\right), \quad i, j = 1, 2, \ldots, N \text{ and } j \neq i \tag{2.393}$$

$$C_{ii}^{(n)} = -\sum_{\substack{j=1 \\ j \neq i}}^{N} C_{ij}^{(n)}, \quad \begin{cases} i = 1, 2, \ldots, N \\ n = 1, 2, \ldots, N-1 \end{cases} \tag{2.394}$$

Once the above instructions are hired in order to compute the derivatives of the bending deflection of the conical shell with respect to x and inserting them into Eq. (2.388), the following relation will be obtained:

$$\begin{bmatrix} A_{bb} & A_{bd} \\ A_{db} & A_{dd} \end{bmatrix}\begin{Bmatrix} x_b \\ x_d \end{Bmatrix} = \omega^2\begin{bmatrix} 0 & 0 \\ B_{db} & B_{dd} \end{bmatrix}\begin{Bmatrix} x_b \\ x_d \end{Bmatrix} \tag{2.395}$$

where A and B are from type of stiffness and mass, respectively. The subscripts "b" and "d" indicate on the boundary and domain points, respectively. In order to find the final relation, it is needed to omit the boundary points by inserting the BCs corresponding with the ends of the truncated conical shell into problem. Once the above manipulation is completely done, the below eigenvalue problem can be gathered:

$$[K - \omega^2 M]x_d = 0 \qquad (2.396)$$

where K and M denote stiffness and mass matrices, respectively. Also, x_d reveals the amplitude vector which is constructed from the domain grid points. Solving the above eigenvalue equation for ω, the natural frequency of the conical shell can be achieved. It is worth noting that the BCs must be applied on the points $x = x_0$ and $x = x_0 + L$. For simply supported edges, the below mathematical constraint must be applied:

$$w = 0 \quad \text{and} \quad \frac{\partial^2 w}{\partial x^2} = 0 \qquad (2.397)$$

Also, the below constraints can satisfy the clamped condition:

$$w = 0 \quad \text{and} \quad \frac{\partial w}{\partial x} = 0 \qquad (2.398)$$

2.2.3.2 First-order shear deformation shell theory

Present section is dedicated to the introduction of the FSDT of the shells. In general, FSDTs are better than classical ones because of the fact that they cover the shear deformations up to the first order. Although this approximation suffers from completeness, it provides better answers and enables the user to investigate thicker structures in comparison with the case of using the classical theory. In this theory, in addition to the longitudinal, circumferential, and flexural deformations of the shell-type elements, two rotation parameters will be implemented, too. According to this theory, the displacement field of the structure can be presented in the below form [1]:

$$\begin{aligned} u_x(x, \theta, z, t) &= u(x, \theta, t) + z\varphi_x(x, \theta, t), \\ u_\theta(x, \theta, z, t) &= v(x, \theta, t) + z\varphi_\theta(x, \theta, t), \\ u_z(x, \theta, z, t) &= w(x, \theta, t) \end{aligned} \qquad (2.399)$$

in which $u(x, \theta, t)$, $v(x, \theta, t)$, and $w(x, \theta, t)$ stand for the axial, circumferential, and lateral deformations, respectively. Also, the rotations around axial and circumferential directions are shown via $\phi_x(x,\theta, t)$ and $\phi_\theta(x,\theta, t)$, respectively. It is clear that the above relations are expressed in the dynamic state and for the case of analyzing static problems, the time variable must be omitted. Because of the fact that in the following subsections, the main focus will be on the linear behaviors of this type of shells, the components of the strain tensor will be derived for the linear case herein. The nonlinear strain–displacement relationships will be provided in the particular section undergoing with them. Considering the abovementioned displacement field in the polar coordinate system, the below expressions for the nonzero components of the shell's strain tensor can be introduced [1]:

$$\varepsilon_{xx} = \frac{\partial u}{\partial x} + z\frac{\partial \varphi_x}{\partial x},$$

$$\varepsilon_{\theta\theta} = \frac{1}{R}\left(\frac{\partial v}{\partial \theta} + z\frac{\partial \varphi_\theta}{\partial \theta} + w\right),$$

$$\gamma_{x\theta} = \frac{1}{R}\frac{\partial u}{\partial \theta} + \frac{\partial v}{\partial x} + \frac{z}{R}\frac{\partial \varphi_x}{\partial \theta} + z\frac{\partial \varphi_\theta}{\partial x},$$

$$\gamma_{xz} = \varphi_x + \frac{\partial w}{\partial x}, \quad \gamma_{\theta z} = \varphi_\theta + \frac{1}{R}\frac{\partial w}{\partial \theta} - \frac{v}{R}$$

(2.400)

Based on the assumption of analyzing a linearly elastic solid shell, the below expression for the variation of the strain energy of the shell can be written:

$$\delta U = \int_0^L \int_0^{2\pi} \begin{bmatrix} N_{xx}\dfrac{\partial \delta u}{\partial x} + M_{xx}\dfrac{\partial \delta \varphi_x}{\partial x} + \dfrac{N_{\theta\theta}}{R}\left(\dfrac{\partial \delta v}{\partial \theta} + \delta w\right) + \dfrac{M_{\theta\theta}}{R}\dfrac{\partial \delta \varphi_\theta}{\partial \theta} \\ + N_{x\theta}\left(\dfrac{1}{R}\dfrac{\partial \delta u}{\partial \theta} + \dfrac{\partial \delta v}{\partial x}\right) + M_{x\theta}\left(\dfrac{1}{R}\dfrac{\partial \delta \varphi_x}{\partial \theta} + \dfrac{\partial \delta \varphi_\theta}{\partial x}\right) + \\ Q_{xz}\left(\delta \varphi_x + \dfrac{\partial \delta w}{\partial x}\right) + Q_{\theta z}\left(\delta \varphi_\theta + \dfrac{1}{R}\dfrac{\partial \delta w}{\partial \theta} - \dfrac{\delta v}{R}\right) \end{bmatrix} R d\theta dx$$

(2.401)

In the above relation, the forces and moments can be expressed in the following form:

$$\begin{Bmatrix} N_{xx} & M_{xx} \\ N_{\theta\theta} & M_{\theta\theta} \\ N_{x\theta} & M_{x\theta} \end{Bmatrix} = \int_{-h/2}^{h/2} \begin{Bmatrix} \sigma_{xx} & z\sigma_{xx} \\ \sigma_{\theta\theta} & z\sigma_{\theta\theta} \\ \sigma_{x\theta} & z\sigma_{x\theta} \end{Bmatrix} dz,$$

$$\begin{Bmatrix} Q_{xz} \\ Q_{z\theta} \end{Bmatrix} = \kappa_s \int_{-h/2}^{h/2} \begin{Bmatrix} \sigma_{xz} \\ \sigma_{z\theta} \end{Bmatrix} dz$$

(2.402)

in which κ_s is the shear correction factor and can be considered to be $\kappa_s = 5/6$.

2.2.3.2.1 Buckling problem

Present part will be allocated to the investigation of the buckling characteristics of cylinders surrounded by a two-parameter stiff foundation. The elastic seat of the shell is considered to be consisted of two linear and shear springs, called Winkler and Pasternak springs, respectively. The buckling load will be applied on the shell in the axial direction. Using the above assumptions, the work done on the shell-type element by the axial compression (P) and stiff springs (k_W and k_P) can be formulated as below

[1, 586]:

$$\delta V = \int_0^L \int_0^{2\pi} \int_{-h/2}^{h/2} \left[k_W w - k_P \left(\frac{\partial^2 w}{\partial x^2} + \frac{1}{R^2} \frac{\partial^2 w}{\partial \theta^2} \right) + P \frac{\partial^2 w}{\partial x^2} \right] R \, dz \, d\theta \, dx \quad (2.403)$$

Now, all of the material required to derive the motion equations describing the buckling problem of the cylindrical shells are in hand. Thus, these relations can be attained by simple substitution of Eqs. (2.401) and (2.403) in Eq. (2.91). Doing so, we have:

$$\frac{\partial N_{xx}}{\partial x} + \frac{1}{R} \frac{\partial N_{x\theta}}{\partial \theta} = 0 \quad (2.404)$$

$$\frac{\partial N_{x\theta}}{\partial x} + \frac{1}{R} \frac{\partial N_{\theta\theta}}{\partial \theta} - \frac{1}{R} Q_{z\theta} = 0 \quad (2.405)$$

$$\frac{\partial Q_{xz}}{\partial x} + \frac{1}{R} \frac{\partial Q_{z\theta}}{\partial \theta} + \frac{N_{\theta\theta}}{R} - k_W w + k_P \left(\frac{\partial^2 w}{\partial x^2} + \frac{1}{R^2} \frac{\partial^2 w}{\partial \theta^2} \right) - P \frac{\partial^2 w}{\partial x^2} = 0 \quad (2.406)$$

$$\frac{\partial M_{xx}}{\partial x} + \frac{1}{R} \frac{\partial M_{x\theta}}{\partial \theta} + Q_{xz} = 0 \quad (2.407)$$

$$\frac{\partial M_{x\theta}}{\partial x} + \frac{1}{R} \frac{\partial M_{\theta\theta}}{\partial \theta} + Q_{z\theta} = 0 \quad (2.408)$$

Now, the above equations must be transferred to a set of equations which are expressed in terms of the displacement field variables. In order to do this job, the stress-resultants must be presented in terms of the components of the shell's displacement field. Consider the following constitutive equation [1, 586]:

$$\begin{Bmatrix} \sigma_{xx} \\ \sigma_{\theta\theta} \\ \sigma_{x\theta} \\ \sigma_{xz} \\ \sigma_{z\theta} \end{Bmatrix} = \begin{bmatrix} Q_{11} & Q_{12} & 0 & 0 & 0 \\ Q_{21} & Q_{22} & 0 & 0 & 0 \\ 0 & 0 & Q_{66} & 0 & 0 \\ 0 & 0 & 0 & Q_{55} & 0 \\ 0 & 0 & 0 & 0 & Q_{44} \end{bmatrix} \begin{Bmatrix} \varepsilon_{xx} \\ \varepsilon_{\theta\theta} \\ \gamma_{x\theta} \\ \gamma_{xz} \\ \gamma_{z\theta} \end{Bmatrix} \quad (2.409)$$

In the above equation, the components of the elasticity tensor are as same as those previously introduced in Eq. (2.314). Integrating from the above equation over the shell's thickness by considering the definitions presented in Eq. (2.402) reveals:

$$N_{xx} = A_{11} \frac{\partial u}{\partial x} + B_{11} \frac{\partial \varphi_x}{\partial x} + \frac{A_{12}}{R} \left(\frac{\partial v}{\partial \theta} + w \right) + \frac{B_{12}}{R} \frac{\partial \varphi_\theta}{\partial \theta} \quad (2.410)$$

$$N_{\theta\theta} = A_{12} \frac{\partial u}{\partial x} + B_{12} \frac{\partial \varphi_x}{\partial x} + \frac{A_{11}}{R} \left(\frac{\partial v}{\partial \theta} + w \right) + \frac{B_{11}}{R} \frac{\partial \varphi_\theta}{\partial \theta} \quad (2.411)$$

$$N_{x\theta} = A_{66} \left(\frac{1}{R} \frac{\partial u}{\partial \theta} + \frac{\partial v}{\partial x} \right) + B_{66} \left(\frac{1}{R} \frac{\partial \varphi_x}{\partial \theta} + \frac{\partial \varphi_\theta}{\partial x} \right) \quad (2.412)$$

2. Micromechanical homogenization and kinematic relations

$$M_{xx} = B_{11}\frac{\partial u}{\partial x} + D_{11}\frac{\partial \varphi_x}{\partial x} + \frac{B_{12}}{R}\left(\frac{\partial v}{\partial \theta} + w\right) + \frac{D_{12}}{R}\frac{\partial \varphi_\theta}{\partial \theta} \quad (2.413)$$

$$M_{\theta\theta} = B_{12}\frac{\partial u}{\partial x} + D_{12}\frac{\partial \varphi_x}{\partial x} + \frac{B_{11}}{R}\left(\frac{\partial v}{\partial \theta} + w\right) + \frac{D_{11}}{R}\frac{\partial \varphi_\theta}{\partial \theta} \quad (2.414)$$

$$M_{x\theta} = B_{66}\left(\frac{1}{R}\frac{\partial u}{\partial \theta} + \frac{\partial v}{\partial x}\right) + D_{66}\left(\frac{1}{R}\frac{\partial \varphi_x}{\partial \theta} + \frac{\partial \varphi_\theta}{\partial x}\right) \quad (2.415)$$

$$Q_{xz} = A_{55}^s\left(\varphi_x + \frac{\partial w}{\partial x}\right) \quad (2.416)$$

$$Q_{z\theta} = A_{55}^s\left(\varphi_\theta + \frac{1}{R}\frac{\partial w}{\partial \theta} - \frac{v}{R}\right) \quad (2.417)$$

In the above relations, the through-the-thickness rigidities of the shell can be formulated in the following form:

$$\left\{\begin{matrix} A_{11} & B_{11} & D_{11} \\ A_{12} & B_{12} & D_{12} \\ A_{66} & B_{66} & D_{66} \end{matrix}\right\} = \int_{-h/2}^{h/2}\left\{\begin{matrix} Q_{11} & zQ_{11} & z^2Q_{11} \\ Q_{12} & zQ_{12} & z^2Q_{12} \\ Q_{66} & zQ_{66} & z^2Q_{66} \end{matrix}\right\}dz, \quad A_{55}^s = \kappa_s\int_{-h/2}^{h/2}Q_{55}dz \quad (2.418)$$

By substituting for the stress-resultants from Eqs. (2.410) to (2.417) into Eqs. (2.404)–(2.408) gives the governing equations of the problem in the following form:

$$A_{11}\frac{\partial^2 u}{\partial x^2} + \frac{A_{66}}{R^2}\frac{\partial^2 u}{\partial \theta^2} + \frac{A_{12}+A_{66}}{R}\frac{\partial^2 v}{\partial x\partial \theta} + \frac{A_{12}}{R}\frac{\partial w}{\partial x} +$$
$$B_{11}\frac{\partial^2 \varphi_x}{\partial x^2} + \frac{B_{66}}{R^2}\frac{\partial^2 \varphi_x}{\partial \theta^2} + \frac{B_{12}+B_{66}}{R}\frac{\partial^2 \varphi_\theta}{\partial x\partial \theta} = 0 \quad (2.419)$$

$$\frac{A_{12}+A_{66}}{R}\frac{\partial^2 u}{\partial x\partial \theta} + A_{66}\frac{\partial^2 v}{\partial x^2} + \frac{A_{11}}{R^2}\frac{\partial^2 v}{\partial \theta^2} + A_{55}^s\frac{v}{R^2} + \frac{A_{11}-A_{55}^s}{R^2}\frac{\partial w}{\partial \theta}$$
$$+\frac{B_{12}+B_{66}}{R}\frac{\partial^2 \varphi_x}{\partial x\partial \theta} + B_{66}\frac{\partial^2 \varphi_\theta}{\partial x^2} + \frac{B_{11}}{R^2}\frac{\partial^2 \varphi_\theta}{\partial \theta^2} - \frac{A_{55}^s}{R}\varphi_\theta = 0 \quad (2.420)$$

$$\frac{A_{12}}{R}\frac{\partial u}{\partial x} + \frac{A_{11}-A_{55}^s}{R^2}\frac{\partial v}{\partial \theta} + \frac{A_{11}}{R^2}w + A_{55}^s\left(\frac{\partial^2 w}{\partial x^2} + \frac{1}{R^2}\frac{\partial^2 w}{\partial \theta^2}\right) - k_W w +$$
$$k_P\left(\frac{\partial^2 w}{\partial x^2} + \frac{1}{R^2}\frac{\partial^2 w}{\partial \theta^2}\right) - P\frac{\partial^2 w}{\partial x^2} + \left(A_{55}^s + \frac{B_{12}}{R}\right)\frac{\partial \varphi_x}{\partial x} + \left(\frac{B_{11}}{R^2} + \frac{A_{55}^s}{R}\right)\frac{\partial \varphi_\theta}{\partial \theta} = 0 \quad (2.421)$$

$$B_{11}\frac{\partial^2 u}{\partial x^2} + \frac{B_{66}}{R^2}\frac{\partial^2 u}{\partial \theta^2} + \frac{B_{12}+B_{66}}{R}\frac{\partial^2 v}{\partial x\partial \theta} + \left(\frac{B_{12}}{R} + A_{55}^s\right)\frac{\partial w}{\partial x}$$
$$+D_{11}\frac{\partial^2 \varphi_x}{\partial x^2} + \frac{D_{66}}{R^2}\frac{\partial^2 \varphi_x}{\partial \theta^2} + A_{55}^s\varphi_x + \frac{D_{12}+D_{66}}{R}\frac{\partial^2 \varphi_\theta}{\partial x\partial \theta} = 0 \quad (2.422)$$

$$\frac{B_{12} + B_{66}}{R} \frac{\partial^2 u}{\partial x \partial \theta} + B_{66} \frac{\partial^2 v}{\partial x^2} + \frac{B_{11}}{R^2} \frac{\partial^2 v}{\partial \theta^2} - \frac{A_{55}^s}{R} v + \left(\frac{B_{11}}{R^2} + \frac{A_{55}^s}{R}\right) \frac{\partial w}{\partial \theta}$$
$$+ \frac{D_{12} + D_{66}}{R} \frac{\partial^2 \varphi_x}{\partial x \partial \theta} + D_{66} \frac{\partial^2 \varphi_\theta}{\partial x^2} + \frac{D_{11}}{R^2} \frac{\partial^2 \varphi_\theta}{\partial \theta^2} + A_{55}^s \varphi_\theta = 0$$

(2.423)

Now, the above set of the governing equations must be solved in order to find the buckling load of the shell. To this goal, the solution function corresponding with the components of the displacement field of the shell can be considered to be in the following form [1]:

$$u(x, \theta) = \sum_{m=1}^{\infty} \sum_{n=1}^{\infty} U_{mn} \cos(\alpha x) \cos(n\theta),$$

$$v(x, \theta) = \sum_{m=1}^{\infty} \sum_{n=1}^{\infty} V_{mn} \sin(\alpha x) \sin(n\theta),$$

$$w(x, \theta) = \sum_{m=1}^{\infty} \sum_{n=1}^{\infty} W_{mn} \sin(\alpha x) \cos(n\theta), \qquad (2.424)$$

$$\varphi_x(x, \theta) = \sum_{m=1}^{\infty} \sum_{n=1}^{\infty} \phi_{xmn} \cos(\alpha x) \cos(n\theta),$$

$$\varphi_\theta(x, \theta) = \sum_{m=1}^{\infty} \sum_{n=1}^{\infty} \phi_{\theta mn} \sin(\alpha x) \sin(n\theta)$$

where U_{mn}, V_{mn}, W_{mn}, φ_{xmn}, and $\varphi_{\theta mn}$ are the unknown amplitudes of the static motion of the shell. In the above solution, the BC at both ends of the shell is considered to be from simply supported type. Also, it must be considered that $\alpha = m\pi/L$. By substituting Eq. (2.424) in Eqs. (2.419)–(2.423) the following expression can be achieved:

$$[\mathbf{K} - \mathbf{K_b}]\mathbf{d} = 0 \qquad (2.425)$$

in which $\mathbf{K}$ and $\mathbf{K_b}$ are the stiffness and buckling load symmetric matrices, respectively. Also, the column vector $\mathbf{d}$ is the vector of unknown amplitudes. In order to find the buckling load of the shell, the below identity must be satisfied:

$$|\mathbf{K} - \mathbf{K_b}| = 0 \qquad (2.426)$$

Solving the above eigenvalue equation for the buckling load P, the response to the problem is achieved. The nonzero arrays of the stiffness and buckling load matrices can be expressed in the below form [1]:

$$k_{11} = -\left(A_{11}\alpha^2 + \frac{A_{66}}{R^2}n^2\right), \quad k_{12} = \frac{A_{12} + A_{66}}{R}\alpha n, \quad k_{13} = \frac{A_{12}}{R}\alpha,$$

$$k_{14} = -\left(B_{11}\alpha^2 + \frac{B_{66}}{R^2}n^2\right), \quad k_{15} = \frac{B_{12} + B_{66}}{R}\alpha n,$$

$$k_{22} = -\left(A_{66}\alpha^2 + \frac{A_{11}}{R^2}n^2 - \frac{A_{55}^s}{R^2}\right), \quad k_{23} = -\frac{A_{11} - A_{55}^s}{R^2}n,$$

$$k_{24} = \frac{B_{12} + B_{66}}{R}\alpha n, \quad k_{25} = -\left(B_{66}\alpha^2 + \frac{B_{11}}{R^2}n^2 - \frac{A_{55}^s}{R}\right),$$

$$k_{33} = \frac{A_{11}}{R^2} - A_{55}^s\left(\alpha^2 + \frac{n^2}{R^2}\right) - k_W - k_P\left(\alpha^2 + \frac{n^2}{R^2}\right),$$

$$k_{34} = -\left(A_{55}^s + \frac{B_{12}}{R}\right)\alpha, \quad k_{35} = \left(\frac{B_{11}}{R^2} + \frac{A_{55}^s}{R}\right)n,$$

$$k_{44} = -\left(D_{11}\alpha^2 + \frac{D_{66}}{R^2}n^2 - A_{55}^s\right), \quad k_{45} = \frac{D_{12} + D_{66}}{R}\alpha n$$

$$k_{55} = -\left(D_{66}\alpha^2 + \frac{D_{11}}{R^2}n^2 - A_{55}^s\right)$$

(2.427)

and

$$k_{b,33} = P\alpha^2 \tag{2.428}$$

2.2.3.2.2 Wave propagation problem

In this subsection, the problem of dispersion of elastic waves in cylindrical shells will be formulated by the means of the FSDT. In order to formulate the aforesaid problem, the energy functionals of the shell are required among them the variations of the strain energy and work done by the elastic substrate are already known to us. Therefore, it is enough to derive an appropriate expression for the variation of the shell's kinetic energy to complete the derivation procedure. The primary definition for the variation of the kinetic energy in the $x - \theta - z$ coordinate system is [392]:

$$\delta T = \int_0^L \int_0^{2\pi} \int_{-h/2}^{h/2} \rho(z)\left[\frac{\partial u_x}{\partial t}\frac{\partial \delta u_x}{\partial t} + \frac{\partial u_\theta}{\partial t}\frac{\partial \delta u_\theta}{\partial t} + \frac{\partial u_z}{\partial t}\frac{\partial \delta u_z}{\partial t}\right]Rdzd\theta dx$$

(2.429)

By substituting for the components of the displacement field of the shell from Eq. (2.399) in the above definition, the variation of the kinetic energy of the shell-type elements can be re-written in the following form:

$$\delta T = \int_0^L \int_0^{2\pi} \left[\begin{array}{c} I_0 \left(\dfrac{\partial u}{\partial t} \dfrac{\partial \delta u}{\partial t} + \dfrac{\partial v}{\partial t} \dfrac{\partial \delta v}{\partial t} + \dfrac{\partial w}{\partial t} \dfrac{\partial \delta w}{\partial t} \right) + \\ I_1 \left(\dfrac{\partial \varphi_x}{\partial t} \dfrac{\partial \delta u}{\partial t} + \dfrac{\partial u}{\partial t} \dfrac{\partial \delta \varphi_x}{\partial t} + \dfrac{\partial \varphi_\theta}{\partial t} \dfrac{\partial \delta v}{\partial t} + \dfrac{\partial v}{\partial t} \dfrac{\partial \delta \varphi_\theta}{\partial t} \right) \\ + I_2 \left(\dfrac{\partial \varphi_x}{\partial t} \dfrac{\partial \delta \varphi_x}{\partial t} + \dfrac{\partial \varphi_\theta}{\partial t} \dfrac{\partial \delta \varphi_\theta}{\partial t} \right) \end{array} \right] R d\theta dx \quad (2.430)$$

In the above relation, the mass moments of inertia can be defined as below:

$$\left\{ \begin{array}{c} I_0 \\ I_1 \\ I_2 \end{array} \right\} = \int_{-h/2}^{h/2} \rho(z) \left\{ \begin{array}{c} 1 \\ z \\ z^2 \end{array} \right\} dz \quad (2.431)$$

It is worth regarding that in the present case, there exists no axial compression acting on the shell. Therefore, the variation of the work done by external loading presented in Eq. (2.403) must be reduced to the below expression [586]:

$$\delta V = \int_0^L \int_0^{2\pi} \int_{-h/2}^{h/2} \left[k_W w - k_P \left(\dfrac{\partial^2 w}{\partial x^2} + \dfrac{1}{R^2} \dfrac{\partial^2 w}{\partial \theta^2} \right) \right] R dz d\theta dx \quad (2.432)$$

Now, the Euler–Lagrange equations can be gathered by inserting Eqs. (2.401), (2.430), and (2.432) into Eq. (2.92). After doing the above substitution, the motion equations of the shell in the dynamic form can be expressed as below:

$$\dfrac{\partial N_{xx}}{\partial x} + \dfrac{1}{R} \dfrac{\partial N_{x\theta}}{\partial \theta} = I_0 \dfrac{\partial^2 u}{\partial t^2} + I_1 \dfrac{\partial^2 \varphi_x}{\partial t^2} \quad (2.433)$$

$$\dfrac{\partial N_{x\theta}}{\partial x} + \dfrac{1}{R} \dfrac{\partial N_{\theta\theta}}{\partial \theta} - \dfrac{1}{R} Q_{z\theta} = I_0 \dfrac{\partial^2 v}{\partial t^2} + I_1 \dfrac{\partial^2 \varphi_\theta}{\partial t^2} \quad (2.434)$$

$$\dfrac{\partial Q_{xz}}{\partial x} + \dfrac{1}{R} \dfrac{\partial Q_{z\theta}}{\partial \theta} + \dfrac{N_{\theta\theta}}{R} - k_W w + k_P \left(\dfrac{\partial^2 w}{\partial x^2} + \dfrac{1}{R^2} \dfrac{\partial^2 w}{\partial \theta^2} \right) = I_0 \dfrac{\partial^2 w}{\partial t^2} \quad (2.435)$$

$$\dfrac{\partial M_{xx}}{\partial x} + \dfrac{1}{R} \dfrac{\partial M_{x\theta}}{\partial \theta} + Q_{xz} = I_1 \dfrac{\partial^2 u}{\partial t^2} + I_2 \dfrac{\partial^2 \varphi_x}{\partial t^2} \quad (2.436)$$

$$\dfrac{\partial M_{x\theta}}{\partial x} + \dfrac{1}{R} \dfrac{\partial M_{\theta\theta}}{\partial \theta} + Q_{z\theta} = I_1 \dfrac{\partial^2 v}{\partial t^2} + I_2 \dfrac{\partial^2 \varphi_\theta}{\partial t^2} \quad (2.437)$$

In order to find the final governing equations of the problem, the stress-resultants available in Eqs. (2.433)–(2.437) must be replaced with their equivalent expressions in terms of the components of the shell's displacement field. Therefore, Eqs. (2.410)–(2.417) must be substituted in

Eqs. (2.433)–(2.437). Once the above substitution is done, the governing equations of the shell can be expressed in the following form:

$$A_{11}\frac{\partial^2 u}{\partial x^2} + \frac{A_{66}}{R^2}\frac{\partial^2 u}{\partial \theta^2} + \frac{A_{12}+A_{66}}{R}\frac{\partial^2 v}{\partial x \partial \theta} + \frac{A_{12}}{R}\frac{\partial w}{\partial x} + B_{11}\frac{\partial^2 \varphi_x}{\partial x^2}$$
$$+ \frac{B_{66}}{R^2}\frac{\partial^2 \varphi_x}{\partial \theta^2} + \frac{B_{12}+B_{66}}{R}\frac{\partial^2 \varphi_\theta}{\partial x \partial \theta} = I_0\frac{\partial^2 u}{\partial t^2} + I_1\frac{\partial^2 \varphi_x}{\partial t^2} \quad (2.438)$$

$$\frac{A_{12}+A_{66}}{R}\frac{\partial^2 u}{\partial x \partial \theta} + A_{66}\frac{\partial^2 v}{\partial x^2} + \frac{A_{11}}{R^2}\frac{\partial^2 v}{\partial \theta^2} + A_{55}^s\frac{v}{R^2} + \frac{A_{11}-A_{55}^s}{R^2}\frac{\partial w}{\partial \theta} +$$
$$\frac{B_{12}+B_{66}}{R}\frac{\partial^2 \varphi_x}{\partial x \partial \theta} + B_{66}\frac{\partial^2 \varphi_\theta}{\partial x^2} + \frac{B_{11}}{R^2}\frac{\partial^2 \varphi_\theta}{\partial \theta^2} - \frac{A_{55}^s}{R}\varphi_\theta = I_0\frac{\partial^2 v}{\partial t^2} + I_1\frac{\partial^2 \varphi_\theta}{\partial t^2}$$
$$(2.439)$$

$$\frac{A_{12}}{R}\frac{\partial u}{\partial x} + \frac{A_{11}-A_{55}^s}{R^2}\frac{\partial v}{\partial \theta} + \frac{A_{11}}{R^2}w + A_{55}^s\left(\frac{\partial^2 w}{\partial x^2} + \frac{1}{R^2}\frac{\partial^2 w}{\partial \theta^2}\right) - k_W w +$$
$$k_P\left(\frac{\partial^2 w}{\partial x^2} + \frac{1}{R^2}\frac{\partial^2 w}{\partial \theta^2}\right) + \left(A_{55}^s + \frac{B_{12}}{R}\right)\frac{\partial \varphi_x}{\partial x} + \left(\frac{B_{11}}{R^2} + \frac{A_{55}^s}{R}\right)\frac{\partial \varphi_\theta}{\partial \theta} = I_0\frac{\partial^2 w}{\partial t^2}$$
$$(2.440)$$

$$B_{11}\frac{\partial^2 u}{\partial x^2} + \frac{B_{66}}{R^2}\frac{\partial^2 u}{\partial \theta^2} + \frac{B_{12}+B_{66}}{R}\frac{\partial^2 v}{\partial x \partial \theta} + \left(\frac{B_{12}}{R} + A_{55}^s\right)\frac{\partial w}{\partial x} + D_{11}\frac{\partial^2 \varphi_x}{\partial x^2}$$
$$+ \frac{D_{66}}{R^2}\frac{\partial^2 \varphi_x}{\partial \theta^2} + A_{55}^s\varphi_x + \frac{D_{12}+D_{66}}{R}\frac{\partial^2 \varphi_\theta}{\partial x \partial \theta} = I_1\frac{\partial^2 u}{\partial t^2} + I_2\frac{\partial^2 \varphi_x}{\partial t^2}$$
$$(2.441)$$

$$\frac{B_{12}+B_{66}}{R}\frac{\partial^2 u}{\partial x \partial \theta} + B_{66}\frac{\partial^2 v}{\partial x^2} + \frac{B_{11}}{R^2}\frac{\partial^2 v}{\partial \theta^2} - \frac{A_{55}^s}{R}v + \left(\frac{B_{11}}{R^2} + \frac{A_{55}^s}{R}\right)\frac{\partial w}{\partial \theta} +$$
$$\frac{D_{12}+D_{66}}{R}\frac{\partial^2 \varphi_x}{\partial x \partial \theta} + D_{66}\frac{\partial^2 \varphi_\theta}{\partial x^2} + \frac{D_{11}}{R^2}\frac{\partial^2 \varphi_\theta}{\partial \theta^2} + A_{55}^s\varphi_\theta = I_1\frac{\partial^2 v}{\partial t^2} + I_2\frac{\partial^2 \varphi_\theta}{\partial t^2}$$
$$(2.442)$$

Now, the above set of coupled governing equations must be solved to obtain the dynamic characteristics of the scattered waves. In order to do so, the recommended solution for the components of the displacement field of the shell can be written in the following form [511]:

$$u(x,\theta,t) = U\exp[i(\beta x + m\theta - \omega t)],$$
$$v(x,\theta,t) = V\exp[i(\beta x + m\theta - \omega t)],$$
$$w(x,\theta,t) = W\exp[i(\beta x + m\theta - \omega t)], \quad (2.443)$$
$$\varphi_x(x,\theta,t) = \phi_x\exp[i(\beta x + m\theta - \omega t)],$$
$$\varphi_\theta(x,\theta,t) = \phi_\theta\exp[i(\beta x + m\theta - \omega t)]$$

where U, V, W, ϕ_x, and ϕ_θ are the unknown amplitudes of the displacements. The wave numbers in axial and circumferential directions are shown with β and m, respectively. Also, ω denotes the frequency of the propagated waves. Whenever the analytical solutions of Eq. (2.443) are inserted into Eqs. (2.438)–(2.442), the below eigenvalue equation can be achieved:

$$[\mathbf{K} - \omega^2 \mathbf{M}]\mathbf{d} = 0 \tag{2.444}$$

where **K** and **M** are the symmetric matrices of stiffness and mass, respectively. In addition, **d** is a column vector including the unknown amplitudes of the displacements. By setting the determinant of the coefficient behind the amplitude vector to zero, the frequency of the propagated waves can be derived:

$$|\mathbf{K} - \omega^2 \mathbf{M}| = 0 \tag{2.445}$$

The nonzero components of the stiffness and mass matrices introduced in Eq. (2.444) are as follows [511]:

$$k_{11} = -\left(A_{11}\beta^2 + \frac{A_{66}}{R^2}m^2\right), \quad k_{12} = -\frac{A_{12}+A_{66}}{R}m\beta, \quad k_{13} = i\beta\frac{A_{12}}{R},$$

$$k_{14} = -\left(B_{11}\beta^2 + \frac{B_{66}}{R^2}m^2\right), \quad k_{15} = -\frac{B_{12}+B_{66}}{R}m\beta, \quad k_{22} = -\left(A_{66}\beta^2 + \frac{A_{11}}{R^2}m^2 - \frac{A^s_{55}}{R^2}\right),$$

$$k_{23} = im\frac{A_{11}-A^s_{55}}{R^2}, \quad k_{24} = -\frac{B_{12}+B_{66}}{R}m\beta, \quad k_{25} = -\left(B_{66}\beta^2 + \frac{B_{11}}{R^2}m^2 + \frac{A^s_{55}}{R^2}\right),$$

$$k_{33} = \frac{A_{11}}{R^2} - A^s_{55}\left(\beta^2 + \frac{m^2}{R^2}\right) - k_W - k_P\left(\beta^2 + \frac{m^2}{R^2}\right), \quad k_{34} = i\beta\left(A^s_{55} + \frac{B_{12}}{R}\right),$$

$$k_{35} = im\left(\frac{B_{11}}{R^2} + \frac{A^s_{55}}{R}\right), \quad k_{44} = -\left(D_{11}\beta^2 + \frac{D_{66}}{R^2}m^2 - A^s_{55}\right),$$

$$k_{45} = -\frac{D_{12}+D_{66}}{R}m\beta, \quad k_{55} = -\left(D_{66}\beta^2 + \frac{D_{11}}{R^2}m^2 - A^s_{55}\right) \tag{2.446}$$

and

$$m_{11} = m_{22} = m_{33} = -I_0, \quad m_{14} = m_{25} = -I_1, \quad m_{44} = m_{55} = -I_2 \tag{2.447}$$

2.2.3.2.3 Nonlinear forced vibration problem

The nonlinear forced vibration characteristics of shell-type elements in the context of the shell's FSDT will be monitored in the present subsection. The shell will be considered to be surrounded by a viscoelastic substrate consisted of linear, Pasternak, and nonlinear springs as well as a damper. The strain–displacement relations introduced in Eq. (2.400) are not valid in the nonlinear analysis. Implementing the displacement field expressed in Eq. (2.399) incorporated with the von Kármán geometrical nonlinearity, the following strains can be introduced [400]:

$$\varepsilon_{xx} = \frac{\partial u}{\partial x} + \frac{1}{2}\left(\frac{\partial w}{\partial x}\right)^2 + z\frac{\partial \varphi_x}{\partial x},$$

$$\varepsilon_{\theta\theta} = \frac{1}{R}\left[\frac{\partial v}{\partial \theta} + w + \frac{1}{2R}\left(\frac{\partial w}{\partial \theta}\right)^2 + z\frac{\partial \varphi_\theta}{\partial \theta}\right], \quad (2.448)$$

$$\gamma_{x\theta} = \frac{1}{R}\frac{\partial u}{\partial \theta} + \frac{\partial v}{\partial x} + \frac{1}{R}\frac{\partial w}{\partial x}\frac{\partial w}{\partial \theta} + \frac{z}{R}\frac{\partial \varphi_x}{\partial \theta} + z\frac{\partial \varphi_\theta}{\partial x},$$

$$\gamma_{xz} = \varphi_x + \frac{\partial w}{\partial x}, \quad \gamma_{\theta z} = \varphi_\theta + \frac{1}{R}\frac{\partial w}{\partial \theta} - \frac{v}{R}$$

Based on the above strain–displacement relationship and with regard to the definition of the strain energy of the linearly elastic solids, the variation of the strain energy of the shells in the nonlinear domain can be presented in the following form:

$$\delta U = \int_0^L \int_0^{2\pi} \begin{bmatrix} N_{xx}\left(\dfrac{\partial \delta u}{\partial x} + \dfrac{\partial w}{\partial x}\dfrac{\partial \delta w}{\partial x}\right) + M_{xx}\dfrac{\partial \delta \varphi_x}{\partial x} + \\[6pt] \dfrac{N_{\theta\theta}}{R}\left(\dfrac{\partial \delta v}{\partial \theta} + \delta w + \dfrac{1}{R}\dfrac{\partial w}{\partial \theta}\dfrac{\partial \delta w}{\partial \theta}\right) + \dfrac{M_{\theta\theta}}{R}\dfrac{\partial \delta \varphi_\theta}{\partial \theta} + \\[6pt] N_{x\theta}\left(\dfrac{1}{R}\dfrac{\partial \delta u}{\partial \theta} + \dfrac{\partial \delta v}{\partial x} + \dfrac{1}{R}\dfrac{\partial \delta w}{\partial x}\dfrac{\partial w}{\partial \theta} + \dfrac{1}{R}\dfrac{\partial w}{\partial x}\dfrac{\partial \delta w}{\partial \theta}\right) \\[6pt] +M_{x\theta}\left(\dfrac{1}{R}\dfrac{\partial \delta \varphi_x}{\partial \theta} + \dfrac{\partial \delta \varphi_\theta}{\partial x}\right) + Q_{xz}\left(\delta\varphi_x + \dfrac{\partial \delta w}{\partial x}\right) + \\[6pt] Q_{\theta z}\left(\delta\varphi_\theta + \dfrac{1}{R}\dfrac{\partial \delta w}{\partial \theta} - \dfrac{\delta v}{R}\right) \end{bmatrix} Rd\theta dx$$

(2.449)

In the above relation, the stress-resultants can be defined similar to what previously introduced in Eq. (2.402). It can be claimed that we are close to the Euler–Lagrange equations because the variation of the kinetic energy is previously obtained in Eq. (2.430). Now, it is enough to introduce an expression for the variation of the work done on the shell by the foundation parameters. So, the variation of the work done on the shell by the viscoelastic medium can be expressed as below [1, 390, 515]:

$$\delta V = \int_0^L \int_0^{2\pi} \int_{-h/2}^{h/2} \left[k_L w - k_P\left(\frac{\partial^2 w}{\partial x^2} + \frac{1}{R^2}\frac{\partial^2 w}{\partial \theta^2}\right) + k_{NL}w^3 + c_d\frac{\partial w}{\partial t}\right] Rdzd\theta dx$$

(2.450)

where k_L, k_P, k_{NL}, and c_d are linear, Pasternak, nonlinear, and damping coefficients of the viscoelastic medium, respectively. Substitution of

2.2 Kinematic relations

Eqs. (2.449), (2.430), and (2.450) in Eq. (2.92) results in finding the below expressions:

$$\frac{\partial N_{xx}}{\partial x} + \frac{1}{R}\frac{\partial N_{x\theta}}{\partial \theta} = I_0 \frac{\partial^2 u}{\partial t^2} + I_1 \frac{\partial^2 \varphi_x}{\partial t^2} \quad (2.451)$$

$$\frac{\partial N_{x\theta}}{\partial x} + \frac{1}{R}\frac{\partial N_{\theta\theta}}{\partial \theta} - \frac{1}{R}Q_{z\theta} = I_0 \frac{\partial^2 v}{\partial t^2} + I_1 \frac{\partial^2 \varphi_\theta}{\partial t^2} \quad (2.452)$$

$$\frac{\partial Q_{xz}}{\partial x} + \frac{1}{R}\frac{\partial Q_{z\theta}}{\partial \theta} + \frac{N_{\theta\theta}}{R} + \frac{\partial}{\partial x}\left\{N_{xx}\frac{\partial w}{\partial x} + \frac{1}{R}N_{x\theta}\frac{\partial w}{\partial \theta}\right\}$$
$$+ \frac{1}{R}\frac{\partial}{\partial \theta}\left\{N_{x\theta}\frac{\partial w}{\partial x} + \frac{1}{R}N_{\theta\theta}\frac{\partial w}{\partial \theta}\right\} - k_L w + k_P\left(\frac{\partial^2 w}{\partial x^2} + \frac{1}{R^2}\frac{\partial^2 w}{\partial \theta^2}\right) \quad (2.453)$$
$$- k_{NL}w^3 - c_d\frac{\partial w}{\partial t} = I_0\frac{\partial^2 w}{\partial t^2}$$

$$\frac{\partial M_{xx}}{\partial x} + \frac{1}{R}\frac{\partial M_{x\theta}}{\partial \theta} + Q_{xz} = I_1\frac{\partial^2 u}{\partial t^2} + I_2\frac{\partial^2 \varphi_x}{\partial t^2} \quad (2.454)$$

$$\frac{\partial M_{x\theta}}{\partial x} + \frac{1}{R}\frac{\partial M_{\theta\theta}}{\partial \theta} + Q_{z\theta} = I_1\frac{\partial^2 v}{\partial t^2} + I_2\frac{\partial^2 \varphi_\theta}{\partial t^2} \quad (2.455)$$

where stress-resultants are similar to what we introduced in Eq. (2.402) previously. In the present problem, the in-plane inertia can be neglected in comparison with the inertia of the shell in the thickness direction. Also, I_1 and I_2 are higher-order functions of the thickness and are negligible compared with I_0. So, Eqs. (2.451), (2.452), (2.454), and (2.455) can be re-written in the following form:

$$\frac{\partial N_{xx}}{\partial x} + \frac{1}{R}\frac{\partial N_{x\theta}}{\partial \theta} = 0 \quad (2.456)$$

$$\frac{\partial N_{x\theta}}{\partial x} + \frac{1}{R}\frac{\partial N_{\theta\theta}}{\partial \theta} - \frac{1}{R}Q_{z\theta} = 0 \quad (2.457)$$

$$\frac{\partial M_{xx}}{\partial x} + \frac{1}{R}\frac{\partial M_{x\theta}}{\partial \theta} + Q_{xz} = 0 \quad (2.458)$$

$$\frac{\partial M_{x\theta}}{\partial x} + \frac{1}{R}\frac{\partial M_{\theta\theta}}{\partial \theta} + Q_{z\theta} = 0 \quad (2.459)$$

By considering the available identity in Eq. (2.456), Eq. (2.453) can be re-written in the following form:

$$\frac{\partial Q_{xz}}{\partial x} + \frac{1}{R}\frac{\partial Q_{z\theta}}{\partial \theta} + \frac{N_{\theta\theta}}{R} + N_{xx}\frac{\partial^2 w}{\partial x^2} + \frac{2}{R}N_{x\theta}\frac{\partial^2 w}{\partial x \partial \theta} + \frac{1}{R^2}N_{\theta\theta}\frac{\partial^2 w}{\partial \theta^2}$$
$$+ \frac{1}{R}\left\{\frac{\partial N_{x\theta}}{\partial x} + \frac{1}{R}\frac{\partial N_{\theta\theta}}{\partial \theta}\right\}\frac{\partial w}{\partial \theta} - k_L w + k_P\left(\frac{\partial^2 w}{\partial x^2} + \frac{1}{R^2}\frac{\partial^2 w}{\partial \theta^2}\right) \quad (2.460)$$
$$- k_{NL}w^3 - c_d\frac{\partial w}{\partial t} = I_0\frac{\partial^2 w}{\partial t^2}$$

By considering Eq. (2.457), the above relation can be re-written as:

$$\frac{\partial Q_{xz}}{\partial x} + \frac{1}{R}\frac{\partial Q_{z\theta}}{\partial \theta} + \frac{N_{\theta\theta}}{R} + N_{xx}\frac{\partial^2 w}{\partial x^2} + \frac{2}{R}N_{x\theta}\frac{\partial^2 w}{\partial x\partial\theta} + \frac{1}{R^2}N_{\theta\theta}\frac{\partial^2 w}{\partial \theta^2}$$
$$+\frac{1}{R^2}Q_{z\theta}\frac{\partial w}{\partial \theta} - k_L w + k_P\left(\frac{\partial^2 w}{\partial x^2} + \frac{1}{R^2}\frac{\partial^2 w}{\partial \theta^2}\right) - k_{NL}w^3 - c_d\frac{\partial w}{\partial t} = I_0\frac{\partial^2 w}{\partial t^2}$$
(2.461)

Now, the Euler–Lagrange equations must be expressed in terms of the components of the displacement field of the shell. To this end, the stress-resultants should be expressed in terms of the displacement field of the shell and therefore, substituted in the motion equations. By integrating from Eq. (2.409) over the thickness of the shell with regard to the definitions provided in Eq. (2.402), the following relations can be derived:

$$N_{xx} = A_{11}\left[\frac{\partial u}{\partial x} + \frac{1}{2}\left(\frac{\partial w}{\partial x}\right)^2\right] + B_{11}\frac{\partial \varphi_x}{\partial x} + \frac{A_{12}}{R}\left[\frac{\partial v}{\partial \theta} + w + \frac{1}{2R}\left(\frac{\partial w}{\partial \theta}\right)^2\right] + \frac{B_{12}}{R}\frac{\partial \varphi_\theta}{\partial \theta}$$
(2.462)

$$N_{\theta\theta} = A_{12}\left[\frac{\partial u}{\partial x} + \frac{1}{2}\left(\frac{\partial w}{\partial x}\right)^2\right] + B_{12}\frac{\partial \varphi_x}{\partial x} + \frac{A_{11}}{R}\left[\frac{\partial v}{\partial \theta} + w + \frac{1}{2R}\left(\frac{\partial w}{\partial \theta}\right)^2\right] + \frac{B_{11}}{R}\frac{\partial \varphi_\theta}{\partial \theta}$$
(2.463)

$$N_{x\theta} = A_{66}\left(\frac{1}{R}\frac{\partial u}{\partial \theta} + \frac{\partial v}{\partial x} + \frac{1}{R}\frac{\partial w}{\partial x}\frac{\partial w}{\partial \theta}\right) + B_{66}\left(\frac{1}{R}\frac{\partial \varphi_x}{\partial \theta} + \frac{\partial \varphi_\theta}{\partial x}\right) \quad (2.464)$$

$$M_{xx} = B_{11}\left[\frac{\partial u}{\partial x} + \frac{1}{2}\left(\frac{\partial w}{\partial x}\right)^2\right] + D_{11}\frac{\partial \varphi_x}{\partial x} + \frac{B_{12}}{R}\left[\frac{\partial v}{\partial \theta} + w + \frac{1}{2R}\left(\frac{\partial w}{\partial \theta}\right)^2\right] + \frac{D_{12}}{R}\frac{\partial \varphi_\theta}{\partial \theta}$$
(2.465)

$$M_{\theta\theta} = B_{12}\left[\frac{\partial u}{\partial x} + \frac{1}{2}\left(\frac{\partial w}{\partial x}\right)^2\right] + D_{12}\frac{\partial \varphi_x}{\partial x} + \frac{B_{11}}{R}\left[\frac{\partial v}{\partial \theta} + w + \frac{1}{2R}\left(\frac{\partial w}{\partial \theta}\right)^2\right] + \frac{D_{11}}{R}\frac{\partial \varphi_\theta}{\partial \theta}$$
(2.466)

$$M_{x\theta} = B_{66}\left(\frac{1}{R}\frac{\partial u}{\partial \theta} + \frac{\partial v}{\partial x} + \frac{1}{R}\frac{\partial w}{\partial x}\frac{\partial w}{\partial \theta}\right) + D_{66}\left(\frac{1}{R}\frac{\partial \varphi_x}{\partial \theta} + \frac{\partial \varphi_\theta}{\partial x}\right) \quad (2.467)$$

$$Q_{xz} = A_{55}^s\left(\varphi_x + \frac{\partial w}{\partial x}\right) \quad (2.468)$$

$$Q_{z\theta} = A_{55}^s\left(\varphi_\theta + \frac{1}{R}\frac{\partial w}{\partial \theta} - \frac{v}{R}\right) \quad (2.469)$$

in which through-the-thickness rigidities of the shell are as same as those introduced in Eq. (2.418). Substitution of Eqs. (2.462)–(2.469) in Eqs. (2.456)–(2.459) and (2.461) results in the governing equations in terms of the components of the shell's displacement field. In order to conserve

2.2 Kinematic relations

the initial form of the motion equations, the equation related to the variation of the shell's bending deflection (δw) is placed between equations related to the deformations (δu and δv) and rotations ($\delta\phi_x$ and $\delta\phi_\theta$). So, the following equations can be written for the nonlinear vibrations of a shell:

$$A_{11}\frac{\partial^2 u}{\partial x^2} + \frac{A_{66}}{R^2}\frac{\partial^2 u}{\partial \theta^2} + \frac{A_{12}+A_{66}}{R}\frac{\partial^2 v}{\partial x\partial\theta} + \frac{A_{12}}{R}\frac{\partial w}{\partial x} + A_{11}\frac{\partial w}{\partial x}\frac{\partial^2 w}{\partial x^2} + \frac{A_{12}+A_{66}}{R^2}$$

$$\frac{\partial w}{\partial \theta}\frac{\partial^2 w}{\partial x\partial\theta} + \frac{A_{66}}{R^2}\frac{\partial w}{\partial x}\frac{\partial^2 w}{\partial \theta^2} + B_{11}\frac{\partial^2 \varphi_x}{\partial x^2} + \frac{B_{66}}{R^2}\frac{\partial^2 \varphi_x}{\partial \theta^2} + \frac{B_{12}+B_{66}}{R}\frac{\partial^2 \varphi_\theta}{\partial x\partial\theta} = 0$$

(2.470)

$$\frac{A_{12}+A_{66}}{R}\frac{\partial^2 u}{\partial x\partial\theta} + A_{66}\frac{\partial^2 v}{\partial x^2} + \frac{A_{11}}{R^2}\frac{\partial^2 v}{\partial \theta^2} + \frac{A^s_{55}}{R^2}v + \frac{A_{11}-A^s_{55}}{R^2}\frac{\partial w}{\partial \theta}$$

$$+\frac{A_{66}}{R}\frac{\partial^2 w}{\partial x^2}\frac{\partial w}{\partial \theta} + \frac{A_{12}+A_{66}}{R}\frac{\partial w}{\partial x}\frac{\partial^2 w}{\partial x\partial\theta} + \frac{A_{11}}{R^3}\frac{\partial w}{\partial \theta}\frac{\partial^2 w}{\partial \theta^2} + \frac{B_{12}+B_{66}}{R}$$

(2.471)

$$\frac{\partial^2 \varphi_x}{\partial x\partial\theta} + B_{66}\frac{\partial^2 \varphi_\theta}{\partial x^2} + \frac{B_{11}}{R^2}\frac{\partial^2 \varphi_\theta}{\partial \theta^2} - \frac{A^s_{55}}{R}\varphi_\theta = 0$$

$$\frac{A_{12}}{R}\frac{\partial u}{\partial x} + \frac{A_{11}-A^s_{55}}{R^2}\frac{\partial v}{\partial \theta} + \frac{A_{11}}{R^2}w + A^s_{55}\left(\frac{\partial^2 w}{\partial x^2} + \frac{1}{R^2}\frac{\partial^2 w}{\partial \theta^2}\right) + \left(A^s_{55} + \frac{B_{12}}{R}\right)\frac{\partial \varphi_x}{\partial x}$$

$$+\left(\frac{A^s_{55}}{R} + \frac{B_{11}}{R^2}\right)\frac{\partial \varphi_\theta}{\partial \theta} + \frac{A_{12}}{2R}\left(\frac{\partial w}{\partial x}\right)^2 + \frac{A_{11}+2A^s_{55}}{2R^3}\left(\frac{\partial w}{\partial \theta}\right)^2 + \frac{A_{12}}{R}w\frac{\partial^2 w}{\partial x^2} + \frac{A_{11}}{R^3}w\frac{\partial^2 w}{\partial \theta^2}$$

$$+\frac{A_{11}}{2}\left(\frac{\partial w}{\partial x}\right)^2\frac{\partial^2 w}{\partial x^2} + \frac{A_{12}}{2R^2}\left(\frac{\partial w}{\partial \theta}\right)^2\frac{\partial^2 w}{\partial x^2} + \frac{2A_{66}}{R^2}\frac{\partial w}{\partial x}\frac{\partial w}{\partial \theta}\frac{\partial^2 w}{\partial x\partial\theta} + \frac{A_{12}}{2R^2}\left(\frac{\partial w}{\partial x}\right)^2\frac{\partial^2 w}{\partial \theta^2}$$

$$+\frac{A_{11}}{2R^4}\left(\frac{\partial w}{\partial \theta}\right)^2\frac{\partial^2 w}{\partial \theta^2} + A_{11}\frac{\partial u}{\partial x}\frac{\partial^2 w}{\partial x^2} + \frac{A_{12}}{R^2}\frac{\partial u}{\partial x}\frac{\partial^2 w}{\partial \theta^2} + \frac{2A_{66}}{R^2}\frac{\partial u}{\partial \theta}\frac{\partial^2 w}{\partial x\partial\theta} + \frac{A_{12}}{R}\frac{\partial v}{\partial \theta}\frac{\partial^2 w}{\partial x^2}$$

$$+\frac{2A_{66}}{R}\frac{\partial v}{\partial x}\frac{\partial^2 w}{\partial x\partial\theta} + \frac{A_{11}}{R^3}\frac{\partial v}{\partial \theta}\frac{\partial^2 w}{\partial \theta^2} - \frac{A^s_{55}}{R^3}v\frac{\partial w}{\partial \theta} + B_{11}\frac{\partial \varphi_x}{\partial x}\frac{\partial^2 w}{\partial x^2} + \frac{2B_{66}}{R^2}\frac{\partial \varphi_x}{\partial \theta}\frac{\partial^2 w}{\partial x\partial\theta}$$

$$+\frac{B_{12}}{R^2}\frac{\partial \varphi_x}{\partial x}\frac{\partial^2 w}{\partial \theta^2} + \frac{B_{12}}{R}\frac{\partial \varphi_\theta}{\partial \theta}\frac{\partial^2 w}{\partial x^2} + \frac{2B_{66}}{R}\frac{\partial \varphi_\theta}{\partial x}\frac{\partial^2 w}{\partial x\partial\theta} + \frac{B_{11}}{R^3}\frac{\partial \varphi_\theta}{\partial \theta}\frac{\partial^2 w}{\partial \theta^2} + \frac{A^s_{55}}{R^2}\varphi_\theta\frac{\partial w}{\partial \theta}$$

$$-k_L w + k_P\left(\frac{\partial^2 w}{\partial x^2} + \frac{1}{R^2}\frac{\partial^2 w}{\partial \theta^2}\right) - k_{NL}w^3 - c_d\frac{\partial w}{\partial t} - I_0\frac{\partial^2 w}{\partial t^2} = 0$$

(2.472)

$$B_{11}\frac{\partial^2 u}{\partial x^2} + \frac{B_{66}}{R^2}\frac{\partial^2 u}{\partial \theta^2} + \frac{B_{12}+B_{66}}{R}\frac{\partial^2 v}{\partial x\partial\theta} + \left(A^s_{55} + \frac{B_{12}}{R}\right)\frac{\partial w}{\partial x}$$

$$+B_{11}\frac{\partial w}{\partial x}\frac{\partial^2 w}{\partial x^2} + \frac{B_{12}+B_{66}}{R^2}\frac{\partial w}{\partial \theta}\frac{\partial^2 w}{\partial x\partial\theta} + \frac{B_{66}}{R^2}\frac{\partial w}{\partial x}\frac{\partial^2 w}{\partial \theta^2} + D_{11}\frac{\partial^2 \varphi_x}{\partial x^2} + \frac{D_{66}}{R^2}\frac{\partial^2 \varphi_x}{\partial \theta^2}$$

(2.473)

$$+A^s_{55}\varphi_x + \frac{D_{12}+D_{66}}{R}\frac{\partial^2 \varphi_\theta}{\partial x\partial\theta} = 0$$

$$\frac{B_{12}+B_{66}}{R}\frac{\partial^2 u}{\partial x \partial \theta}+B_{66}\frac{\partial^2 v}{\partial x^2}+\frac{B_{11}}{R^2}\frac{\partial^2 v}{\partial \theta^2}-\frac{A^s_{55}}{R}v+\left(\frac{A^s_{55}}{R}+\frac{B_{11}}{R^2}\right)\frac{\partial w}{\partial \theta}+\frac{B_{66}}{R}$$
$$\frac{\partial^2 w}{\partial x^2}\frac{\partial w}{\partial \theta}+\frac{B_{12}+B_{66}}{R}\frac{\partial w}{\partial x}\frac{\partial^2 w}{\partial x \partial \theta}+\frac{B_{11}}{R^3}\frac{\partial w}{\partial \theta}\frac{\partial^2 w}{\partial \theta^2}+\frac{D_{12}+D_{66}}{R}\frac{\partial^2 \varphi_x}{\partial x \partial \theta} \quad (2.474)$$
$$+D_{66}\frac{\partial^2 \varphi_\theta}{\partial x^2}+\frac{D_{11}}{R^2}\frac{\partial^2 \varphi_\theta}{\partial \theta^2}+A^s_{55}\varphi_\theta=0$$

Now, the above governing equations must be solved to find the answer of the problem. To do so, the solution functions related to each component of the displacement field of the shell can be considered to be in the following form based on the well-known Navier's analytical solution [1]:

$$u(x,\theta,t)=\sum_{m=1}^{\infty}\sum_{n=1}^{\infty}U_{mn}(t)\cos(\alpha x)\cos(n\theta),$$

$$v(x,\theta,t)=\sum_{m=1}^{\infty}\sum_{n=1}^{\infty}V_{mn}(t)\sin(\alpha x)\sin(n\theta),$$

$$w(x,\theta,t)=\sum_{m=1}^{\infty}\sum_{n=1}^{\infty}W_{mn}(t)\sin(\alpha x)\cos(n\theta), \quad (2.475)$$

$$\varphi_x(x,\theta,t)=\sum_{m=1}^{\infty}\sum_{n=1}^{\infty}\phi_{xmn}(t)\cos(\alpha x)\cos(n\theta),$$

$$\varphi_\theta(x,\theta,t)=\sum_{m=1}^{\infty}\sum_{n=1}^{\infty}\phi_{\theta mn}(t)\sin(\alpha x)\sin(n\theta)$$

where $U_{mn}(t)$, $V_{mn}(t)$, $W_{mn}(t)$, $\phi_{xmn}(t)$, and $\phi_{\theta mn}(t)$ are the dynamic amplitudes of the shell's oscillations. Also, $\alpha = m\pi/L$. It is worth regarding that m and n are the longitudinal mode number and circumferential wave number, respectively. The above solutions are introduced by considering both ends of the cylindrical shell to be simply supported. Now, by inserting analytical solution introduced in Eq. (2.475) into Eqs. (2.470)–(2.474), the below relations can be achieved:

$$C_{11}U_{mn}(t)+C_{12}V_{mn}(t)+C_{14}\phi_{xmn}(t)+C_{15}\phi_{\theta mn}(t)=C_{13}W_{mn}(t)+\bar{C}_{13}W_{mn}^2(t) \quad (2.476)$$

$$C_{21}U_{mn}(t)+C_{22}V_{mn}(t)+C_{24}\phi_{xmn}(t)+C_{25}\phi_{\theta mn}(t)=C_{23}W_{mn}(t)+\bar{C}_{23}W_{mn}^2(t) \quad (2.477)$$

$$C_{31}U_{mn}(t)+\eta_{13}U_{mn}(t)W_{mn}(t)+C_{32}V_{mn}(t)+\eta_{23}V_{mn}(t)W_{mn}(t)+C_{33}W_{mn}(t)$$
$$+C_{34}\phi_{xmn}(t)+\eta_{43}\phi_{xmn}(t)W_{mn}(t)+C_{35}\phi_{\theta mn}(t)+\eta_{53}\phi_{\theta mn}(t)W_{mn}(t)$$
$$+\bar{C}_{33}W_{mn}^2(t)+\hat{C}_{33}W_{mn}^3(t)+\mu\dot{W}_{mn}(t)+M\ddot{W}_{mn}(t)=0$$
$$(2.478)$$

$$C_{41}U_{mn}(t) + C_{42}V_{mn}(t) + C_{44}\phi_{xmn}(t) + C_{45}\phi_{\theta mn}(t) = C_{43}W_{mn}(t) + \bar{C}_{43}W_{mn}^2(t) \tag{2.479}$$

$$C_{51}U_{mn}(t) + C_{52}V_{mn}(t) + C_{54}\phi_{xmn}(t) + C_{55}\phi_{\theta mn}(t) = C_{53}W_{mn}(t) + \bar{C}_{53}W_{mn}^2(t) \tag{2.480}$$

in which

$$C_{11} = -\left(A_{11}\alpha^2 + \frac{A_{66}}{R^2}n^2\right), \quad C_{12} = \frac{A_{12}+A_{66}}{R}\alpha n, \quad C_{14} = -\left(B_{11}\alpha^2 + \frac{B_{66}}{R^2}n^2\right),$$
$$C_{15} = \frac{B_{12}+B_{66}}{R}\alpha n, \quad C_{13} = -\frac{A_{12}}{R}\alpha, \quad \bar{C}_{13} = A_{11}\alpha^3 - \frac{A_{12}}{R^2}\alpha n^2 \tag{2.481}$$

$$C_{21} = \frac{A_{12}+A_{66}}{R}\alpha n, \quad C_{22} = -\left(A_{66}\alpha^2 + \frac{A_{11}}{R^2}n^2 - \frac{A_{55}^s}{R^2}\right), \quad C_{24} = \frac{B_{12}+B_{66}}{R}\alpha n,$$
$$C_{25} = -\left(B_{66}\alpha^2 + \frac{B_{11}}{R^2}n^2 + \frac{A_{55}^s}{R}\right), \quad C_{23} = \frac{A_{11}-A_{55}^s}{R^2}\alpha, \quad \bar{C}_{23} = \frac{A_{12}}{R}\alpha^2 n - \frac{A_{11}}{R^3}n^3 \tag{2.482}$$

$$C_{31} = -\frac{A_{12}}{R}\alpha, \quad C_{32} = \frac{A_{11}-A_{55}^s}{R^2}n, \quad C_{33} = \frac{A_{11}}{R^2} - A_{55}^s\left(\alpha^2 + \frac{n^2}{R^2}\right) - k_L - k_P\left(\alpha^2 + \frac{n^2}{R^2}\right),$$
$$C_{34} = -\left(A_{55}^s + \frac{B_{12}}{R}\right)\alpha, \quad C_{35} = \left(\frac{A_{55}^s}{R} + \frac{B_{11}}{R^2}\right)n, \quad \eta_{13} = A_{11}\alpha^3 + \frac{A_{12}+2A_{66}}{R^2}\alpha n^2,$$
$$\eta_{23} = -\frac{A_{12}+2A_{66}}{R}\alpha^2 n - \frac{A_{11}}{R^3}n^3 + \frac{A_{55}^s}{R^3}n, \quad \eta_{43} = B_{11}\alpha^3 + \frac{B_{12}+2B_{66}}{R^2}\alpha n^2,$$
$$\eta_{53} = -\left(\frac{B_{12}+2B_{66}}{R}\alpha^2 n + \frac{B_{11}}{R^3}n^3 + \frac{A_{55}^s}{R^2}n\right), \quad \bar{C}_{33} = -\frac{A_{12}}{2R}\alpha^2 + \frac{2A_{55}^s - A_{11}}{2R^3}n^2,$$
$$\hat{C}_{33} = -\left(\frac{A_{11}}{2}\alpha^4 + \frac{A_{12}-2A_{66}}{R^2}\alpha^2 n^2 + \frac{A_{11}}{2R^4}n^4 + k_{NL}\right), \quad \mu = -c_d, \quad M = -I_0 \tag{2.483}$$

$$C_{41} = -\left(B_{11}\alpha^2 + \frac{B_{66}}{R^2}n^2\right), \quad C_{42} = \frac{B_{12}+B_{66}}{R}\alpha n, \quad C_{44} = -\left(D_{11}\alpha^2 + \frac{D_{66}}{R^2}n^2 - A_{55}^s\right),$$
$$C_{45} = \frac{D_{12}+D_{66}}{R}\alpha n, \quad C_{43} = -\left(A_{55}^s + \frac{B_{12}}{R}\right)\alpha, \quad \bar{C}_{43} = B_{11}\alpha^3 - \frac{B_{12}}{R^2}\alpha n^2 \tag{2.484}$$

$$C_{51} = \frac{B_{12}+B_{66}}{R}\alpha n, \quad C_{52} = -\left(B_{66}\alpha^2 + \frac{B_{11}}{R^2}n^2 + \frac{A_{55}^s}{R}\right), \quad C_{54} = \frac{D_{12}+D_{66}}{R}\alpha n,$$
$$C_{55} = -\left(D_{66}\alpha^2 + \frac{D_{11}}{R^2}n^2 - A_{55}^s\right), \quad C_{53} = \left(\frac{A_{55}^s}{R} + \frac{B_{11}}{R^2}\right)n, \quad \bar{C}_{53} = \frac{B_{12}}{R}\alpha^2 n - \frac{B_{11}}{R^3}n^3 \tag{2.485}$$

Now, coupled relations introduced in Eqs. (2.476)–(2.480) are aimed to be reduced to a single equation describing the dynamics of the system. In order to satisfy this desire, Eqs. (2.476), (2.477), (2.479), and (2.480) can be considered to be in the following set:

$$\begin{bmatrix} C_{11} & C_{12} & C_{14} & C_{15} \\ C_{21} & C_{22} & C_{24} & C_{25} \\ C_{41} & C_{42} & C_{44} & C_{45} \\ C_{51} & C_{52} & C_{54} & C_{55} \end{bmatrix} \begin{Bmatrix} U_{mn}(t) \\ V_{mn}(t) \\ \phi_{xmn}(t) \\ \phi_{\theta mn}(t) \end{Bmatrix} = \begin{Bmatrix} C_{13} \\ C_{23} \\ C_{43} \\ C_{53} \end{Bmatrix} W_{mn}(t) + \begin{Bmatrix} \bar{C}_{13} \\ \bar{C}_{23} \\ \bar{C}_{43} \\ \bar{C}_{53} \end{Bmatrix} W_{mn}^2(t) \tag{2.486}$$

Now, the above set of four-unknown four-equation algebraic relation can be solved for the oscillation amplitudes $U_{mn}(t)$, $V_{mn}(t)$, $\varphi_{xmn}(t)$, and $\varphi_{\theta mn}(t)$. Once the above set of equation is solved, the below relation for each of the aforementioned amplitudes can be expressed:

$$\begin{aligned} U_{mn}(t) &= \lambda_1 W_{mn}(t) + \psi_1 W_{mn}^2(t), \\ V_{mn}(t) &= \lambda_2 W_{mn}(t) + \psi_2 W_{mn}^2(t), \\ \phi_{xmn}(t) &= \lambda_3 W_{mn}(t) + \psi_3 W_{mn}^2(t), \\ \phi_{\theta mn}(t) &= \lambda_4 W_{mn}(t) + \psi_4 W_{mn}^2(t) \end{aligned} \quad (2.487)$$

Now, the final equation can be derived by substituting for the time-dependent amplitudes from the above relation in Eq. (2.478). Once the aforementioned substitution is done, the following relation will be attained:

$$M\ddot{W}_{mn}(t) + \mu \dot{W}_{mn}(t) + \gamma_1 W_{mn}(t) + \gamma_2 W_{mn}^2(t) + \gamma_3 W_{mn}^3(t) = 0 \quad (2.488)$$

in which

$$\gamma_1 = C_{31}\lambda_1 + C_{32}\lambda_2 + C_{33} + C_{34}\lambda_3 + C_{35}\lambda_4,$$
$$\gamma_2 = C_{31}\psi_1 + \eta_{13}\lambda_1 + C_{32}\psi_2 + \eta_{23}\lambda_2 + C_{34}\psi_3 + \eta_{43}\lambda_3 + C_{35}\psi_4 + \eta_{53}\lambda_4 + \bar{C}_{33},$$
$$\gamma_3 = \eta_{13}\psi_1 + \eta_{23}\psi_2 + \eta_{43}\psi_3 + \eta_{53}\psi_4 + \hat{C}_{33}$$

$$(2.489)$$

It is worth mentioning that Eq. (2.488) describes the nonlinear free vibrations of the shells and the effect of the external loading is not included in it. In order to show the influence of the external harmonic excitation ($f \cos \Omega t$), the above relation can be re-written in the below form:

$$\ddot{W}_{mn}(t) + \varsigma_1 W_{mn}(t) + \hat{\mu}\dot{W}_{mn}(t) + \varsigma_2 W_{mn}^2(t) + \varsigma_3 W_{mn}^3(t) = \hat{f} \cos \Omega t \quad (2.490)$$

where

$$\hat{\mu} = \frac{\mu}{M}, \quad \varsigma_1 = \frac{\gamma_1}{M}, \quad \varsigma_2 = \frac{\gamma_2}{M}, \quad \varsigma_3 = \frac{\gamma_3}{M}, \quad \hat{f} = \frac{f}{M} \quad (2.491)$$

By considering the fact that the nonlinearities and damping term are small in comparison with the linear term and with regard to the fact that we are about to monitor the primary resonance of the system (i.e., corresponding with soft excitation), Eq. (2.490) can be re-written in the following scaled format:

$$\ddot{W}_{mn}(t) + \varsigma_1 W_{mn}(t) + 2\varepsilon^2 \bar{\mu} \dot{W}_{mn}(t) + \varepsilon \varsigma_2 W_{mn}^2(t) + \varepsilon^2 \varsigma_3 W_{mn}^3(t) = 2\varepsilon^2 \bar{f} \cos \Omega t \quad (2.492)$$

where

$$\bar{\mu} = \hat{\mu}/2, \quad \bar{f} = \hat{f}/2 \quad (2.493)$$

2.2 Kinematic relations

The coefficient "2" behind damping and excitation load is provided for mathematical simplification and does not relate to any physical phenomenon. Just like Section 2.1.1.2, herein, the linear natural frequency of the system is $\omega_L = \omega_0 = \sqrt{\varsigma_1}$. Now, the final equation must be solved in the time domain in order to achieve the steady-state response of the nonlinear system under action of a soft excitation. Due to the fact that there exists a quadratic nonlinear term in the system as well as cubic nonlinearity, implementation of the first-order expansion of the multiple scales method is not allowed. The reason is that by doing so, the troublesome terms cannot be omitted completely. Therefore, the second-order expansion must be hired. So, the approximate solution will be assumed to be in the following form [722]:

$$W(t) = w_0(T_0, T_1, T_2) + \varepsilon w_1(T_0, T_1, T_2) + \varepsilon^2 w_2(T_0, T_1, T_2) + \cdots \quad (2.494)$$

By inserting the above solution into Eq. (2.492) and following the instructions mentioned in Section 2.1.1.2, the below relations can be achieved [722]:

$$\varepsilon^0 \to D_0^2 w_0 + \varsigma_1 w_0 = 0 \quad (2.495)$$

$$\varepsilon^1 \to D_0^2 w_1 + \varsigma_1 w_1 = -2 D_0 D_1 w_0 - \varsigma_2 w_0^2 \quad (2.496)$$

$$\varepsilon^2 \to D_0^2 w_2 + \varsigma_1 w_2 = -2 D_0 D_1 w_1 - 2 D_0 D_2 w_1 - D_1^2 w_0 \\ - 2\varsigma_2 w_0 w_1 - \varsigma_3 w_0^3 + \bar{f} \exp(i\Omega T_0) + CC \quad (2.497)$$

Solving Eq. (2.495) for w_0, we have [722]:

$$w_0 = A(T_1, T_2) \exp(i\omega_0 T_0) + CC \quad (2.498)$$

By substituting for the w_0 from Eq. (2.498) in Eq. (2.496), the following expression will be obtained [722]:

$$D_0^2 w_1 + \varsigma_1 w_1 = -2i\omega_0 D_1 A \exp(i\omega_0 T_0) - \varsigma_2 \left[A^2 \exp(2i\omega_0 T_0) + A\bar{A}\right] + CC \quad (2.499)$$

In order to eliminate the secular terms, the following constraint must be valid:

$$D_1 A = 0 \quad (2.500)$$

So, it can be concluded that A is a function of T_2 only ($A \equiv A(T_2)$). Now, the particular part of w_1 can be determined with the aid of the method of unknown coefficients in the following form:

$$w_1 = \frac{\varsigma_2}{\omega_0^2} \left[\frac{1}{3} A^2 \exp(2i\omega_0 T_0) - A\bar{A}\right] + CC \quad (2.501)$$

Now, by inserting Eqs. (2.498) and (2.501) into Eq. (2.497), we have:

$$D_0^2 w_2 + \varsigma_1 w_2 = -\left[2i\omega_0(A' + \bar{\mu}A) + \left(3\varsigma_3 - \frac{10\varsigma_2^2}{3\omega_0^2}\right)A^2\bar{A}\right]\exp(i\omega_0 T_0) + \bar{f}\exp(i\Omega T_0) + CC + NST$$

(2.502)

where NST denotes the nonsecular terms and $(.)' = D_2(.)$. In this problem, we are about to probe the primary resonance of the system ($\Omega \simeq \omega_0$) and therefore, the detuning parameter (σ) can be defined in the below form [722]:

$$\Omega \simeq \omega_0 \rightarrow \Omega = \omega_0 + \varepsilon^2\sigma \qquad (2.503)$$

Once Eq. (2.502) is re-written by the means of the above definition of the detuning parameter, the below expression will be attained:

$$D_0^2 w_2 + \varsigma_1 w_2 = -\left[2i\omega_0(A' + \bar{\mu}A) + \left(3\varsigma_3 - \frac{10\varsigma_2^2}{3\omega_0^2}\right)A^2\bar{A} - \bar{f}\exp(i\sigma T_2)\right]\exp(i\omega_0 T_0) + CC + NST$$

(2.504)

Similar to any other relation in the perturbation theory, the troublesome terms must be omitted from the above identity via algebraic constraints. Hence:

$$2i\omega_0(A' + \bar{\mu}A) + \left(3\varsigma_3 - \frac{10\varsigma_2^2}{3\omega_0^2}\right)A^2\bar{A} - \bar{f}\exp(i\sigma T_2) = 0 \qquad (2.505)$$

Now, the following polar form for A can be implemented to solve the above equation [722]:

$$A = \frac{1}{2}a(T_2)\exp(i\beta(T_2)) \qquad (2.506)$$

Once the above solution for $A(T_2)$ is substituted in Eq. (2.505) and mathematical simplifications are procured, the following modulation equations can be achieved:

$$\begin{aligned} a' + \bar{\mu}a &= \frac{\bar{f}}{\omega_0}\sin\gamma \\ a\gamma' &= a\sigma - \frac{9\varsigma_1\varsigma_3 - 10\varsigma_2^2}{24\omega_0^3}a^3 + \frac{\bar{f}}{\omega_0}\cos\gamma \end{aligned} \qquad (2.507)$$

in which $\gamma = \sigma T_2 - \beta$. To track the steady-state response of the oscillating system, we will be about to find the fixed points of this system. To this end, the following identities must be valid simultaneously:

$$\begin{aligned} a' &= 0, \\ a\gamma' &= 0 \end{aligned} \qquad (2.508)$$

So, one can write the following steady-state modulation equations:

$$\begin{aligned} \bar{\mu} &= \frac{\bar{f}}{\omega_0 a}\sin\gamma \\ \sigma - \frac{9\varsigma_1\varsigma_3 - 10\varsigma_2^2}{24\omega_0^3}a^2 &= -\frac{\bar{f}}{\omega_0 a}\cos\gamma \end{aligned} \qquad (2.509)$$

By computing the square of the both sides of the above relations and summing the result, the below frequency–response equation can be extracted:

$$\sigma = \frac{9\varsigma_1\varsigma_3 - 10\varsigma_2^2}{24\omega_0^3}a^2 \pm \sqrt{\frac{\bar{f}^2}{\omega_0^2 a^2} - \bar{\mu}^2} \qquad (2.510)$$

Therefore, the frequency–response curves of the shell's forced oscillations can be plotted simply with the aid of the above formula. Again, it must be mentioned that the above formula describes the nonlinear primary resonance of the system and the secondary resonances and combination-mode resonances are not included in it. However, due to the great importance of the primary resonance of the system, studying the above relation can be led to have a mentionable understanding from the system's dynamics. It is obvious that all of the qualitative issues instructed in Section 2.1.1.2 for the purpose of analyzing the system's nonlinear dynamic behavior can be hired here, too. However, it must be kept in mind that the results of the linearization via Jacobian matrix are not permanently valid due to the existence of damping in the underobservation system. In fact, the Hartman–Grobman theorem must be implemented in order to analyze the nonlinear dynamic behaviors of the analyzed oscillator through linearization.

CHAPTER 3

Static analysis of multiscale hybrid nanocomposite structures

3.1 Bending, buckling, and postbuckling of beams

In this section, the bending behavior and static stability margin of the beam-type structures manufactured from multiscale hybrid (MSH) polymer nanocomposite (PNC) materials will be monitored. To this purpose, the static responses achieved in Sections 2.2.1.1.1, 2.2.1.2.1, and 2.2.1.2.2 of Chapter 2 will be kept in mind to discuss about both bending and buckling behaviors of PNC beams.

3.1.1 Bending analysis of MSH PNC beams

This subsection undergoes with the bending behaviors of PNC beams reinforced via multiscale reinforcing elements. Illustrative examples will be presented in this part to show the influence of different variants on the variation of the deflection of the continua in the framework of the refined-type higher-order shear deformation theory. For the sake of simplicity in the investigation of the bending response of MSH PNC beams, the following dimensionless variables are employed:

$$W = \frac{100 E_m h^3 \sum_{m=1}^{\infty} (W_{bm} + W_{sm}) \sin(\alpha x)}{q_0 L^4}, \quad X = \frac{x}{L} \quad (3.1)$$

Before going through detailed discussion about deflection response of the MSH PNC beams, the accuracy of the presented method is shown in Table 3.1. In this table, the dimensionless deflection of composite beams with small slenderness ratios is calculated and it is revealed that the present modeling can regenerate the data provided in former studies

TABLE 3.1 Comparison of the dimensionless deflection of thick shear deformable composite beams subjected to uniform loading ($L/h = 5$).

Gradient index, p	Source	Dimensionless deflection, w
0.0	Ref. [731]	3.1657
	Ref. [727]	3.1654
	Present	3.1990
0.5	Ref. [731]	4.8292
	Ref. [727]	4.8285
	Present	4.8679
1.0	Ref. [731]	6.2599
	Ref. [727]	6.2594
	Present	6.2327
2.0	Ref. [731]	8.0602
	Ref. [727]	8.0677
	Present	7.8407

[727, 731]. The small difference between the results of present study and those reported in Refs. [731] and [727] is because of the difference in the utilized shape functions.

As the first illustration, the influence of gradient index introduced in Eq. (2.44) on the bent shape of the beam under uniform loading is drawn in Fig. 3.1. It is worth mentioning that the second type of thickness-dependent $F(z)$ function (i.e., $\left(\frac{1}{2} - \frac{z}{h}\right)^p$) is considered in this diagram. Also, the inputs of the problem show the full agglomeration case like what explained in Section 2.1.1.5 of Chapter 2. Clearly, the amplitude of the PNC beam's deflection is aggrandized by increasing the gradient index of the nanofillers' volume fraction. By remembering the negative influence of the gradient index on the Young's modulus of the nanomaterial (refer to Fig. 2.6), it is simple to justify the observed trend. Indeed, the deflection amplitude becomes greater because the stiffness of the MSH PNC will be reduced by adding the gradient index. So, based upon the inverse relation between deflection and stiffness, it is natural to see such a phenomenon. Moreover, it is shown that implementation of a higher content of the reinforcing nanofillers in the composition of the nanomaterial results in lower deflection of the structure due to the same reason.

In order to capture the most critical case in the subsequent examples, the maximum deflection (i.e., corresponding with the deflection of the beam's center) will be tracked. Fig. 3.2 deals with the monitoring of the coupled

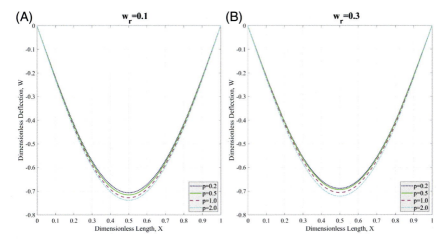

FIGURE 3.1 Variation of the dimensionless deflection of the MSH PNC beam subjected to uniform bending load against dimensionless length for various gradient indices ($L/h = 10$, $\eta = 1$, $\mu = 0.5$, $V_F = 0.2$).

influences of the volume fraction of the clusters and volume fraction of the carbon nanotubes (CNTs) inside the clusters on the maximum amount of the beam's deflection. According to this plot, it can be observed that the absolute value of the dimensionless deflection of the beam grows nonlinearly as the volume fraction of the CNTs inside the clusters tends to one. This means that the MSH PNC beam manifests a softer behavior which leads to an increase in the deflection due to the inverse relation between deflection and stiffness. The aforesaid softening behavior of the continuous system is attributed to the negative influence of the availability of the CNT agglomerates on the total stiffness of the constituent nanomaterial. In reverse, the deflection of the beam will be reduced in magnitude when the volume fraction of the cluster becomes greater. This is because of the fact that by adding this term, many small clusters will be replaced with limited big ones which is proven to have lower destructive impact on the total stiffness of the MSH PNC.

Furthermore, the effects of both partial and full patterns of agglomeration, volume fraction of the clusters, and mass fraction of the nanofillers on the deflection amplitude of the MSH PNC beams subjected to a uniform-type bending load are covered in Fig. 3.3. In this figure, the outcomes of former illustrations can be observed again. In fact, addition of the content of the CNTs in the composition of the MSH PNC results in a reciprocal enhancement in the stiffness of the nanomaterial which shows its effect on the absolute value of the deflection amplitude in a decreasing manner. The same trend can be observed as the volume fraction of the clusters will be added. It can be figured out that existence of a limited number

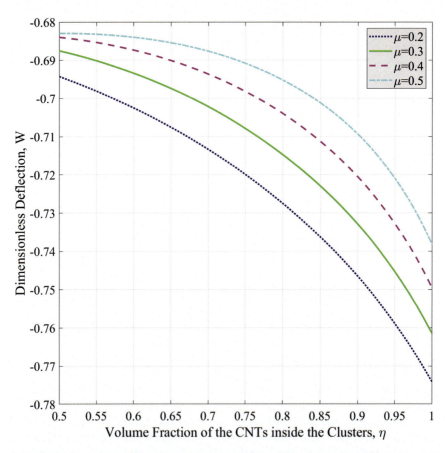

FIGURE 3.2 Variation of the dimensionless maximum deflection of the MSH PNC beam subjected to uniform bending load against volume fraction of the CNTs inside the clusters for various volume fractions of the clusters ($L/h = 10, p = 2, w_r = 0.1, V_F = 0.2$).

of big clusters corresponds with lower deflection of the MSH PNC beam in comparison with the case of existence of a large number of locally distributed small clusters in the media. On the other hand, it is illustrated that full agglomeration case leads to higher deflection amplitude compared with the partial case. The reason of this phenomenon is that nonaggregated CNTs can act as efficient reinforcing nanofillers in the media if the partial agglomeration is happened.

In Fig. 3.4, the impacts of wavy shape and content of the reinforcing CNTs on the deflection amplitude of the MSH PNC beams subjected to uniform loading are shown. Clearly, the waviness phenomenon can play a crucial role in the determination of the static response of the MSH PNC beam. Based on this figure, using volume fraction of $V_r = 1\%$ for the CNTs

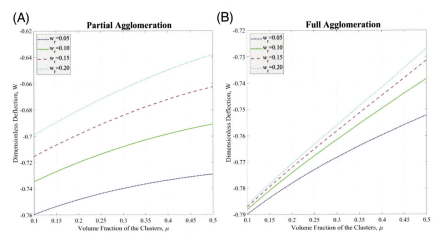

FIGURE 3.3 Variation of the dimensionless maximum deflection of the MSH PNC beam subjected to uniform bending load against volume fraction of the clusters for various mass fractions of the CNTs ($L/h = 10$, $p = 2$, $V_F = 0.2$).

with straight shape can intensify the stiffness of the MSH PNC more than the case of implementing wavy CNTs with volume fraction of $V_r = 2\%$. It is noteworthy that the aforementioned trend can be sensed easily in low contents of the macroscale carbon fibers (CFs). Indeed, the deflection amplitudes corresponding with each of the considered cases will tend to a steady value as the volume fraction of the CFs in the MSH PNC grows. In addition, it is worth regarding that implementation of a limited content of the CFs can be resulted in a remarkable jump in the static response of the system. Quantitatively, a rough 90% reduction in the deflection amplitude of the beam can be attained by changing the volume fraction of the CFs from zero to $V_F = 0.1$.

In the next illustration, the influence of the waviness coefficient on the deflection amplitude of the MSH PNC beam is shown in Fig. 3.5. According to this figure, the reinforcing efficiency of the CNTs implemented in the composition of the MSH PNC can be manipulated by changing the waviness coefficient. As explained in Section 2.1.1.3 of Chapter 2, high values of the waviness coefficient represent nearly straight CNTs. Hence, it is not strange that the deflection amplitude of the beam is decreased whenever great values are assigned to the waviness coefficient. This trend can be physically justified by remembering the inverse relationship between deflection amplitude and stiffness. Based on the instructions presented in the aforesaid section of Chapter 2, it is suitable to use waviness coefficients like $C_w = 0.35$ to provide data close to the experimental ones.

In all of the above examples, the structure was presumed to be subjected to uniform loading and the bending characteristics of continuous systems

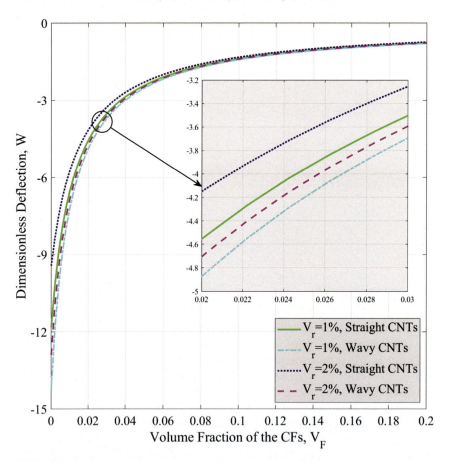

FIGURE 3.4 Variation of the dimensionless maximum deflection of the MSH PNC beam subjected to uniform bending load against volume fraction of the CFs for various loadings of both straight and wavy CNTs ($L/h = 10$).

subjected to sinusoidal and point loadings were dismissed. To compensate the above lack in the former results and also capture the impact of the great length-to-diameter ratio of the reinforcing CNTs in the framework of the 3D Mori–Tanaka method, Fig. 3.6 is depicted herein. It can be well observed that the deflection amplitude of the MSH PNC beams will tend to a steady amount once the length-to-diameter ratio of the CNTs grows. Prior to reach its steady value, the deflection amplitude will be increased at first. This is in agreement with the fact that slender CNTs can help to the reinforcement of the PNCs in a better way. So, it is natural to see an upward shift in the deflection curves of the system. Besides, it is shown that beams subjected to point loading experience the greatest deflection amplitudes followed by those subjected to uniform and sinusoidal loadings, respectively. This

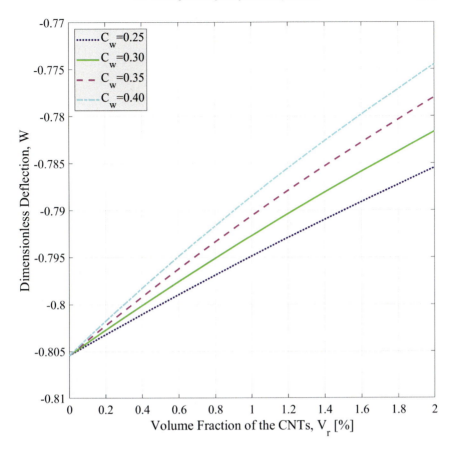

FIGURE 3.5 Variation of the dimensionless maximum deflection of the MSH PNC beam subjected to uniform bending load against volume fraction of the CNTs for various values of waviness coefficient ($L/h = 10$, $V_F = 0.2$).

can be simply understood by taking a look at the definitions presented in Eq. (2.220) for the amplitude of the bending load applied on the MSH PNC beam.

3.1.2 Buckling analysis of MSH PNC beams

Herein, the main concern is to exhibit numerical results for the purpose of tracking the critical buckling load of the MSH PNC beams with both S–S and C–C boundary conditions (BCs) while issues such as aggregation of the nanofillers, the curvy shape of the reinforcing CNTs, and the 3D random orientation of the CNTs in the matrix are taken into consideration. Similar to Section 3.1.1, the shear deformation effects on the approximation of the

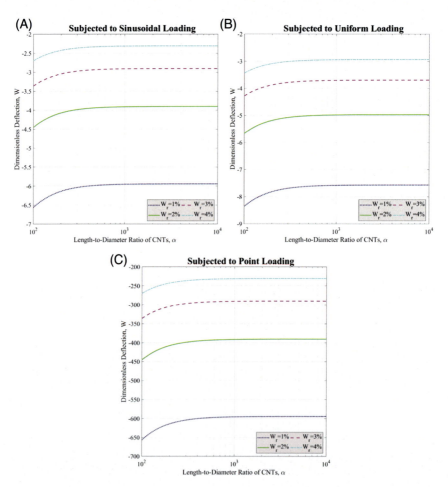

FIGURE 3.6 Variation of the dimensionless maximum deflection of the MSH PNC beam subjected to (A) sinusoidal, (B) uniform, and (C) point load against CNTs' length-to-diameter ratio for various weight fractions of the nanofillers ($L/h = 10$, $V_F = 0.2$).

system's buckling load are captured and therefore, thick PNC structures are analyzed. Like previous subsection, the following dimensionless form of the foundation springs and buckling load are used herein:

$$P_{b,cr} = \frac{PL^2}{D_{11}}, \quad K_W = \frac{k_W L^4}{D_{11}}, \quad K_P = \frac{k_P L^2}{D_{11}} \quad (3.2)$$

The validity of the methodology implemented in this study can be recognized by referring to the data tabulated in Table 3.2. In this table, the buckling loads of CNTR PNC beams are obtained and compared with those reported in Ref. [284]. It is clear that the presented method is able to

TABLE 3.2 Comparison of the dimensionless buckling load of moderately thick CNTR PNC beams ($L/h = 15$).

Type of BC	Volume fraction of the CNTs	Ref. [284]	Present
S–S	0.12	0.098597	0.098316
	0.17	0.150559	0.150019
	0.28	0.220904	0.219274
C–C	0.12	0.213958	0.212638
	0.17	0.344251	0.341437
	0.28	0.455602	0.448723

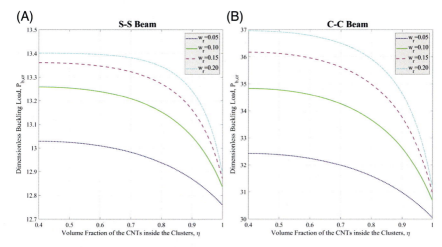

FIGURE 3.7 Variation of the dimensionless buckling load of the MSH PNC beam with (A) S–S and (B) C–C BCs against volume fraction of the CNTs inside the clusters for various mass fraction of the nanofillers ($L/h = 10$, $\mu = 0.4$, $p = 2$, $V_F = 0.2$, $K_W = 20$, $K_P = 2$).

estimate the buckling behaviors of both S–S and C–C beams manufactured from PNCs as well.

Starting with the impact of the availability of the CNT agglomerates in the composition of the MSH PNC on the buckling load of the system, the variation of the dimensionless buckling load versus volume fraction of the CNTs inside the clusters is plotted in Fig. 3.7 once the mass fraction of the CNTs is changed. It is obvious that the beam can be buckled via smaller compressive stimulations if the volume fraction of the CNTs entangled in the clusters grows. This is due to the negative effect of the agglomeration phenomenon on the effective stiffness of the three-phase PNC beam. Because of the fact that the stability load is proportional to the stiffness, it is natural to see such a lessening trend. Besides, the beam

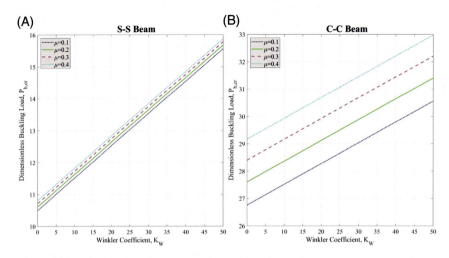

FIGURE 3.8 Variation of the dimensionless buckling load of the MSH PNC beam with (A) S–S and (B) C–C BCs against Winkler coefficient for various volume fractions of the clusters ($L/h = 10$, $\eta = 1$, $w_r = 0.1$, $p = 2$, $V_F = 0.2$, $K_P = 2$).

will be able to tolerate greater buckling loads by adding the mass fraction of the nanofillers because of the marvelous stiffness of the CNTs which leads to a mentionable improvement in the equivalent stiffness of the PNC structure. Meanwhile, beams with C–C BCs support extremer buckling loads according to higher structural stiffness of this type of BC compared with the S–S one.

Moreover, the effect of the volume fraction of the clusters on the determination of the critical buckling load of both S–S and C–C beams is illustrated in Figs. 3.8 and 3.9. In the first figure, the Winkler coefficient is changed and, in the latter, the Pasternak coefficient is varied. According to these figures, the beam will be able to tolerate greater buckling excitations once a bigger value is assigned to the volume fraction of the clusters. As explained in Section 2.1.1.5 of Chapter 2, high volume fractions of the clusters correspond with the state of having greater clusters and moving toward construction of a big unique cluster which contains all of the agglomerated CNTs. Thus, the negative influence of agglomeration phenomenon on the stiffness of the continuous system will be decreased which results in a consequent enhancement in the critical buckling load of the MSH PNC beam. This trend can be observed in any desired value of foundation coefficients and is free from the type of the considered BC. Based on the abovementioned figures, it can be perceived that increment of both of the medium's stiff coefficients can improve the compressive resistance of the PNC beam against buckling excitations. Obviously, the Pasternak spring can do the aforesaid strengthening task better than the Winkler one.

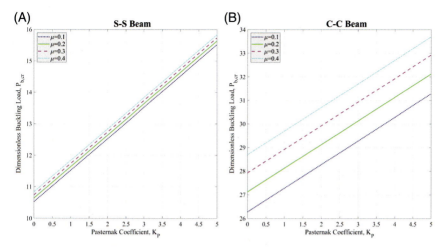

FIGURE 3.9 Variation of the dimensionless buckling load of the MSH PNC beam with (A) S–S and (B) C–C BCs against Pasternak coefficient for various volume fractions of the clusters ($L/h = 10$, $\eta = 1$, $w_r = 0.1$, $p = 2$, $V_F = 0.2$, $K_W = 20$).

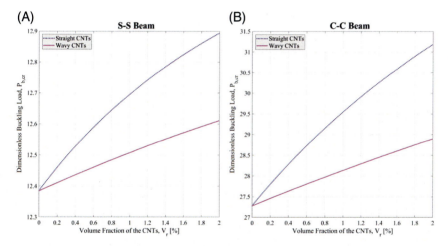

FIGURE 3.10 Variation of the dimensionless buckling load of the MSH PNC beam with (A) S–S and (B) C–C BCs against volume fraction of the CNTs for PNC structures reinforced with both straight and wavy nanofillers ($L/h = 10$, $V_F = 0.2$, $K_W = 20$, $K_P = 2$).

Next, the influence of waviness phenomenon on the buckling performance of the MSH PNC beams will be monitored. To do so, in the first numerical example, this issue is studied qualitatively in the framework of Fig. 3.10. Based on this figure, it can be realized that existence of a curve in the chord of the reinforcing nanofillers can affect the endurable buckling load of the beam negatively. The physical reason of this issue is that the reinforcing efficiency of the CNTs will be decreased due to the available

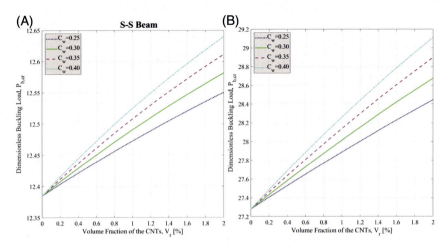

FIGURE 3.11 Variation of the dimensionless buckling load of the MSH PNC beam with (A) S–S and (B) C–C BCs against volume fraction of the CNTs for various waviness coefficients ($L/h = 10$, $V_F = 0.2$, $K_W = 20$, $K_P = 2$).

curve in their shape. Indeed, the entire length of the CNT cannot be aligned in the desired direction in the case of using wavy CNTs and this results in a reduced stiffness in the purposed direction. Hence, it is not strange to see such a trend because of the direct relationship between buckling load and stiffness. Therefore, maybe a higher CNT loading is required to design a powerful element so that bears under axial compression as well. However, this solution cannot be assumed as a permanent alternative because in the case of using a great amount of the CNTs in the composition of the MSH PNC structure, the possibility of the construction of CNT agglomerates in the PNC's microstructure increases. So, the necessity of consideration of the waviness phenomenon in the design of three-phase PNC beams that are supposed to be subjected to compressive loads can be sensed based on this figure. In addition to the above qualitative investigation, another example is depicted in Fig. 3.11 indicating on the role of the waviness coefficient on the quantitative determination of the stability limit of the PNC beams. Based upon this illustration, it is clear that the critical buckling load of the MSH PNC beam will be intensified as higher values are assigned to the waviness coefficient. From Section 2.1.1.3 of Chapter 2, we know that bigger waviness coefficient stands for the case of using CNTs with smaller curvature. Therefore, our estimation will be closer to the state of implementing ideal nanofillers. Based on the above explanation, it can be physically figured out why such a trend exists in the underobservation illustration. It is worth mentioning that C–C beams buckle in bigger compressive loads in comparison with S–S ones.

On the other hand, the impact of the length-to-diameter ratio of the CNTs, distributed in the media in a 3D form, on the buckling performance

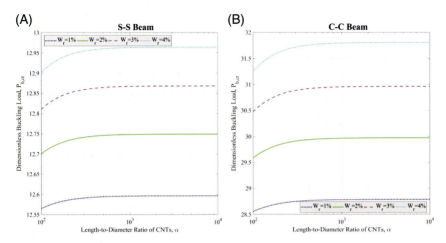

FIGURE 3.12 Variation of the dimensionless buckling load of the MSH PNC beam with (A) S–S and (B) C–C BCs against length-to-diameter ratio of CNTs for various weight fractions of the nanofillers ($L/h = 10$, $V_F = 0.2$, $K_W = 20$, $K_P = 2$).

of the MSH PNC beams rested on Winkler–Pasternak elastic seat is plotted in Fig. 3.12. In this diagram, the plot of critical buckling load versus volume fraction of the CNTs is presented for both S–S and C–C BCs. It can be revealed that utilization of longer CNTs can be resulted in stiffer MSH PNC structures which are able to endure greater buckling loads thanks to their enhanced stiffness. However, it must be considered that the increasing effect of the length-to-diameter ratio on the buckling performance of the beam is obvious in the cases that CNTs with $\alpha = 100$ are replaced with those possessing $\alpha = 1000$. Also, the designer must notice that limitless enhancement of the length of the selected CNTs can be resulted in an increase in the possibility of the appearance of a curve in the shape of the reinforcing nanofillers.

3.1.3 Postbuckling analysis of MSH PNC beams

Within this subsection, the nonlinear postbuckling characteristics of MSH PNC beams with both S–S and C–C BCs will be investigated in the context of the Euler–Bernoulli beam theorem. The impacts of agglomeration and waviness phenomena in addition to the 3D random distribution of the nanofillers in the matrix on the static stability of the continuous PNC system will be included herein. Also, the influence of temperature gradient on the postbuckling path of the MSH PNC beams will be monitored, too. Due to the consideration of the thermal effects on the nonlinear buckling behaviors of the PNC media, the values of coefficient of thermal expansions (CTEs) related to each phase are required to be known.

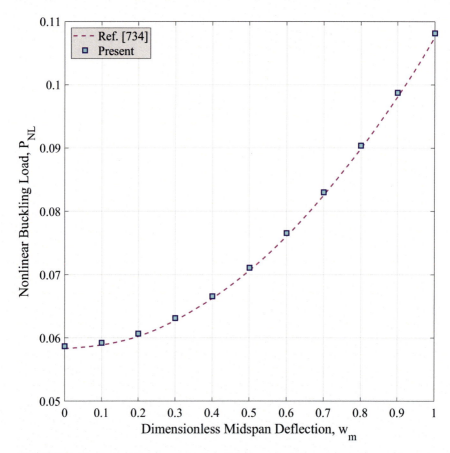

FIGURE 3.13 Postbuckling path of C–C PNC beams reinforced with GPLs ($L/h = 10$, $W_r = 0.3\%$).

The CTE of polymer matrix introduced in Section 2.1.1.2 of Chapter 2 is $\alpha_m = 66 \times 10^{-6}$ (K^{-1}) [373]. Also, the SWCNT(10,10) used there possesses a CTE of $\alpha_r = 2.4 \times 10^{-6}$ (K^{-1}) [732,733]. In the following numerical results, the following dimensionless form of the foundation parameters and postbuckling load will be used:

$$P_{NL} = \frac{P_b L^2}{E_m h^3}, \quad W = \frac{W_m}{h}, \quad K_L = \frac{k_L L^4}{D_{11}}, \quad K_P = \frac{k_P L^2}{D_{11}}, \quad K_{NL} = \frac{k_{NL} L^4}{A_{11}} \quad (3.3)$$

To be confident about the correctness of the results which are going to be provided in the following examples, the postbuckling path of the PNC beams reinforced via graphene platelets are generated and compared with that recommended by researchers in Ref. [734]. The results of this comparison, illustrated in Fig. 3.13, state the accuracy of the presented

3.1 Bending, buckling, and postbuckling of beams

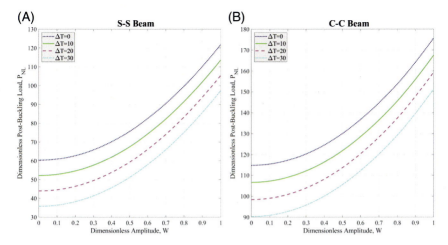

FIGURE 3.14 Thermally influenced postbuckling path of the MSH PNC beams with (A) S–S and (B) C–C BCs ($L/h = 100$, $V_r = 2\%$, $K_L = 100$, $K_P = 10$, $K_{NL} = 1$).

modeling. There exist differences between the predicted results which can be attributed to the difference between the kinematic theories employed herein and in Ref. [734] to derive the governing equations of the problem.

To show the effect of existence of a temperature gradient in the environment that the MSH PNC beam is located in, Fig. 3.14 is depicted. In this diagram, the postbuckling path of the MSH PNC beam is plotted for the cases of having no change in the temperature profile ($\Delta T = 0$) and existence of a temperature gradient in the environment ($\Delta T \neq 0$). It can be well observed that by adding the temperature change, the stability load which can be endured by the PNC beam will be decreased. This reduction can be simply attributed to the negative effect of addition of the local temperature on the stiffness of the material. Indeed, because of the fact that the total CTE of the MSH PNC beam in the studied state is positive and based on our previous insight about the influence of the temperature on the stiffness and compliance of the linearly elastic solids, this trend can be justified physically. However, it must be declared that this result is related to the present case and it is not permanent for all PNC beams. In other words, if another chirality of the CNTs is chosen or an alternative macroscale fiber with negative CTE is implemented in the fabrication of the MSH PNC beam, it is possible to see even an improvement in the postbuckling load of the continua by increasing the temperature. So, such issues must be kept in mind while trying to design MSH PNC elements so that the manufactured device be able to tolerate desired compressive stimulations in thermal environments.

After observing the stability margins of the MSH PNC beams in thermal environments, it is turn to probe the effect of CNT agglomerates on the

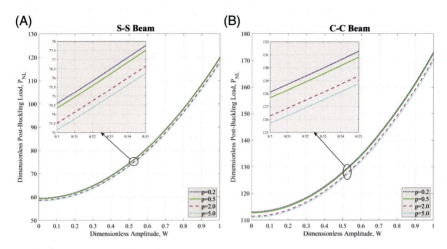

FIGURE 3.15 Postbuckling path of the MSH PNC beams with (A) S–S and (B) C–C BCs for various gradient indices ($L/h = 100$, $\eta = 1$, $\mu = 0.4$, $w_r = 0.1$, $V_F = 0.2$, $K_L = 100$, $K_P = 10$, $K_{NL} = 1$).

postbuckling performance of the system. In Sections 3.1.1 and 3.1.2 of present chapter, the second type of $F(z)$ function introduced in Eq. (2.44) was selected. In the following numerical examples however, the first type will be utilized for the sake of generality. In Fig. 3.15, the influence of gradient index on the postbuckling path of the MSH PNC beams containing aggregated CNTs is presented. According to this figure, it can be realized that the static response of the PNC structure will be shifted downward while a greater value is assigned to the gradient index. This trend is similar to what we previously observed in Sections 3.1.1 and 3.1.2. So, it can be deduced that the type of the $F(z)$ function cannot affect the physics of the problem and only is able to determine the location of the agglomerated nanofillers. The physical reason of the reducing impact of the gradient index on the postbuckling load of the beam is the lessening effect of enhancement of this term on the total stiffness of the MSH PNC material (refer to Fig. 2.6 for detailed explanations). So, based on the direct relation between the static response and stiffness of the system, it is clear that the PNC beam will be postbuckled in smaller excitations as gradient index becomes larger. Like Fig. 3.14, in the present illustration the effect of the BC on the postbuckling path of the beam is shown. As predicted before, beams with C–C BC bear larger buckling excitations rather than those with S–S ends.

The next diagram is allocated to the investigation of the coupled influences of partial/full agglomeration of the nanotubes and mass fraction of the reinforcing CNTs on the postbuckling path of both S–S and C–C beams. Based on Fig. 3.16, the worst situation corresponds with full agglomeration

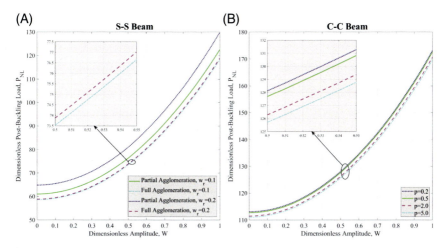

FIGURE 3.16 Postbuckling path of the MSH PNC beams with (A) S–S and (B) C–C BCs for various mass fractions of both partially and fully agglomerated nanofillers ($L/h = 100$, $\mu = 0.4$, $p = 2$, $V_F = 0.2$, $K_L = 100$, $K_P = 10$, $K_{NL} = 1$).

of the CNTs. In this condition, all of the reinforcing nanoparticles will be entangled in the clusters and cannot accomplish their reinforcing task as well. Actually, the destroying impact of the agglomeration phenomenon on the stiffness of the PNC material will be in its utmost degree in this case and thus, the postbuckling load will be decreased proportionally. Obviously, MSH PNC beams containing a greater amount of reinforcements can endure greater postbuckling loads thanks to the positive effect of the nanofillers' extraordinary modulus on the total stiffness of the continua. It is noteworthy that the positive impact of the mass fraction of the CNTs on the postbuckling endurance of the MSH PNC beams will be weakened in the full agglomeration state. So, the agglomeration state can also affect the other aspects of CNTs' reinforcing mechanism, too. Again, it is observed that the branches related to the postbuckling path of the C–C beams are located above those concerned with S–S beams.

Moreover, the effect of volume fraction of the clusters on the determination of the postbuckling path of the MSH PNC beams with various BCs is plotted in Fig. 3.17. By taking a brief look at this figure, it can be easily inferred that if the volume fraction of the clusters is increased, the stability limit of the MSH PNC beam will be aggrandized. In fact, the PNC system can be considered like a unified cluster including all of the CNTs inside whenever the volume fraction of clusters is added. Therefore, the destroying impact of the aggregation of the CNTs on the equivalent stiffness of the nanomaterial will be reduced. So, with regard to the direct relation between postbuckling response and stiffness of the MSH PNC material, it is logical to see such an enhancement in the

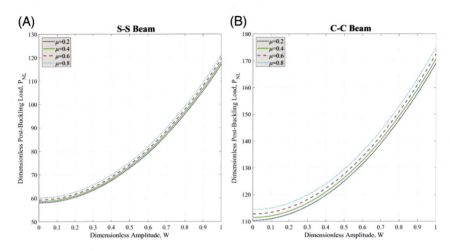

FIGURE 3.17 Postbuckling path of the MSH PNC beams with (A) S–S and (B) C–C BCs for various volume fractions of the clusters ($L/h = 100$, $\eta = 1$, $w_r = 0.1$, $p = 2$, $V_F = 0.2$, $K_L = 100$, $K_P = 10$, $K_{NL} = 1$).

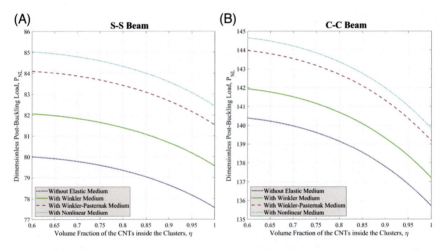

FIGURE 3.18 Variation of the postbuckling load of the MSH PNC beams with (A) S–S and (B) C–C BCs against various volume fractions of the CNTs inside the clusters once the elastic seat of the continuous system is changed ($L/h = 100$, $W = 1$, $\mu = 0.4$, $w_r = 0.1$, $p = 2$, $V_F = 0.2$).

postbuckling path of the beam if volume fraction of the clusters is intensified. Based on the above, it can be concluded that consideration of small values for the volume fraction of the clusters in the theoretical estimations can be resulted to extract a safer nonlinear stability region for the designed elements.

On the other hand, Fig. 3.18 exhibits the variation of the postbuckling load at $W = 1$ versus volume fraction of the CNTs inside the clusters for S–S

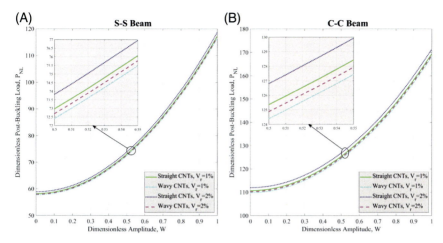

FIGURE 3.19 Postbuckling path of the MSH PNC beams with (A) S–S and (B) C–C BCs for various volume fractions of both straight and wavy CNTs ($L/h = 100$, $V_F = 0.2$, $K_L = 100$, $K_P = 10$, $K_{NL} = 1$).

and C–C beams with/without elastic foundation. It is clear that by placing the beam on the elastic medium, the resistance of the continua against compressive stimulations will be enhanced. The reason is the positive influence of the stiff parameters of the elastic substrate on the total stiffness of the MSH PNC system. In addition, the postbuckling load of the beam will be decreased gradually whenever the volume fraction of the CNTs inside the clusters grows. This trend can be physically justified by pointing the fact that entanglement of the CNTs in the clusters leads to a remarkable reduction in the positive effect of the nanoparticles on the improvement of the total stiffness of the system. Once again, the greatest postbuckling load in any desired η corresponds with C–C beam followed by S–S one.

After a wide discussion about the influence of the agglomeration parameters on the postbuckling path of the beams, Fig. 3.19 shows the nonlinear stability characteristics of MSH PNC beams consisted of different contents of both straight and wavy CNTs. Although ideal estimations recommend a reciprocal improvement in the static response of the system once the CNT loading is increased, the prediction achieved from modified Halpin–Tsai is in accordance with the experiments. Based on the figure, it is obvious that addition of the volume fraction of the CNTs cannot certify the improvement of the postbuckling load of the beam because of the existence of a curve in the shape of the reinforcements. In other words, wavy shape of the CNTs decreases the reinforcing efficiency of the CNTs and thus, the postbuckling load of the beam will be lessened because of the direct relation between stiffness and mechanical response.

In another example, wavy CNTs with various wave amplitudes are implemented to reinforce MSH PNC beams and the postbuckling path

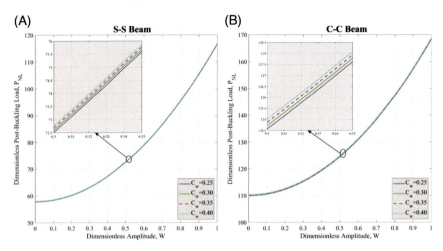

FIGURE 3.20 Postbuckling path of the MSH PNC beams with (A) S–S and (B) C–C BCs for various waviness coefficients ($L/h = 100$, $V_r = 2\%$, $V_F = 0.2$, $K_L = 100$, $K_P = 10$, $K_{NL} = 1$).

of such PNC structures are drawn in Fig. 3.20. In this diagram, different values are assigned to the waviness coefficient and it is observed that MSH PNC beams with bigger waviness coefficients provide greater postbuckling loads. This trend is originated from the fact that utilization of high waviness coefficients corresponds with low amplitudes for the wave existing in the shape of the nanofillers. So, the modeled CNTs will be closer to the ideal ones and an overestimation from the postbuckling performance will be attained. According to the modified Halpin–Tsai model, implementation of $C_w = 0.35$ can be an appropriate selection to reflect the realistic postbuckling path of the MSH PNC beam. In the present illustration, the influence of the BC of the beam on the static stability behaviors of the system is covered, too. Like former examples, herein, C–C beams manifest stiffer behavior from themselves in comparison with S–S ones.

At the end of the numerical examples in this subsection, the variation of postbuckling load of the MSH PNC beams at $W = 1$ is plotted against length-to-diameter ratio of the CNTs for different weight fractions of the nanofillers (see Fig. 3.21). It is revealed that using long CNTs can be resulted in greater postbuckling loads. This trend is logical because of the direct relation between the length of the reinforcing nanofillers and the improvement of the stiffness of the PNC material. It is worth regarding that this effect can be well observed in length-to-diameter ratios smaller than $\beta = 1000$. Indeed, from the referenced value on, the effect of changing the length cannot be sensed clearly. In addition, this figure represents that by adding the value that is assigned to the weight fraction of the CNTs, the postbuckling load of the system will be increased continuously.

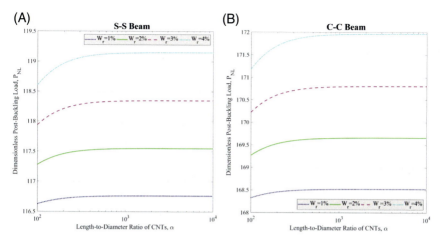

FIGURE 3.21 Variation of the postbuckling load of the MSH PNC beams with (A) S–S and (B) C–C BCs against length-to-diameter ratio of the CNTs for different weight fractions of the nanofillers ($L/h = 100$, $V_F = 0.2$, $K_L = 100$, $K_P = 10$, $K_{NL} = 1$).

3.2 Bending and buckling of plates

In this section, the concentration will be on the investigation of both bending and buckling performances of rectangular plates consisted of MSH PNC materials. Herein, the instructions presented in Sections 2.2.2.2.1 and 2.2.2.2.2 of the previous chapter will be utilized to derive the deflection amplitude and buckling load of the plate, respectively. Like former section about beam-type elements, herein, the homogenization schemes addressing agglomeration of the CNTs, their wavy shape, and random 3D distribution in the polymer will be hired.

3.2.1 Bending analysis of MSH PNC plates

Within this subsection, the deflection amplitude of the MSH PNC plates subjected to various types of bending stimulations will be monitored. The underobservation plate will be modeled with the aid of the refined higher-order shear deformation theory as explained in Section 2.2.2.2.1 of Chapter 2. In some cases, the mode shape will be shown in a 3D plot and the rest of the results will be dedicated to the 2D plots addressing the variation of the deflection amplitude versus contributing parameters. For the sake of simplicity, the below dimensionless parameters are considered in what follows:

$$W = \frac{100 E_m h^3 \sum_{m=1}^{\infty} (W_{bm} + W_{sm}) \sin(\alpha x) \sin(\beta y)}{q_0 a^4}, \quad X = \frac{x}{a}, \quad Y = \frac{y}{b}$$

(3.4)

TABLE 3.3 Comparison of the dimensionless deflection of SSSS plates consisted of CNTR PNCs ($V_{CNT} = 0.11$, $q = 0.1$ MPa).

a/h	Distribution pattern through thickness	Ref. [285]	Present
10	Uniform	3.739e−3	3.717e−3
	FG-V	4.466e−3	4.463e−3
	FG-O	5.230e−3	5.248e−3
	FG-X	3.177e−3	3.150e−3
20	Uniform	3.628e−2	3.644e−2
	FG-V	4.879e−2	4.874e−2
	FG-O	6.155e−2	6.135e−2
	FG-X	2.701e−2	2.713e−2

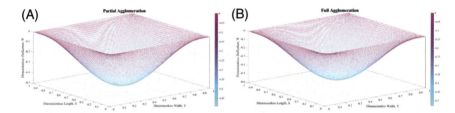

FIGURE 3.22 Fundamental mode shape of MSH PNC square plates subjected to sinusoidal load for (A) partial agglomeration and (B) full agglomeration of the CNTs ($a/h = 10$, $\mu = 0.4$, $w_r = 0.1$, $p = 2$, $V_F = 0.2$).

In Table 3.3, the validity of the results obtained by present methodology is checked by comparing them with those reported in Ref. [285]. It can be realized that the dimensionless deflection of CNTR PNC plates can be accurately predicted by the formulation presented in this text book.

Fig. 3.22 illustrates the mode shape of the MSH PNC plates subjected to sinusoidal loading while the state of the CNT agglomerates is changed. In this figure, both partial and full agglomeration cases are taken into consideration and it is demonstrated that the deflection amplitude of the plate can be monotonically increased if the full agglomeration case is considered. Indeed, this trend is because of the negative influence of the full agglomeration condition on the total stiffness of the MSH PNC material. On the other hand, we are aware of the inverse relation between stiffness and deflection of the structure. So, the observed trend can be justified based on the above explanations easily.

In the next example, the variation of the plate's deflection amplitude against volume fraction of the CNTs inside the clusters is depicted in Fig. 3.23 for the case of changing the mass fraction of the reinforcing

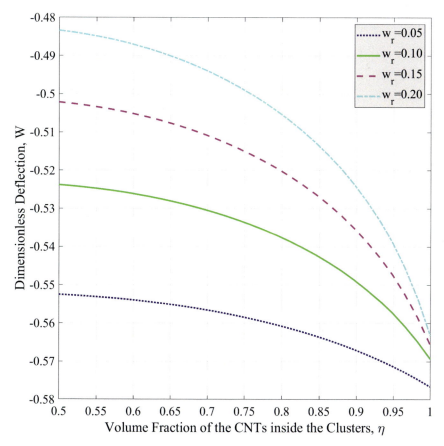

FIGURE 3.23 Variation of the deflection of the MSH PNC plates subjected to sinusoidal load against volume fraction of the CNTs inside the clusters for various values of mass fraction of the nanofillers ($a/h = 10$, $b/a = 2$, $\mu = 0.4$, $p = 2$, $V_F = 0.2$).

nanofillers. It is observed that an increase in the agglomeration parameter η leads to a reciprocal enhancement in the absolute value of the plate's deflection. According to the negative impact of term η on the stiffness of the nanomaterial and with regard to the relation between stiffness and deflection in a bending problem, it is not strange to observe such a trend. On the other hand, it can be inferred that the deflection amplitude of the plate can be reduced in magnitude by increasing the mass fraction of the CNTs. It is obvious that existence of higher amounts of nanofillers in the composition of the nanomaterial can help to the improvement of the equivalent stiffness of the PNC whose consequent result will be a reduction in the absolute value of the deflection amplitude thanks to the inverse relation between deflection and stiffness. It is worth mentioning that the positive effect of the mass fraction of the nanofillers on the bending

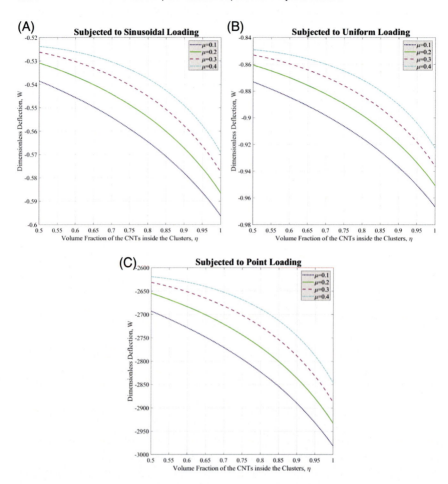

FIGURE 3.24 Variation of the deflection of the MSH PNC plates subjected to (A) sinusoidal, (B) uniform, and (C) point loading against volume fraction of the CNTs inside the clusters for various volume fractions of the clusters ($a/h = 10$, $b/a = 2$, $w_r = 0.1$, $p = 2$, $V_F = 0.2$).

characteristics of the MSH PNC plate can be better observed in lower values of the agglomeration parameter η. So, it can be concluded that the intensity of the efficiency of nanofillers' mass fraction can be influenced by the volume fraction of the CNTs inside the clusters.

In the last two examples, the bending behaviors of plates subjected to sinusoidal loading were tracked and the influence of the loading type on the plate's deflection amplitude was ignored. To cover this issue and with regard to the impact of the agglomeration parameters η and μ on the bending response of the system, Fig. 3.24 is plotted in three subfigures

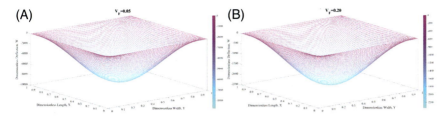

FIGURE 3.25 Fundamental mode shape of rectangular MSH PNC plates subjected to point load for (A) $V_F = 0.05$ and (B) $V_F = 0.20$ ($a/h = 10$, $b/a = 2$, $\eta = 1$, $\mu = 0.4$, $w_r = 0.1$, $p = 2$).

related to sinusoidal, uniform, and point loads, respectively. Again, it can be observed that the absolute value of the deflection amplitude increases as higher content of the reinforcing nanofillers will be entangled in the clusters. This reason of this issue was mentioned in the previous paragraphs and herein, it will not be repeated. However, another trend can be observed in this figure that is related to the effect of the volume fraction of the clusters on the deflection amplitude of the plate. According to this figure, the deflection curve will be shifted upward if a greater value is assigned to the term μ. This trend happens due to the positive influence of adding the volume fraction of the clusters on the reinforcing efficiency of the CNTs which results in a higher stiffness for the material. Based on this and by recalling the relationship between stiffness and deflection, it is clear to see such a positive shift in the deflection diagram. On the other hand, it is clear that the maximum value of deflection amplitude in any arbitrary η and μ corresponds with the case of subjecting the plate to a point load followed by uniform and sinusoidal loads, respectively. Therefore, it is recommended to design MSH PNC plates for the case of being subjected to point load to capture the critical condition of loading.

Based on the last sentence of the previous paragraph, the effect of the volume fraction of the CFs on the fundamental mode shape of a MSH PNC plate subjected to a point load is shown in Fig. 3.25. Based on this figure, it can be seen that an increase in the volume fraction of the CFs results in a limitation in the deflection amplitude of the plate thanks to the positive effect of the CFs on the effective stiffness of the MSH PNC material. In this case study, it is tried to capture a critical condition. In other words, the plate is subjected to point load and the full agglomeration case is taken into account by choosing $\eta = 1$ and $\mu = 0.4$. So, it can be claimed that the provided data are related to at least one of the worst cases if not the worst one.

The agglomeration phenomenon was extensively discussed in Figs. 3.22–3.25. Now, the effect of curvy shape of the CNTs on the deflection

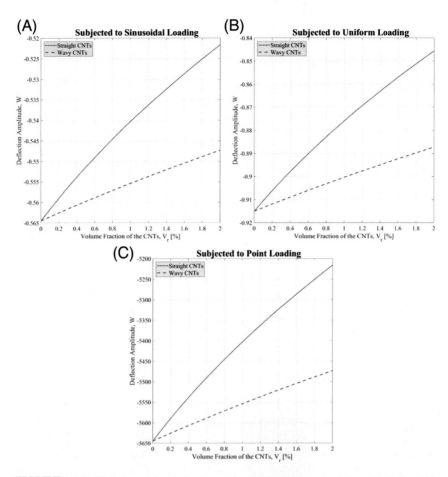

FIGURE 3.26 Variation of the deflection of the MSH PNC square plates subjected to (A) sinusoidal, (B) uniform, and (C) point loading against volume fraction of the CNTs for both straight and wavy nanofillers ($a/h = 10$, $V_F = 0.2$).

amplitude of the MSH PNC plate will be monitored. As the first example, consider the qualitative comparison depicted in Fig. 3.26 for MSH PNC plates reinforced with low contents of the nanofillers. It is obvious that the static response of the problem with be smaller in magnitude if the utilized CNTs are supposed to be ideal ones with straight shape. The physical reason of this trend is the negative influence of waviness phenomenon on the stiffness of the nanomaterial. In other words, because of the lower stiffness of the MSH PNCs reinforced with wavy nanofillers and according to the inverse relation between stiffness and deflection amplitude, PNC structures reinforced with such nonideal nanofillers experience greater deflections. Once again, it can be perceived that plates subjected to

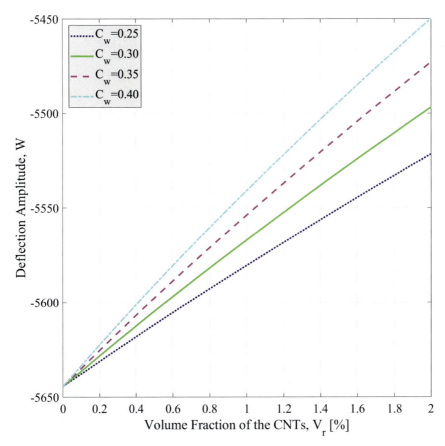

FIGURE 3.27 Variation of the deflection of the MSH PNC square plates subjected to point loading against volume fraction of the CNTs for different waviness coefficients ($a/h = 10$, $V_F = 0.2$).

point load will possess greater deflection amplitudes followed by those subjected to uniform and sinusoidal loads, respectively.

Moreover, the impact of waviness coefficient on the determination of the deflection amplitude of the MSH PNC plates subjected to point load is illustrated in Fig. 3.27. Based on this diagram, the absolute value of the deflection amplitude in any desired volume fraction of the CNTs will be decreased as the waviness coefficient increases. This is because of the fact that if a high value is assigned to the waviness coefficient, the amplitude of the curve which exists in the CNT will be decreased and the CNT will be closer to an ideal one. Consequently, the stiffness of the nanomaterial will be greater in this case and therefore, the plate will experience lower deflections. Free from the abovementioned explanations, it is proven that implementation of waviness coefficient of

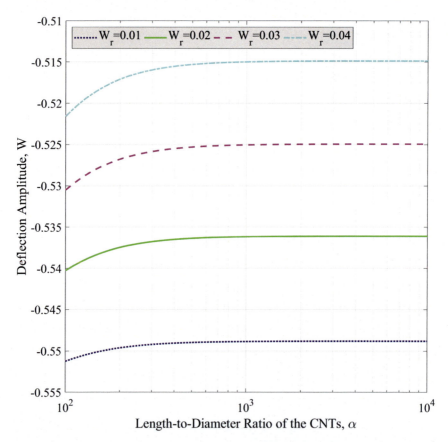

FIGURE 3.28 Variation of the deflection of the MSH PNC square plates subjected to sinusoidal loading against length-to-diameter ratio of the CNTs for different weight fractions of the nanofillers ($a/h = 10$, $V_F = 0.2$).

equal with 0.35 results in a reliable numerical estimation close to experimental measurements.

As the final illustration in this section, the combined impacts of length-to-diameter ratio of the CNTs and weight fraction of the nanofillers on the deflection amplitude of the MSH PNC plate were covered in the context of Fig. 3.28. This figure reveals that slender nanotubes are better candidates for reinforcing MSH PNCs. However, there exists an optimum state for selection of slender CNTs in the composition of the MSH PNC materials. On one hand, this figure exhibits that using CNTs whose length-to-diameter ratio is greater than $\alpha = 1000$ does not have a remarkable influence on the stiffness of the system and thus, it cannot make special changes in the deflection amplitude of the plate. On the other hand, extraordinary values cannot be assigned to the term α because it increases the probability

3.2.2 Buckling analysis of MSH PNC plates

Like previous section, herein, we are about to track the static performance of the MSH PNC plates whenever subjected to axial compression. In what follows, buckling of thick-type plates will be considered and to this goal, the formulation presented in Section 2.2.2.2.2 of Chapter 2 will be employed. Practical issues like aggregation of the nanofillers, existence of a curve in the shape of the CNTs, and 3D random distribution of the nanofillers in the matrix will be covered in the following case studies. To make it easier understand the trends, the below dimensionless terms will be utilized in the future examples:

$$P_{b,cr} = \frac{Pa^2}{E_m h^3}, \quad K_W = \frac{k_W a^4}{E_m h^3}, \quad K_P = \frac{k_P a^2}{E_m h^3} \tag{3.5}$$

To examine the accuracy of the presented static stability results, the buckling loads of present text are compared with those presented in Ref. [377]. The outcome of this comparison is illustrated in Fig. 3.29 for both SSSS and CCCC plates. It can be well observed that the present modeling can regenerate the valid data reported in the former studies.

Starting with the crucial impact of the agglomeration parameters on the buckling behaviors of MSH PNC plates, Fig. 3.30 is plotted for plates with all edges simply supported (SSSS) and clamped (CCCC). Based on this figure, the plate can withstand smaller buckling excitations whenever the volume fraction of the CNTs inside the clusters tends to its maximum possible value. This trend originates from the negative effect of the aforesaid term on the total stiffness of the nanomaterial. In contrast, increment of the volume fraction of the clusters helps the structure to be able to support larger buckling loads due to the positive effect of this parameter on the reinforcing efficiency of the nanofillers. So, as a brief conclusion, it can be stated that consideration of high and low values for η and μ, respectively, leads to a safe design of MSH PNC plates which are aimed to be subjected to buckling stimulations. Furthermore, it is obvious that CCCC plates can tolerate extremer buckling excitations compared with SSSS ones.

The variation of buckling load of MSH PNC plates versus volume fraction of the CNTs inside the clusters for different values of mass fraction of the CNTs is plotted in Fig. 3.31. According to this diagram, it can be figured out that an addition in the mass fraction of the nanofillers can be resulted in an improvement in the axial compressive load which can be endured by the plate-type element. This positive shift of the curve is due to the positive effect of the content of the available CNTs in the composition of hybrid nanomaterial on its stiffness which can be led to a reciprocal increase in

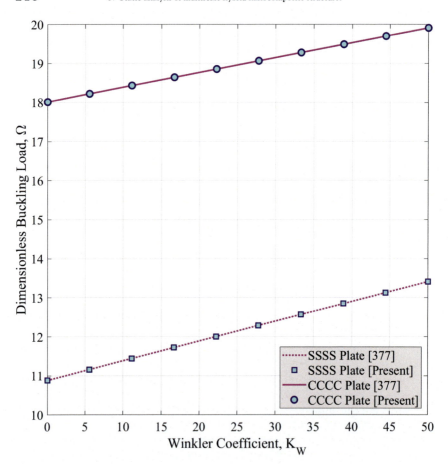

FIGURE 3.29 Comparison of the variation of dimensionless buckling load of MSH PNC square plates rested on two-parameter elastic seat versus Winkler coefficient for both SSSS and CCCC plates ($a/h = 50$, $W_r = 1\%$, $V_F = 0.1$, $\theta = \pi/4$).

the buckling performance of the plate. Again, it is shown that CCCC plates are able to withstand under more critical bidirectional buckling loads in comparison with the SSSS ones. The reason of this trend is the greater structural stiffness of the clamped BC compared with the simply supported one.

In both of the abovementioned examples, there was no stiff medium under the plate. In practice, however, the continuous systems are usually supported by elastic mediums. To give insight about the buckling performance of embedded plates to the readers, Fig. 3.32 illustrates the variation of the buckling load of the MSH PNC plate against Winkler coefficient while different values are assigned to the CFs' volume fraction. It can be

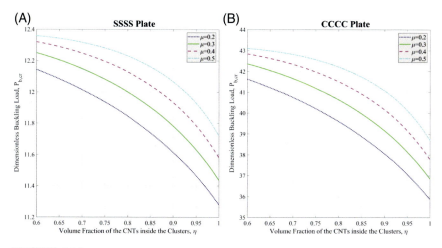

FIGURE 3.30 Variation of the dimensionless buckling load of MSH PNC square plates with (A) SSSS and (B) CCCC BCs against volume fraction of the CNTs inside the clusters for different volume fractions of the clusters ($a/h = 10$, $w_r = 0.1$, $p = 2$, $V_F = 0.25$, $K_W = 0$, $K_P = 0$).

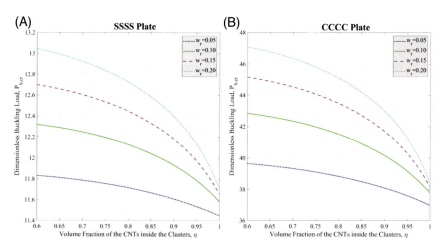

FIGURE 3.31 Variation of the dimensionless buckling load of MSH PNC square plates with (A) SSSS and (B) CCCC BCs against volume fraction of the CNTs inside the clusters for different mass fractions of the nanofillers ($a/h = 10$, $\mu = 0.4$, $p = 2$, $V_F = 0.25$, $K_W = 0$, $K_P = 0$).

simply realized that an increase in the stiff coefficient of the Winkler spring can be resulted in an enhancement in the buckling load of the plate. In other words, it can be inferred that if the plate is rested on a stiff substrate, it will be stiffer and due to the direct relation between stiffness and buckling response, the resistance of the continua against buckling-type failure will be increased. Besides, it is clear that an increase in the volume fraction of

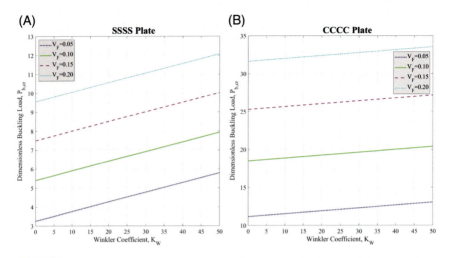

FIGURE 3.32 Variation of the dimensionless buckling load of MSH PNC square plates with (A) SSSS and (B) CCCC BCs against Winkler coefficient for different volume fractions of the CFs ($a/h = 10$, $\eta = 1$, $\mu = 0.4$, $w_r = 0.1$, $p = 2$, $K_P = 0$).

the CFs leads to a reciprocal improvement in the endurable buckling load of the MSH PNC plate thanks to the positive influence of the CFs on the total stiffness of the PNC system.

In the above diagram, the Pasternak spring was assumed to be absent. To measure the effect of this spring on the buckling performance of both SSSS and CCCC plates, the variation of buckling load versus Pasternak coefficient is illustrated in Fig. 3.33 when the gradient index is changed from 0.2 to 5.0. It can be observed that a small increase in the value of Pasternak coefficient can be resulted in a noticeable improvement in the compressive load which can be tolerated by the MSH PNC plate. It is clear that the origin of this positive effect is the direct relation between stiff coefficients of the foundation and equivalent stiffness of the continuous system. Moreover, it is shown that the plate will fail under smaller buckling loads if a great value is assigned to the gradient index. As explained in Fig. 2.6 of Chapter 2, the reason of this trend is the reducing effect of the gradient index on the stiffness of the MSH PNC material whose consequent will be reduction of the buckling endurance of the plate. Similar to Fig. 3.32, herein, CCCC plates manifest stiffer performance compared with SSSS ones. In both of the above figures, the full agglomeration condition is considered to provide safer results for designers of MSH PNC structures subjected to axial excitations.

After a detailed investigation about buckling characteristics of MSH PNC plates with both SSSS and CCCC BCs with regard to the agglomeration phenomenon, it is time to show the negative effect of the curvy-shaped nanofillers on the static stability of the MSH PNC plates.

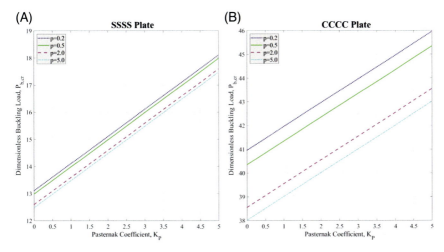

FIGURE 3.33 Variation of the dimensionless buckling load of MSH PNC square plates with (A) SSSS and (B) CCCC BCs against Pasternak coefficient for different gradient indices ($a/h = 10$, $\eta = 1$, $\mu = 0.4$, $w_r = 0.1$, $V_F = 0.25$, $K_W = 20$).

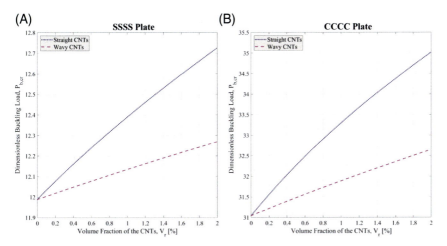

FIGURE 3.34 Variation of the dimensionless buckling load of MSH PNC square plates with (A) SSSS and (B) CCCC BCs against volume fraction of the CNTs for cases of using both straight and wavy nanofillers ($a/h = 10$, $V_F = 0.2$, $K_W = 20$, $K_P = 2$).

To this end and in the first step, the variation of the buckling load of the plate against volume fraction of the reinforcing CNTs is drawn in Fig. 3.34 for MSH PNC materials reinforced via both straight and wavy CNTs. According to this figure, it can be well observed that the stability limit of the MSH PNC plates can be affected dramatically if nonideal wavy CNTs are utilized in the fabrication procedure. This is due to the negative impact of the waviness phenomenon on the stiffness enhancement

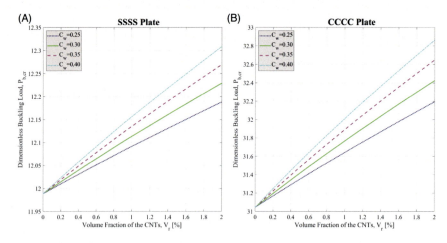

FIGURE 3.35 Variation of the dimensionless buckling load of MSH PNC square plates with (A) SSSS and (B) CCCC BCs against volume fraction of the CNTs for different waviness coefficients ($a/h = 10$, $V_F = 0.2$, $K_W = 20$, $K_P = 2$).

mechanism in the nanomaterial. Like previous examples, in this figure, CCCC plates can support greater compressive loading rather than SSSS ones.

Fig. 3.35 is presented for the goal of showing the influence of waviness coefficient on the buckling performance of the MSH PNC plates with either SSSS or CCCC BCs. It can be easily found that higher values are assigned to the endurable buckling load of the plate if the waviness coefficient is increased. This is due to the fact that higher waviness coefficients are corresponding with CNTs closer to the ideal straight-shaped nanofillers. So, the maximum efficiency for the CNTs will be considered in this condition and therefore, the estimated buckling load will be an overestimated value. It is recommended to choose waviness coefficients between 0.3 and 0.4 to gather numerical evaluation which can be trusted for design of such continuous systems in real-world applications.

Finally, the influence of length-to-diameter ratio of the CNTs on the approximation of the buckling load of the plate is illustrated in Fig. 3.36. In this figure, the buckling loads of both SSSS and CCCC plates are provided and it is observed that the response of the CCCC plates is many times bigger than that of the SSSS ones due to the greater structural stiffness of the clamped edge compared with the simply supported one. Based on this figure, implementation of CNTs with great slenderness ratios can be resulted in an increase in the buckling response of the plate. This reality however is limited to some issues. First of all, it is clear that the length-to-diameter ratio cannot be increased limitless because it increases the possibility of waviness phenomenon. On the other hand, CNTs with length-to-diameter

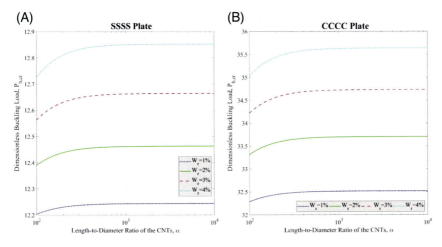

FIGURE 3.36 Variation of the dimensionless buckling load of MSH PNC square plates with (A) SSSS and (B) CCCC BCs against length-to-diameter ratio of the CNTs for different weight fractions of the nanofillers ($a/h = 10$, $V_F = 0.2$, $K_W = 20$, $K_P = 2$).

ratios greater than $\alpha = 1000$ cannot help to the improvement of the stability margin of the structure. So, it is logical to select a mid-range value to satisfy all of the design criteria.

3.3 Buckling of shells

In this section of this textbook, we are about to analyze the static stability characteristics of cylindrical shells manufactured from MSH PNC materials. To this end, the mathematical framework introduced in Section 2.2.3.2.1 of Chapter 2 will be followed to extract the buckling load of the shell whose constitutive modeling will be accomplished like what introduced in Sections 2.1.1.3, 2.1.1.5, and 2.1.1.6 of the same chapter. To go through the details related to the buckling performance of such PNC systems, it is better to introduce the dimensionless forms of the buckling load and foundation parameters which are going to be utilized in the following case studies. So, consider the below dimensionless variables:

$$P_{b,cr} = \frac{PL^2}{1000 E_m h^3}, \quad K_W = \frac{k_W L^4}{E_m h^3}, \quad K_P = \frac{k_P L^2}{E_m h^3} \qquad (3.6)$$

It is worth mentioning that in the following examples, the considered structure is a thin-walled one as the length-to-thickness ratio is fixed on $L/h = 100$ and the shell's radii are considered to be one fifth of its length (i.e., $L/R = 5$). To be sure about the accuracy of the future results, a

TABLE 3.4 Comparison of the critical buckling load of the cylindrical PNC shells reinforced with GPLs.

R/t	L/R	Ref. [735]	Present
30	2	499602	498232
	3	489986	488877
	4	485494	484419
40	2	498381	491878
	3	488982	484389
	4	485396	481341
50	2	496553	488930
	3	488608	482309
	4	485805	479916

comparison between the critical buckling load of PNC shells obtained with the aid of FEM [735] and those achieved from present study is accomplished whose results are appeared in Table 3.4. Based on this table, the validity of this modeling can be guaranteed. It is noteworthy that small differences between the results originate from the fact the FE answers are stiffer compared with analytical ones.

As the first example in this section, the variation of the buckling load against circumferential wave number for shells with or without elastic substrate is drawn in Fig. 3.37. In this figure, an extreme case is considered so that all of the nanofillers are entangled in the local clusters available in the microstructure of the continuous system. According to this illustration, it can be realized that shells rested on an elastic medium are able to provide greater buckling loads compared with bare shells. The reason of this trend is of course the positive effect of the stiffnesses of the elastic springs on the total stiffness of the PNC shell. On the other hand, it is clear that among of the integer circumferential wave numbers, the lowest buckling load is related to the wave number of equal with $n = 2$. So, in what follows, the buckling surveys will be accomplished on shells with mode numbers $m = 1$ and $n = 2$ to capture the most critical condition called fundamental buckling problem.

In Fig. 3.38, the variation of the dimensionless buckling load versus volume fraction of the CNTs inside the clusters is plotted whenever different values are assigned to the mass fraction of the CNTs. It can be deduced that an increase in the mass fraction of the reinforcing nanofillers shows its impact on the buckling load of the shell in a positive manner. In other words, MSH PNC shells with higher content of the CNTs are proven to be better candidates to be employed in conditions that a severe buckling stimulation

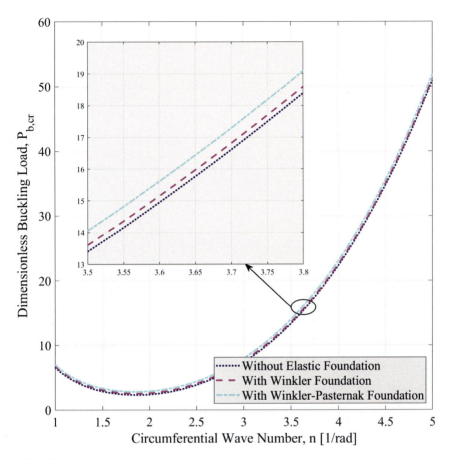

FIGURE 3.37 Variation of the dimensionless buckling load of MSH PNC shells versus circumferential wave number for shells with or without elastic foundation ($m = 1$, $L/h = 100$, $L/R = 5$, $\eta = 1$, $\mu = 0.4$, $w_r = 0.1$, $p = 2$, $V_F = 0.2$).

is going to be applied on the system. This improvement in the buckling performance of the PNC shell is appeared thanks to the extraordinary modulus of the nanotubes compared with the polymer matrix. Moreover, it is shown that the endurable buckling load of the shell will be reduced if a great value is assigned to the agglomeration parameter η. This is because of the negative impact of the aggregation phenomenon on the reinforcing mechanism which leads to a decrease in the equivalent stiffness of the MSH PNC material. So, as a brief conclusion, it is recommended to consider PNC shells with higher degrees of agglomeration that are reinforced with a low content of the nanofillers to achieve safe estimations from the behaviors of the PNC shell while it is subjected to compressive loadings.

Keeping on analyzing the agglomeration-affected buckling characteristics of MSH PNC shells, the influence of the gradient index on the

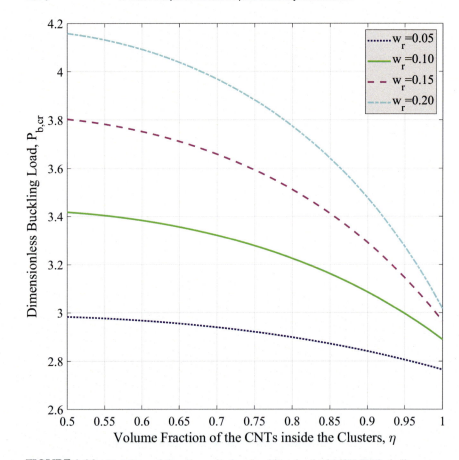

FIGURE 3.38 Variation of the dimensionless buckling load of MSH PNC shells versus volume fraction of the CNTs inside the clusters for different mass fractions of the nanofillers ($m = 1, n = 2, L/h = 100, L/R = 5, \mu = 0.4, p = 2, V_F = 0.2, K_W = 2000, K_P = 200$).

determination of the stability region of the shells is studied in Fig. 3.39. In the aforesaid illustration, the negative effect of the agglomeration parameter on the buckling performance of the shell is observable which was discussed in the former paragraph with physical explanations. The new highlight of this figure is the reducing influence of the gradient index on the buckling load of the MSH PNC shell. Indeed, this decreasing effect is related to the negative impact of this term on the modulus of the MSH PNC material which is in direct relation with the determination of the equivalent stiffness of the nanoengineered system. So, assignment of high values to the gradient index results in reaching a safer theoretical approximation about the limit of the compressive load which can be tolerated by the MSH PNC shell.

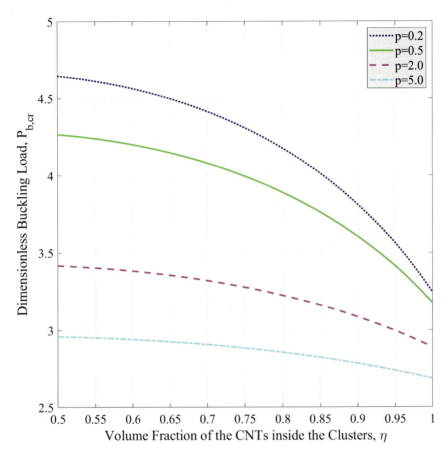

FIGURE 3.39 Variation of the dimensionless buckling load of MSH PNC shells versus volume fraction of the CNTs inside the clusters for different mass fractions of the nanofillers ($m = 1$, $n = 2$, $L/h = 100$, $L/R = 5$, $\mu = 0.4$, $w_r = 0.1$, $V_F = 0.2$, $K_W = 2000$, $K_P = 200$).

As the final example about the influence of agglomeration phenomenon on the buckling performance of the shell, a plot similar to what drawn in Fig. 3.39 is shown in Fig. 3.40 which is mainly depicted to highlight the importance of the agglomeration parameter μ in the approximation of the buckling load of the shell. Based on the latter diagram, it is clear that the shell will be able to endure huge buckling excitations if a great value is assigned to the volume fraction of the clusters. The physical reason of this trend can be stated to be the lower destructive impact of the large clusters on the total stiffness of the nanomaterial in comparison with that of a large number of small clusters. In other words, an increase in the volume fraction of the clusters means a movement toward assumption of having a unified cluster whose space is filled with a large number of nanofillers. In this state,

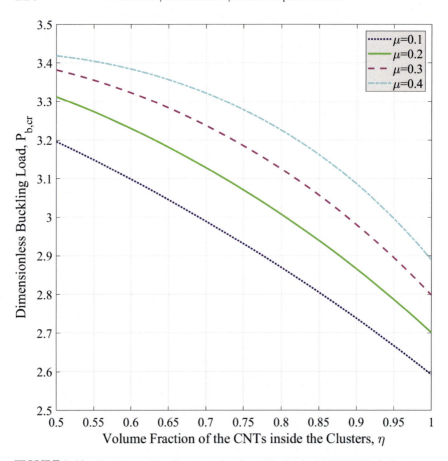

FIGURE 3.40 Variation of the dimensionless buckling load of MSH PNC shells versus volume fraction of the CNTs inside the clusters for different mass fractions of the nanofillers ($m = 1, n = 2, L/h = 100, L/R = 5, p = 2, w_r = 0.1, V_F = 0.2, K_W = 2000, K_P = 200$).

the destroying impact of the agglomeration on the stiffness of the PNC will be reduced in magnitude whose consequence will be an increase in the buckling limit of the cylinder.

Now, a general insight about the agglomeration-affected buckling performance of the MSH PNC shells is given to the readers. The next crucial mechanism that can reduce the reinforcing efficiency of the nanofillers is the possibility of existence of a curve in the shape of the CNTs. In the next two illustrations, this issue is going to be analyzed via qualitative and quantitative viewpoints, respectively. Based on Fig. 3.41, the difference between buckling load supported by MSH PNC cylinders reinforced with straight and wavy CNTs is shown. It can be easily realized that waviness phenomenon can reduce the ability of the continuous system to resist

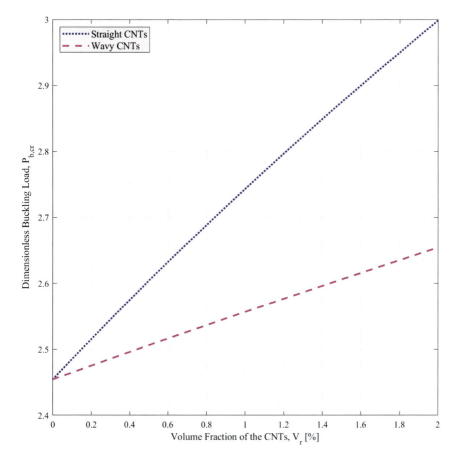

FIGURE 3.41 Variation of the dimensionless buckling load of MSH PNC shells versus volume fraction of the CNTs for PNCs reinforced via straight and wavy nanofillers ($m = 1$, $n = 2$, $L/h = 100$, $L/R = 5$, $V_F = 0.2$, $K_W = 2000$, $K_P = 200$).

against buckling excitations in any desired loading of the nanofillers. However, this effect can be sensed easier if a high content of the CNTs is implemented in the fabrication procedure. The physical reason of this negative effect is that wavy shape of the CNTs avoids from ideal positioning of the nanofillers in the desired direction. So, the stiffness of the PNC will be reduced due to this defect in the CNTs' shape. Therefore, the cylindrical shell will fail under lower compressive loadings.

The emphasis in Fig. 3.42 is on the important role of the waviness coefficient on the buckling load of the shell. As shown in this diagram, higher buckling load corresponds with the cases that a great value is assigned to the waviness coefficient. In such cases, the amplitude of the dome-shaped curve of the CNT will be smaller and hence, the nanofiller

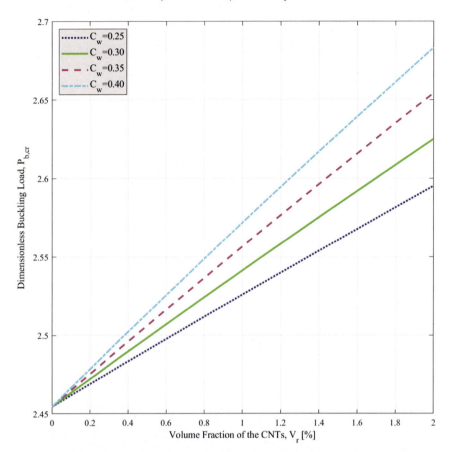

FIGURE 3.42 Variation of the dimensionless buckling load of MSH PNC shells versus volume fraction of the CNTs for different waviness coefficients ($m = 1$, $n = 2$, $L/h = 100$, $L/R = 5$, $V_F = 0.2$, $K_W = 2000$, $K_P = 200$).

will be assumed to be closer to an ideal one. Therefore, its reinforcing efficiency will be higher than the case of assigning a small value to the waviness coefficient. To satisfy the condition close to what takes place in practice, it is recommended to choose waviness coefficients between 0.3 and 0.4. Such a selection results in an estimation of the buckling behaviors of the PNC shell which can be relied on as a guess from the system's reaction if it is subjected to axial compression.

At the end of this section, Fig. 3.43 is presented in order to show the influence of the length-to-diameter ratio of the CNTs on the buckling load which can be supported by the MSH PNC shells. In this diagram, four values are assigned to the weight fraction of the nanofillers and it

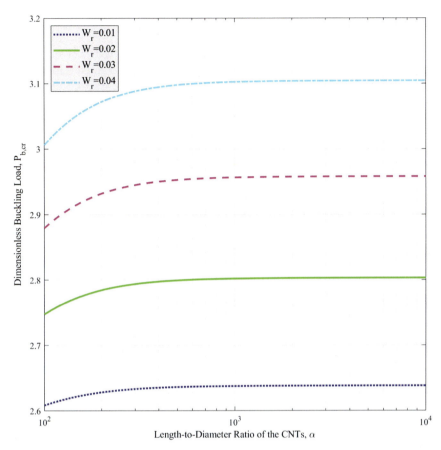

FIGURE 3.43 Variation of the dimensionless buckling load of MSH PNC shells versus length-to-diameter ratio of the CNTs for different weight fractions of the nanofillers ($m = 1$, $n = 2$, $L/h = 100$, $L/R = 5$, $V_F = 0.2$, $K_W = 2000$, $K_P = 200$).

is observed that addition of the CNTs' content in the composition of the nanomaterial leads to an upward shift in the buckling load curve of the shell. This is exactly due to the extraordinary stiffness of the CNTs which leads to a remarkable enhancement in the equivalent stiffness of the PNC system. Furthermore, this figure reveals that the outcome of an increase in the length-to-diameter ratio of the CNTs will be a reciprocal improvement in the compressive load which can be endured by the shell-type element. Indeed, if thinner CNTs are utilized in the fabrication of the MSH PNC systems, they will be able to promote a stiffer behavior from themselves. But this issue has an efficiency range. For example, by taking a brief look at Fig. 3.43 it can be perceived that utilization of CNTs with length-to-diameter ratios bigger than $\alpha = 1000$, cannot help to the improvement of the endurable buckling load of the MSH PNC shell. Free from our observations based on this diagram, it is not logical to use extraordinary long CNTs in the manufacturing of the MSH PNC shells because this decision increases the probability of existence of curvy-shaped nanofillers in the microstructure of the PNC system which is not a desired outcome for the designer engineer at all.

3.4 Concluding remarks

In what we discussed in this chapter, the concentration was on the theoretical micromechanics-assisted analysis of the static behaviors of MSH PNC beams, plates, and shells subjected to different types of loadings. The influence of three major phenomena, namely agglomeration of CNTs, their wavy shape, and their random distribution in the polymer matrix, on the static response of the nanomaterial-made continuous systems was captured. Herein, the authors are aimed to interpret the provided material with an engineering-based point of view.

According to the several case studies presented in this chapter, it can be declared that an improper theoretical estimation from the mechanical behaviors of PNC systems subjected to different types of loadings can be resulted in unbelievable destructions. As the opening example, consider a MSH PNC system designed to be able to resist against buckling-type failure mode. If the probability of existence of CNT agglomerates or wavy shape of the reinforcing nanofillers is dismissed, what may happen? The answer is easy. The overestimation of the designer makes his/her mind to take inaccurate decision. The designer is not aware of the buckling load which can be tolerated by the structure in practice and therefore, the failure takes place in a compressive load smaller than what he/she had in mind.

The same story can be valid for the designer who is aimed to apply a bending load on a PNC system. If an accurate approximation of the constitutive behavior of the utilized material is not determined by the designer engineer, the bending deflection of the structure maybe different from what he/she has estimated before. On the other hand, if a designer relies on the theoretical essays that indicate on the positive role of choosing thin nanotubes as the reinforcing elements in the composition of the nanomaterial on the mechanical response of the system, it is possible for him/her to add the possibility of existence of wavy nanofillers in the continua.

Based on the above, the provided material can help the designers of structural systems manufactured from MSH PNC materials to carry out their design with lower percentages of error. However, they are seriously recommended to accomplish experiments in small scales prior to manufacture their designed device.

CHAPTER 4

Dynamic analysis of multiscale hybrid nanocomposite structures

4.1 Wave propagation, free vibration, and nonlinear forced vibration of beams

In this section of the fourth chapter, the dynamic responses of multiscale hybrid (MSH) polymer nanocomposite (PNC) beams will be monitored in the context of illustrative examples. In what follows, the kinematic formulations formerly introduced in Sections 2.2.1.1.2, 2.2.1.1.3, and 2.2.1.2.3 of Chapter 2 will be followed in order to extract the response to the nonlinear forced vibrations, wave propagation, and free vibrations of MSH PNC beams, respectively. Also, the homogenization schemes mentioned in Sections 2.1.1.3, 2.1.1.5, and 2.1.1.6 will be utilized to capture the influences of wavy shape of the nanotubes, aggregation phenomenon, and 3D random distribution of the nanofillers on the system's dynamics, respectively.

4.1.1 Free vibration analysis of MSH PNC beams

In this subsection, the free vibration problem of MSH PNC beams will be studied in detail. As stated in the general explanations of this section, the free vibration problem will be analyzed with the aid of the refined-type higher-order shear deformation theory (HSDT) whose complete mathematical description was introduced in Section 2.2.1.2.3 of Chapter 2. Because of the fact that the shear deformation effects are included in the future illustrations, it is tried to analyze thick-walled beams in this section. Hence, it can be considered that in all of the following examples, the slenderness ratio of the beam is assumed to be fixed on $L/h = 10$. It is noteworthy that in the following examples, the below dimensionless

TABLE 4.1 Comparison of the dimensionless natural frequencies of CNTR PNC beams ($L/h = 15$).

Volume fraction of the CNTs, V_{CNT}	Ref. [284]	Ref. [288]	Present
0.12	0.9753	0.9745	0.9628
0.17	1.1999	-	1.1574
0.28	1.4401	-	1.4348

parameters are hired for the sake of simplicity:

$$K_W = \frac{k_W L^4}{E_m h^3}, \quad K_P = \frac{k_P L^2}{E_m h^3}, \quad \Omega = \omega h \sqrt{\frac{\rho_m}{E_m}} \quad (4.1)$$

where Ω stands for the dimensionless natural frequency of the beam's free vibrations. Before starting to discuss about the trends available in the diagrams, the validity of the provided data is double-checked in tabulated form. Glancing Table 4.1, it can be figured out that the methodology introduced in this text book is powerful enough to approximate the linear natural frequency of an oscillatory carbon nanotube-reinforced (CNTR) PNC beam. Small differences between the results achieved from our modeling and those reported in Refs. [284] and [288] are due to the differences between theoretical assumptions. So, the accuracy of the reported results can be simply found based on this table. Although Table 4.1 proved the accuracy of the present model, the authors found it interesting to compare their analytical data with those achieved by them with the aid of the well-known finite element method (FEM) [375]. Based on this comparison tabulated in Table 4.2, it can be perceived that the present model can estimate the natural frequency of MSH PNC beams with a marvelous precision. According to Table 4.2, the positive role of the CNTs' mass fraction and volume fraction of the clusters on the natural frequency of the MSH PNC beam can be easily observed. It must be declared that the tiny differences are originated from the fact that FE answers reported in [375] are stiffer than analytical ones.

As the first case study, the combined impacts of the agglomeration parameters (μ and η) on the determination of the linear natural frequency of the MSH PNC beams are shown in Fig. 4.1. Based on this figure, the variations of the natural frequency versus volume fraction of the CNTs inside the clusters are plotted for both S–S and C–C boundary conditions (BCs). The diagrams reveal that the dimensionless frequency of the PNC system tends to lower values as the volume fraction of the CNTs inside the clusters grows. This trend is because of the negative influence of the agglomeration on the stiffness-enhancement mechanism which can be considered in the PNCs in ideal form. So, this reduction in the stiffness shows itself as a reduction in the natural frequency of the continua. On

4.1 Wave propagation, free vibration, and nonlinear forced vibration of beams

TABLE 4.2 Comparison of the dimensionless natural frequency of MSH PNC beams with S–S BC once mass fraction of the CNTs and volume fraction of the clusters are changed ($L/h = 25$, $p = 1$, $\eta = 0.8$, $V_F = 0.2$).

w_r	μ	Ref. [375]	Present
0.1	0.1	0.576259	0.576228
	0.2	0.579449	0.579390
	0.3	0.582515	0.582416
	0.4	0.585394	0.585244
	0.5	0.587962	0.587756
0.2	0.1	0.577706	0.577594
	0.2	0.583512	0.583312
	0.3	0.591587	0.589451
	0.4	0.602247	0.595785
	0.5	0.612573	0.601871
0.3	0.1	0.579340	0.579087
	0.2	0.593044	0.587491
	0.3	0.608610	0.596990
	0.4	0.625353	0.607251
	0.5	0.641980	0.617486

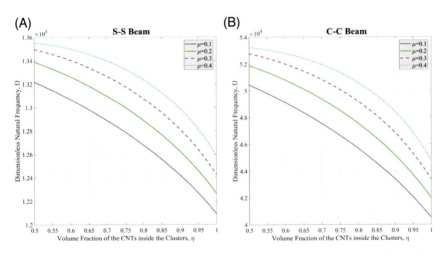

FIGURE 4.1 Variation of the dimensionless natural frequency of MSH PNC beams with (A) S–S and (B) C–C BCs against volume fraction of the CNTs inside the clusters for different volume fractions of the clusters ($L/h = 10$, $w_r = 0.1$, $p = 2$, $V_F = 0.2$, $K_W = 20$, $K_P = 2$).

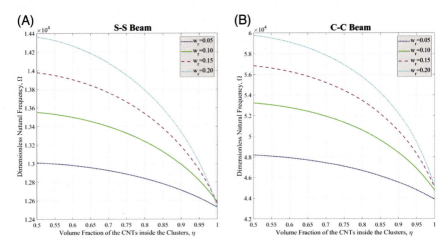

FIGURE 4.2 Variation of the dimensionless natural frequency of MSH PNC beams with (A) S–S and (B) C–C BCs against volume fraction of the CNTs inside the clusters for different mass fractions of the nanofillers ($L/h = 10$, $\mu = 0.4$, $p = 2$, $V_F = 0.2$, $K_W = 20$, $K_P = 2$).

the other hand, this figure denotes that an increase in the volume fraction of the clusters is able to magnify the dimensionless frequency of the MSH PNC beam. The reason of the latter trend is the positive movement of the entire media toward construction of a unified cluster containing all of the entangled CNTs inside. Therefore, the stiffness of the nanomaterial will be lower affected by the agglomeration phenomenon in this circumstance. Furthermore, it is obvious that hybrid PNC beams with C–C supports can provide greater natural frequencies compared with those with S–S BC. The physical justification for this trend is the greater structural stiffness of the clamped support compared with the simply supported one.

In the next example, the effect of mass fraction of the nanofillers on the dynamic responses of the MSH PNC beam is covered for continuous systems with both S–S and C–C BCs. According to Fig. 4.2, it can be perceived that the PNC beam will promote a stiffer behavior if a greater value is assigned to the mass fraction of the CNTs. The reason of this issue is the extraordinary modulus of the nanoengineered CNTs which makes it possible for the PNC system to exhibit stiffer nature if the content of the nanofillers in the composition of the PNC is increased. It is worth regarding that in the cases close to the full agglomeration state, that is, corresponding with $\eta = 1$, the positive impact of the nanofillers' mass fraction on the system's dynamic response can be replaced with a negative influence. This reveals that the agglomeration parameter η cannot only decrease the system's frequency with its increase, but it can also change the effects of other terms on the dynamic response of the continuous system. Like previous diagram, the dimensionless frequency of the beam will be

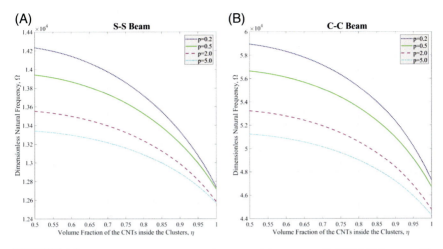

FIGURE 4.3 Variation of the dimensionless natural frequency of MSH PNC beams with (A) S–S and (B) C–C BCs against volume fraction of the CNTs inside the clusters for different gradient indices ($L/h = 10$, $\mu = 0.4$, $w_r = 0.1$, $V_F = 0.2$, $K_W = 20$, $K_P = 2$).

decreased as the volume fraction of the CNTs inside the clusters becomes greater. This issue was justified in the former paragraph in detail. Again, it is observed that in any arbitrary agglomeration parameter η and mass fraction w_r, the peak frequency corresponds with MSH PNC beams with C–C BC.

In addition, the effect of gradient index on the dynamic behaviors of MSH PNC system can be clearly observed in Fig. 4.3. In this diagram, both x- and y-axes of the plot are as same as those appeared in the previous figures. So, it is natural to see the destroying effect of the agglomeration parameter η on the system's dynamic behavior in this figure, too. Based on Fig. 4.3, the natural frequency of the MSH PNC beam will be reduced as a greater value is assigned to the gradient index p. To justify this trend physically, it is enough to recall the trends previously shown in Fig. 2.6. According to the mentioned diagram, free from the type of selected $F(z)$ function, the Young's modulus of the PNC will be decreased all over the structure's thickness if the gradient index is intensified. Therefore, it is not strange to see such a reducing trend in the variations of the natural frequency because of the direct relation between frequency and total stiffness of the PNC system. On the other hand, it is another time certified that thanks to the greater structural stiffness of the clamped BC, the C–C beams are able to support higher frequencies compared with S–S ones.

In the final illustration concerned with the issue of nanofillers' aggregation, the variation of the dimensionless natural frequency of the MSH PNC beams versus volume fraction of the macroscale reinforcing fibers is drawn

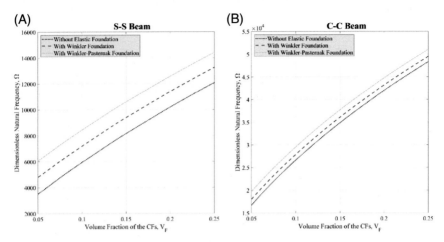

FIGURE 4.4 Variation of the dimensionless natural frequency of MSH PNC beams with (A) S–S and (B) C–C BCs against volume fraction of the CFs for various types of elastic seats for the structure ($L/h = 10$, $\eta = 1$, $\mu = 0.4$, $w_r = 0.1$, $p = 2$).

in Fig. 4.4. In this figure, the effect of existence of elastic substrate under the beam on the dynamic response of the system is covered. According to this example, it can be easily perceived that higher frequencies correspond with MSH PNC beams rested on Winkler–Pasternak medium followed by those rested on Winkler medium and those not rested on a substrate, respectively. The reason of this order is the positive impact of the stiff coefficients related to the foundation's springs on the equivalent stiffness of the MSH PNC beam. In addition to this trend, it can be observed that C–C beams, consisted of PNCs with any desired volume fraction of the carbon fibers (CFs), can support greater frequencies rather than S–S ones. Also, it can be evidently realized that an increase in the volume fraction of the CFs results in a reciprocal enhancement in the natural frequency of the continua. Clearly, the reason is the positive effect of addition of CFs to the composition of the MSH PNC on the total stiffness of the nanomaterial.

Furthermore, the effect of existence of a curve in the chord of the CNTs on the dynamic characteristics of MSH PNC beams is going to be analyzed in Figs. 4.5 and 4.6 from qualitative and quantitative viewpoints, respectively. In Fig. 4.5, the variation of dimensionless natural frequency of the MSH PNC beams against volume fraction of the nanofillers is plotted for hybrid systems consisted of both straight and wavy CNTs. It can be well observed that in both cases, the PNC beam will be able to support higher frequencies whenever the content of the reinforcing CNTs is aggrandized. However, the amplitude of this strengthening mechanism varies for hybrid PNCs reinforced with straight and wavy CNTs. It is clear that implementation of wavy nanofillers for the goal of reinforcing leads

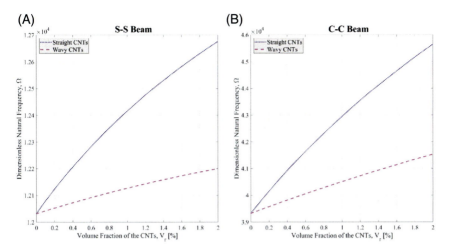

FIGURE 4.5 Variation of the dimensionless natural frequency of MSH PNC beams with (A) S–S and (B) C–C BCs against volume fraction of the CNTs for PNC beams consisted of either straight or wavy nanofillers ($L/h = 10$, $V_F = 0.2$, $K_W = 20$, $K_P = 2$).

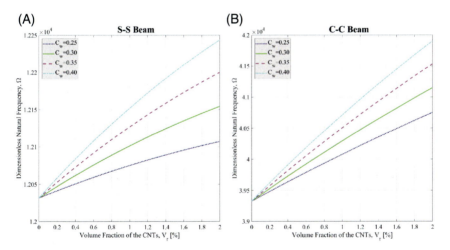

FIGURE 4.6 Variation of the dimensionless natural frequency of MSH PNC beams with (A) S–S and (B) C–C BCs against volume fraction of the CNTs for different waviness coefficients ($L/h = 10$, $V_F = 0.2$, $K_W = 20$, $K_P = 2$).

to a lower slope in the frequency curve of the system. This is completely due to the negative effect of the curvy shape of the CNTs on the stiffness of the MSH PNC material. So, it is strongly recommended to capture the effect of wavy shape of the CNTs in the theoretical calculations to avoid from unpredicted resonances in the structure.

Besides, the influence of the waviness coefficient on the determination of the natural frequency of both S–S and C–C beams manufactured from

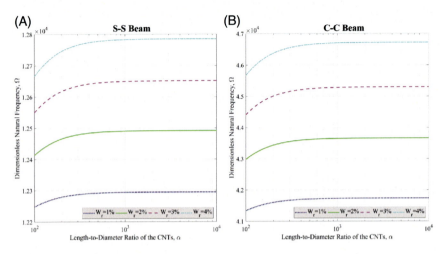

FIGURE 4.7 Variation of the dimensionless natural frequency of MSH PNC beams with (A) S–S and (B) C–C BCs against length-to-diameter ratio of the CNTs for different weight fractions of the CNTs ($L/h = 10$, $V_F = 0.2$, $K_W = 20$, $K_P = 2$).

MSH PNC materials is monitored in the framework of Fig. 4.6. Based on this diagram, it can be recognized that an increase in the value of the waviness coefficient results in aggrandization of the natural frequency of the MSH PNC beam. The reason of this issue is that in the cases of assigning high values to the waviness coefficient, the amplitude of the wave existing in the chord of the CNT will be lessened and thus, the CNTs will be considered to be ideal ones. So, the peak stiffness will be considered for the constituent nanomaterial which leads to an increase in the natural frequency by recalling the direct relation between stiffness and dynamic response of the problem. It is recommended to utilize waviness coefficients between 0.3 and 0.4 to provide reliable data in theoretical calculations. It must be pointed out that like previous diagrams, herein, C–C beams support higher natural frequencies compared with S–S ones.

The concentration of the final case study is on the impact of the length-to-diameter ratio of the CNT on the natural frequency of the MSH PNC beams consisted of randomly oriented nanofillers. According to Fig. 4.7, it is clear that the natural frequency of the MSH PNC beams can be intensified by adding the length-to-diameter ratio of the nanofillers. However, it is obvious that this positive effect is not a limitless one and possesses a range of efficiency. In other words, utilization of CNTs with $\alpha > 1000$ is not a logical choice at all because it does not affect the dynamic response of the nanoengineered system. This selection can just increase the probability of the wavy-shaped being of the nanofillers that is not a desired issue in the designers' point of view. So, it is recommended to choose moderate values and try to synthesize moderately thin nanotubes for the goal of fabricating

MSH PNC materials. Moreover, it is observed that the frequency curve of the beams can be shifted upward as the weight fraction of the CNTs is increased. This is because of the outstanding modulus of the CNTs compared with that of the polymer matrix.

4.1.2 Nonlinear forced vibration analysis of MSH PNC beams

This part is dedicated to the investigation of the nonlinear dynamic response of S–S thin-walled beams fabricated from MSH PNC materials. To this purpose, the aforesaid physical problem is described in mathematical language with the aid of the classical theory of the beams (refer to Section 2.2.1.1.2 of Chapter 2 for details of the derivation procedure). Herein, influences of physical issues like aggregation of the nanoparticles, their curved shape, and the length-to-diameter ratio of the CNTs on the system's nonlinear dynamic characteristics will be discussed in detail. In addition, a geometrical point of view will be utilized incorporated with a linearization, to analyze the phase portraits in some of the case studies. For the sake of simplicity in the calculations, the below dimensionless terms are defined in this subsection:

$$K_L = \frac{k_L L^4}{E_m h^3}, \quad K_P = \frac{k_P L^2}{E_m h^3}, \quad K_{NL} = \frac{k_{NL} L^4}{E_m h}, \quad C_d = \frac{c_d L^2}{h^2 \sqrt{\rho_m E_m}}, \quad F = \frac{\bar{f} M L^2}{E_m h^2} \quad (4.2)$$

In order to be sure that the following illustrations can be trusted, frequency–response curve of CNTR PNC beams, previously reported in Ref. [289], is regenerated and illustrated in Fig. 4.8. According to this diagram, it can be well observed that the frequency–response curve of the CNTR PNC beams can be accurately generated by the present methodology. The differences between the results of our work and those reported in [289] originate from the difference between the kinematic theories utilized to model the problem. However, it is worth mentioning that such differences appear in small slenderness ratios corresponding with thick-walled beams. In other words, it can be guaranteed that in the future case studies, the results of this work are in an excellent agreement with those available in the open literature due to the assumption of thin-walled being of the beam.

As the first example in this subsection, the impact of partial and full agglomeration patterns on the nonlinear dynamic behaviors of MSH PNC beams is studied in Fig. 4.9. The frequency–response and amplitude–response curves of MSH PNC systems with both partial and full agglomeration states can be observed in Fig. 4.9A and B, respectively. According to these subplots, it can be recognized that full agglomeration state reduces the stiffness of the MSH PNC in a worst manner. This issue can be easily

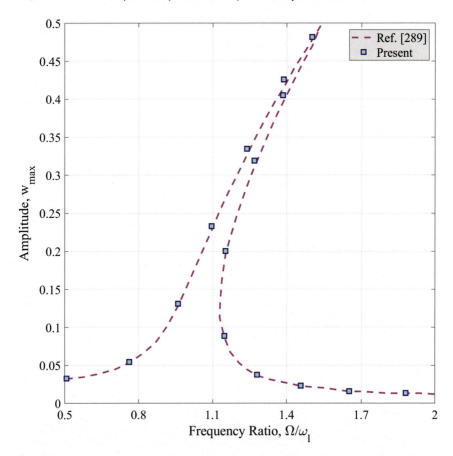

FIGURE 4.8 Comparison of the frequency–response curves of FGΛ-CNTR beams with S–S BC ($L/h = 12$, $c = 0.01$, $f = 0.02$, $\omega_l = 1.3$).

found by looking at Fig. 4.9A simply. Based on this figure, it is clear that the frequency–response curve of the nonlinear system bends to the left in the case of considering the full agglomeration. It means that the system promotes a softer behavior from itself in such a condition. Due to this issue, the unstable region surrounded between dotted vertical lines in both frequency–response and amplitude–response diagrams tends to the left-hand side. This unwanted outcome comes from the destroying impact of the full agglomeration state of the reinforcing CNTs on the total stiffness of the PNC material. Indeed, the full agglomeration case increases the possibility of occurrence of the resonance phenomenon in lower frequencies because it changes the path of the system's backbone curve. In Fig. 4.9C and D, two sample points in the unstable region are chosen to draw the phase portraits corresponding with partial and full

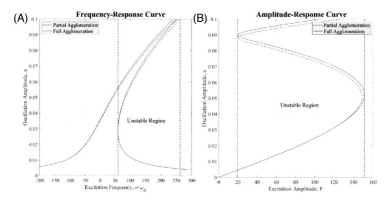

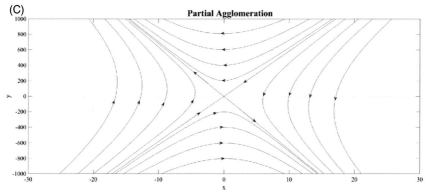

FIGURE 4.9 Effect of both partial and full agglomeration states on the nonlinear forced vibration characteristics of MSH PNC beams ($L/h = 100$, $\mu = 0.2$, $p = 2$, $w_r = 0.1$, $V_F = 0.2$, $C_d = 100$, $K_L = 1000$, $K_P = 100$, $K_{NL} = 10$, and $F = 200$). In subplots (A) and (B), the frequency–response and amplitude–response curves of MSH PNC beams with both partial and full agglomeration states are plotted, respectively. Part (B) is plotted at point $\sigma/\omega_0 = 200$, that is, in the unstable region of the frequency–response diagram. Subplots (C) and (D) correspond with the phase portraits of the linearized system in the unstable region with partial and full agglomeration states, respectively. In the latter subplots, points $(a^*, \sigma/\omega_0) = (0.07328, 150.2)$ and $(a^*, \sigma/\omega_0) = (0.07478, 150.1)$ are chosen, respectively, to plot the trajectories.

agglomeration cases, respectively. To this purpose, the oscillating system is linearized and the Jacobi matrix is constructed by the means of the coordinates of the aforementioned sample points. Due to the fact that the determinant of the enhanced Jacobi matrix is negative, the linearization suggests saddle fixed points in such points. By referring to the Hartman–Grobman theorem, the accomplished linearization in the unstable region can be trusted and considered to be topologically identical with the characteristics of the nonlinear system. According to the phase portraits, it can be realized that the system is consisted of two stable and unstable manifolds related to high- and low-energy limit cycles. In this condition, depending

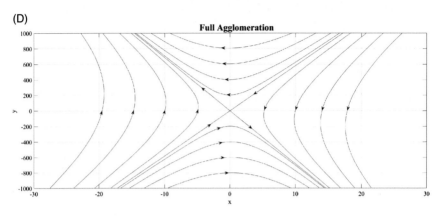

FIGURE 4.9 Continued

on the applied initial condition, the system will tend to the stable limit cycle and approaches to the fixed point or becomes far from it parallel with the unstable manifold.

In the second case study, Fig. 4.10 is depicted to analyze the nonlinear dynamic behaviors of MSH PNC beams whenever different values are assigned to the volume fraction of the clusters. Like previous diagram, both frequency–response and amplitude–response curves of the underobservation system are shown which indicate on the positive influence of aggrandization of the agglomeration parameter μ on the stiffer behavior of the PNC system. The physical reason of this issue is the movement of the nanomaterial's microstructure toward construction of a big cluster containing a large number of aggregated CNTs inside in the case of assigning a big value to the volume fraction of the clusters. Hence, it is natural to see that, for example, the frequency–response is bent to the right in greater values of μ. In subplots Fig. 4.10A and B, it can be simply understood that increment of the volume fraction of the clusters results in enlargement of the unstable region of the system. In the unstable region, more than one value can be assigned to the response of the system in a desired excitation frequency. These distinct values are related to the low- and high-energy limit cycles. Similar to Fig. 4.9, herein, the linearization instructed in Chapter 2 will be employed in order to track the trajectories of the system. Once again, our linearization can be considered to be accurate from qualitative viewpoint because of the fact that linearization suggests a hyperbolic fixed point in the system's unstable region. Even if it is hard for one to recognize the differences between the phase portraits illustrated in Fig. 4.10, it must be kept in mid that the phase portrait related to the case of $\mu = 0.9$ exhibits a stiffer path. In both of these portraits, it can be again mentioned that depending on the location of the initial condition, the trajectories will tend to the stable or unstable limit cycles.

4.1 Wave propagation, free vibration, and nonlinear forced vibration of beams 243

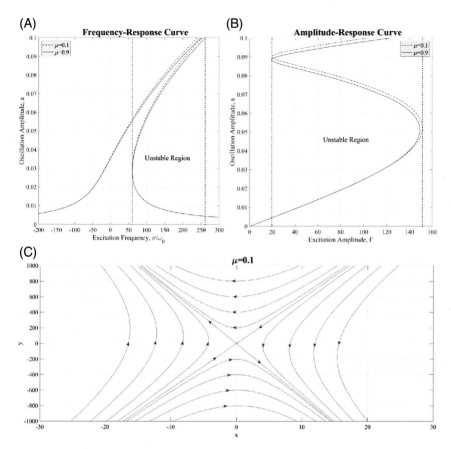

FIGURE 4.10 Effect of volume fraction of the clusters on the nonlinear forced vibration characteristics of MSH PNC beams ($L/h = 100$, $\eta = 1$, $p = 2$, $w_r = 0.1$, $V_F = 0.2$, $C_d = 100$, $K_L = 1000$, $K_P = 100$, $K_{NL} = 10$, and $F = 200$). In subplots (A) and (B), the frequency–response and amplitude–response curves of MSH PNC beams with both partial and full agglomeration states are plotted, respectively. Part (B) is plotted at point $\sigma/\omega_0 = 200$, that is, in the unstable region of the frequency–response diagram. Subplots (C) and (D) correspond with the phase portraits of the linearized system in the unstable region while the volume fraction of the clusters is fixed on 0.1 and 0.9, respectively. In the latter subplots, points $(a^*, \sigma/\omega_0) = (0.07498, 150.2)$ and $(a^*, \sigma/\omega_0) = (0.07348, 150.1)$ are chosen, respectively, to plot the trajectories.

Furthermore, the concentration of Fig. 4.11 is on the determination of the influence of the gradient index on the nonlinear forced oscillation characteristics of MSH PNC beams. Based on the curves drawn in this diagram, it can be perceived that a rise in the value of the gradient index results in a reduction in the total stiffness of the MSH PNC structure. In fact, in high values of gradient index p, frequency–response curve of the system will bend to the left which stands for a softer behavior. This issue can also be certified by taking a look at the amplitude–response curve presented in

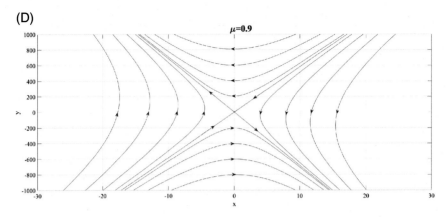

FIGURE 4.10 Continued

Fig. 4.11B. The aforementioned trend could be estimated with the aid of our previous knowledge related to the effect of the gradient index on the stiffness of the PNC material (refer to Fig. 2.6 of Chapter 2). Based on the phase portraits plotted in Fig. 4.11C and D, it can be realized that the system oscillates between two low- and high-energy limit cycles whose connection is a saddle-type fixed point. It must be pointed out that the aforesaid phase portraits are obtained from a linearization procedure which can be trusted because of the fact that it suggests a hyperbolic-type fixed point for the nonlinear system. By considering this issue and recalling our former insight about the physics of such a nonlinear oscillator, the plotted phase portrait can be trusted.

As the final example concerned with both quantitative and qualitative analysis of MSH PNC beams manufactured from agglomerated CNTs, Fig. 4.12 is plotted in order to show the impact of the nanofillers' mass fraction on the system's nonlinear dynamic behaviors. According to Fig. 4.12A, it can be realized that an increase in the mass fraction of the CNTs results in an improvement in the equivalent stiffness of the PNC beam. In other words, the tendency of both high- and low-energy branches of the frequency–response curve to the right-hand side grows as a higher value is assigned to the mass fraction of the CNTs. Therefore, it can be concluded that the nonlinear system behaves stiffer and this is the positive outcome of implementation of ultrastiff CNTs in the composition of the continuous system. All of the abovementioned trends can be observed in Fig. 4.12B, too. Based on the amplitude–response curve, the effect of adding the content of the nanofillers to the composition of the hybrid nanomaterial on the dynamic characteristics of the nonlinear system can be well observed. In addition to frequency–response and amplitude–response curves of the system, the linearization technique is employed in this example in order to find out an equivalent linear system describing the

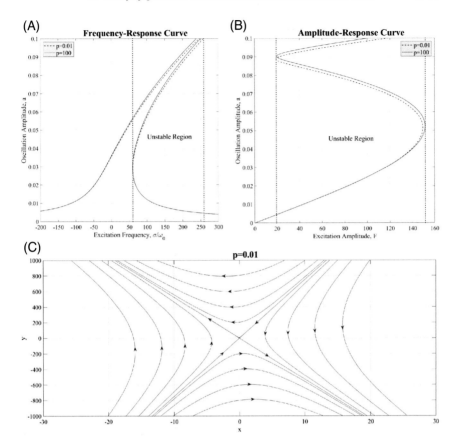

FIGURE 4.11 Effect of gradient index on the nonlinear forced vibration characteristics of MSH PNC beams ($L/h = 100$, $\eta = 1$, $\mu = 0.4$, $w_r = 0.1$, $V_F = 0.2$, $C_d = 100$, $K_L = 1000$, $K_P = 100$, $K_{NL} = 10$, and $F = 200$). In subplots (A) and (B), the frequency–response and amplitude–response curves of MSH PNC beams for different gradient indices are plotted, respectively. Part (B) is plotted at point $\sigma/\omega_0 = 200$, that is, in the unstable region of the frequency–response diagram. Subplots (C) and (D) correspond with the phase portraits of the linearized system in the unstable region while the gradient index is fixed on 0.01 and 100, respectively. In the latter subplots, points $(a^*, \sigma/\omega_0) = (0.07378, 150)$ and $(a^*, \sigma/\omega_0) = (0.07518, 150.2)$ are chosen, respectively, to plot the trajectories.

underobservation system. Again, the linearization can be accepted due to the hyperbolic being of the estimated fixed point. The trajectories of the linearized system indicate on the existence of two stable and unstable limit cycles in the system whose features can be tuned by changing the material properties of the MSH PNC material. It is worth regarding that in the rest of the diagrams, the geometrical viewpoint will be skipped for the sake of brevity.

Investigation of the influence of CNTs' wavy shape on the nonlinear vibrations of harmonically stimulated MSH PNC beams is the main

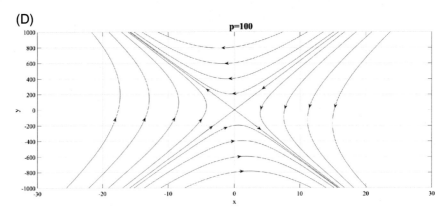

FIGURE 4.11 Continued

objective of Fig. 4.13. It can be observed that utilization of ideal nanofillers results in a stiffer system whose frequency–response and amplitude–response curves promote this issue as well. In other words, the illustrated diagrams indicate on the softening role of the waviness phenomenon on the nonlinear dynamic response of the system. This softening effect originates from the negative influence of the waviness phenomenon on the reinforcing mechanism which is predicted based on the theoretical approximations. Hence, the effect of waviness on the resonance characteristics of the systems must be kept in mind while analyzing MSH PNC systems. The aforementioned destroying effect can be well observed in higher amplitudes in both frequency–response and amplitude–response curves.

Finally, the influence of the 3D distribution of the CNTs in the polymer matrix on the nonlinear characteristics of the oscillating beam is going to be probed in the framework of Fig. 4.14. In this figure, both frequency–response and amplitude–response curves of the MSH PNC beam for different length-to-diameter ratios of the CNTs are drawn. According to this diagram, it can be confirmed that an increase in the length-to-diameter ratio of the CNTs can be resulted in bending of the frequency–response curve of the system to the right-hand side. This consequence corresponds with the better enhancement of the beam's stiffness in the case of utilization of thinner nanofillers. However, it must be mentioned that although this trend is proven in the experimental measurements, it is not logical to increase the length-to-diameter ratio of the CNTs limitless. This is due to the obligation of avoiding from implementation of ultrathin nanofillers in the composition of the nanomaterial which can be resulted in a rise in the probability of existence of wavy CNTs in the nanoengineered system.

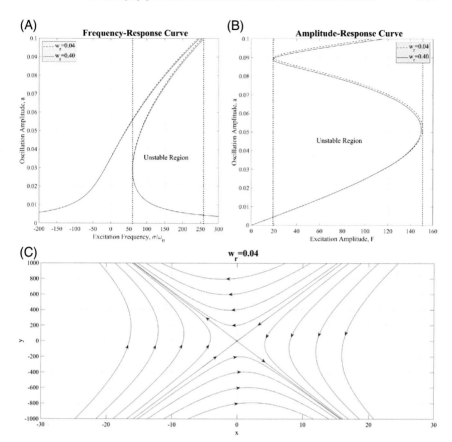

FIGURE 4.12 Effect of nanofillers' mass fraction on the nonlinear forced vibration characteristics of MSH PNC beams ($L/h = 100$, $\eta = 1$, $\mu = 0.4$, $p = 2$, $V_F = 0.2$, $C_d = 100$, $K_L = 1000$, $K_P = 100$, $K_{NL} = 10$, and $F = 200$). In subplots (A) and (B), the frequency–response and amplitude—response curves of MSH PNC beams for different mass fractions of the CNTs are plotted, respectively. Part (B) is plotted at point $\sigma/\omega_0 = 200$, that is, in the unstable region of the frequency–response diagram. Subplots (C) and (D) correspond with the phase portraits of the linearized system in the unstable region while the mass fraction is fixed on 0.04 and 0.4, respectively. In the latter subplots, points $(a^*, \sigma/\omega_0) = (0.07468, 150)$ and $(a^*, \sigma/\omega_0) = (0.07388, 149.9)$ are chosen, respectively, to plot the trajectories.

4.1.3 Wave propagation analysis of MSH PNC beams

In this subsection, the dispersion behaviors of the elastic waves scattered in MSH PNC beams are going to be tracked in the framework of transient analyses. The beam-type element is assumed to be rested on visco-Pasternak medium and subjected to bending force. In order to investigate the propagation characteristics of the PNC beam, the kinematic relations of the Euler–Bernoulli beam hypothesis are employed (refer to

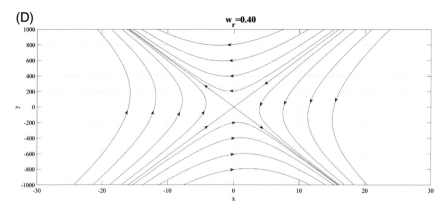

FIGURE 4.12 Continued

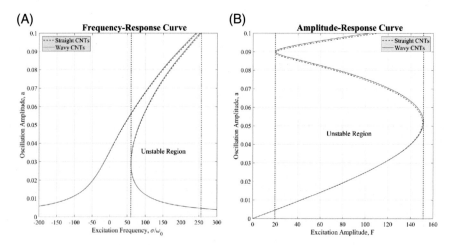

FIGURE 4.13 Qualitative effect of waviness phenomenon on the nonlinear forced vibration characteristics of MSH PNC beams ($L/h = 100$, $V_r = 0.02$, $V_F = 0.2$, $C_d = 100$, $K_L = 1000$, $K_P = 100$, $K_{NL} = 10$, and $F = 200$). In subplots (A) and (B), the frequency–response and amplitude–response curves of MSH PNC beams for PNC beams consisted of straight and wavy CNTs, respectively, are plotted. Part (B) is plotted at point $\sigma/\omega_0 = 200$, that is, in the unstable region of the frequency–response diagram.

Section 2.2.1.1.3 of Chapter 2). The influence of agglomeration and waviness phenomena as well as length-to-diameter ratio of the CNTs on the system's dynamics will be covered in this subsection, too. In the first part of the following case studies, transient response of the system will be studied. Next, the stability analysis will be accomplished with the aid of Bode and Root Locus criteria. In order to be sure from the precision of the provided material, the slenderness ratio of the beam is assumed to be fixed

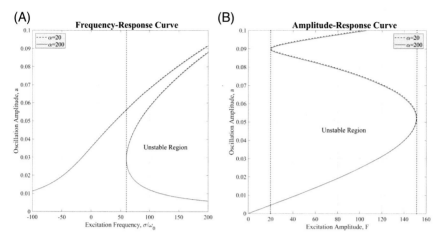

FIGURE 4.14 Effect of nanofillers' length-to-diameter ratio on the nonlinear forced vibration characteristics of MSH PNC beams ($L/h = 100$, $W_r = 0.02$, $V_F = 0.2$, $C_d = 100$, $K_L = 1000$, $K_P = 100$, $K_{NL} = 10$, and $F = 200$). In subplots (A) and (B), the frequency–response and amplitude–response curves of MSH PNC beams for PNC beams consisted of CNTs with $\alpha = 20$ and $\alpha = 200$, respectively, are plotted. Part (B) is plotted at point $\sigma/\omega_0 = 200$, that is, in the unstable region of the frequency–response diagram.

on $L/h = 100$ in the future case studies. Also, the below dimensionless forms of the foundation parameters and deflection will be hired in the following examples for the sake of simplicity:

$$K_W = \frac{k_W L^4}{E_m h^3}, \quad K_P = \frac{k_P L^2}{E_m h^3}, \quad C_d = \frac{c_d L^2}{h^2 \sqrt{\rho_m E_m}}, \quad W = \frac{w(t, L/2) E_m h}{f_0 L^2} \quad (4.3)$$

Prior to discuss about the mechanical behaviors of the scattered waves, it is necessary to prove the accuracy of the presented data. In Fig. 4.15, the variation of wave frequency against volume fraction of the CFs in the composition of the MSH PNC beams is shown. In this diagram, the results obtained from the present study were compared with those formerly reported in Ref. [383]. It can be well observed that present modeling can accurately predict the wave dispersion characteristics of the hybrid NC beams with different contents of the CFs. The small differences are attributed to the difference between the kinematic theories utilized to model the problem. In the present study, classical theory is implemented, whereas, HSDT is employed in Ref. [383].

As the first example in this subsection, the influence of the gradient index on the dispersion characteristics of MSH PNC beams rested on a viscoelastic substrate is depicted in Fig. 4.16. Based on Fig. 4.16A, it can be figured out that implementation of great gradient indices results into a downward shift in the deflection-time response of the scattered waves.

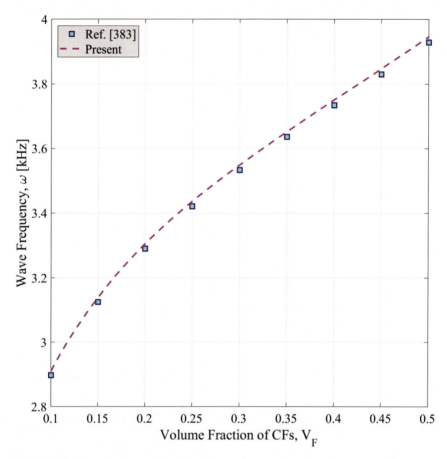

FIGURE 4.15 Comparison of the frequency curve of MSH PNC beam for various volume fractions of the CFs.

In other words, the amplitude of the system's deflection whenever it is subjected to propagation of elastic waves will be enlarged because of the reduction induced in the stiffness of the system due to the increment of the gradient index. To better understand this trend, one can refer to Fig. 2.6 of Chapter 2. It is obvious that in large times, the deflection of the system converges to its steady-state value because of the damping coefficient of the visco-Pasternak medium. In subplots shown in parts B and C of Fig. 4.16, the Root Locus and Bode diagrams of the systems analyzed in Fig. 4.16A are illustrated. The transfer function of the analyzed systems can be considered to be in the general form of $G(s) = \frac{k'}{s(Ms^2+Cs+K)}$; where M, C, and K are mass, damping, and stiffness of the eigenvalue problem, respectively. It is worth regarding that $k' = -k$ and k is the gain of the

4.1 Wave propagation, free vibration, and nonlinear forced vibration of beams   251

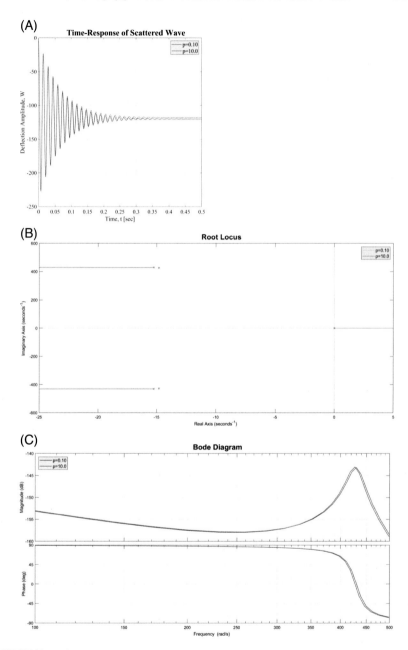

FIGURE 4.16 Effect of the gradient index on the wave dynamics and stability of the acoustical waves dispersed in MSH PNC beams rested on a three-parameter visco-Pasternak medium ($L/h = 100$, $\eta = 1$, $\mu = 0.4$, $w_r = 0.1$, $V_F = 0.2$, $K_W = 10$, $K_P = 1$, $C_d = 0.5$). In this diagram, the full agglomeration state is considered. Subplot (A) illustrates the time–response of the system; whereas, subplots (B) and (C) exhibit the stability of the system in the framework of root locus and bode criteria, respectively.

system. Based on the figures, the present system possesses a negative gain (i.e., corresponding with $k < 0$). It is obvious that a change in the value of the gradient index from 0.10 to 10.0 leads to a general change in the position of both real and imaginary parts of the system's poles. It can be claimed that an increase in the value of the gradient index results in a more stable system because the location of the poles will move toward the left-hand side. Clearly, there exists no intersection between the real axis and Root Locus in the present system because of the negative being of the system's gain. However, in the case of having a system with positive gain, the location of the intersection between the Root Locus and real axis of the system can be calculated via $\delta = \frac{\Sigma P - \Sigma Z}{n-m}$; where ΣP and ΣZ denote the summation of the poles and zeros of the system, respectively. Also, n and m stand for the order of the denominator and numerator of the transfer function $G(s)$. According to the above transfer function, the location of the system's break point (B.P.) can be attained by using the definition of this point [725], that is, $\frac{dG(s)}{ds} = 0 \rightarrow s = $ B.P. From another viewpoint and by considering the characteristics equation of the system to be like $s^2 + 2\xi\omega_n s + \omega_n^2 = 0$ (in which ω_n and ξ are natural frequency and damping ratio, respectively; also, damped frequency and damping coefficient can be defined as $\omega_d = \omega_n\sqrt{1-\xi^2}$ and $\alpha = \xi\omega_n$, respectively), it can be perceived that an increase in the gradient index leads to an increase in the damping coefficient α and consequently $\theta = \arctan\left(\frac{\omega_d}{\alpha}\right)$ will be reduced. Hence, the damping ratio of the system will be lessened by assigning $p = 10.0$ instead of $p = 0.10$. On the basis of the Bode diagram, there exists a pole in the origin and therefore, the phase must begin from $-\pi/2$ radian for type-1 systems. However, the phase of the present system starts from $\pi/2$ radian which certifies the negative being of the system's gain. According to the π radian phase change in the Bode diagram, it can be realized that the system has a pole in the left side of the plane which was previously observed in the Root Locus of the system. Based on the greater damped frequency of the system in the case of $p = 10.0$ and with regard to the definition of the peak time (t_p), it can be conceived that the peak time of the system reduces as a greater value is assigned to the gradient index. Based on the characteristics equations of the systems with $p = 0.10$ and $p = 10.0$, it can be claimed that the peak of overshoot for both of the aforesaid systems is similar. Furthermore, greater θ in the case of $p = 0.10$ results in a longer settling time (t_s) for this system compared with the case of having a system with $p = 10.0$.

In the second case study, the impact of the agglomeration pattern on the dynamic response of the propagated waves is studies in the framework of Fig. 4.17. In the first part of this diagram, the variation of the deflection amplitude of the MSH PNC beam attacked by acoustic waves versus elapsed time is depicted for PNCs with both partial and full agglomeration

4.1 Wave propagation, free vibration, and nonlinear forced vibration of beams   253

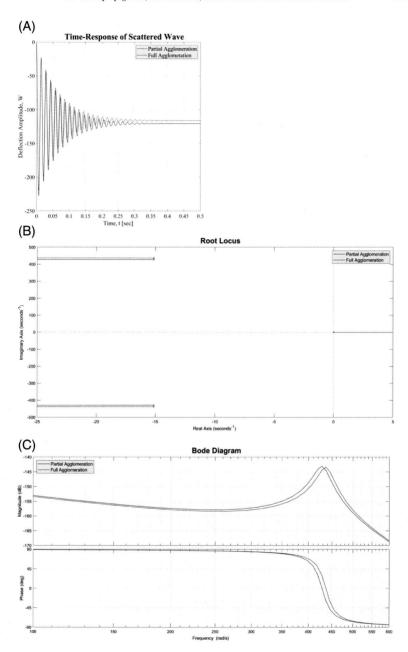

FIGURE 4.17 Effect of the agglomeration pattern on the wave dynamics and stability of the acoustical waves dispersed in MSH PNC beams rested on a three-parameter visco-Pasternak medium ($L/h = 100$, $\mu = 0.2$, $w_r = 0.1$, $p = 2$, $V_F = 0.2$, $K_W = 10$, $K_P = 1$, $C_d = 0.5$). In this diagram, the full agglomeration state is considered. Subplot (A) illustrates the time–response of the system; whereas, subplots (B) and (C) exhibit the stability of the system in the framework of root locus and bode criteria, respectively.

cases. Based on this figure, one can understand the destroying effect of the CNT agglomerates on the dynamic characteristics of the dispersed waves. Clearly, PNC beams consisted of fully aggregated nanofillers experience greater deflections due to the lower stiffness of such nanoengineered systems in comparison with those reinforced with partially agglomerated CNTs. This difference can be better sensed once the oscillating system reaches to its steady-state answer. Like former example, herein, the mass, stiffness, and damping components related to each of the above cases will be utilized in order to construct both characteristics equation and transfer function of the system to analyze the dynamic stability of the system via Bode and Root Locus criteria. With respect to the transfer functions of these two cases, Fig. 4.17B is related to the plot of the system's Root Locus. Again, a system with negative gain is enhanced. The location of the system's poles varies as the agglomeration pattern changes due to the differences between the transfer functions corresponding to these cases. It is worth mentioning that the imaginary part of the poles is decreased in magnitude in the full agglomeration condition compared with partial one. No contact between the real axis and Root Locus of the system is observed due to the negative being of the system's gain. Based on the Root Locus diagram, reduction of the imaginary part of the poles by changing the agglomeration pattern from partial to full results in the reduction of the angle θ. So, by considering the inverse relationship between θ and ξ, it is clear that the damping ratio of the system will be enlarged in the full agglomeration case compared with the partial agglomeration one. On the other hand, Bode diagram states that the underobservation type-1 system contains a pole at the origin which leads to the start of the phase's variation from $-\pi/2$ radian. But this does not take place and the phase starts its decreasing trend from $\pi/2$ instead. This is due to the negative gain of the system. Moreover, difference in the characteristics equation of these systems states that the damped frequency of the MSH PNC beam with partial agglomeration state is larger than that of the system with full agglomeration pattern. Therefore, the peak time of the system containing fully aggregated CNTs will be greater thanks to the inverse relation between damped frequency and peak time. It can be declared that because of the tiny difference between the damping ratio of partial and full patterns (i.e., $\xi_{Full} > \xi_{Partial}$), the peak of overshoot for PNC beam with fully aggregated nanofillers will be smaller than that of PNC beams reinforced via partially aggregated CNTs. According to the same reason, settling time of the system in the partial pattern is greater than that of the system in the full case.

On the other hand, Fig. 4.18 illustrates the effect of the volume fraction of the clusters on the wave propagation behaviors of MSH PNC beams reinforced via agglomerated nanotubes. Based on the first subplot in this illustration, an increase in the volume fraction of the clusters leads to an

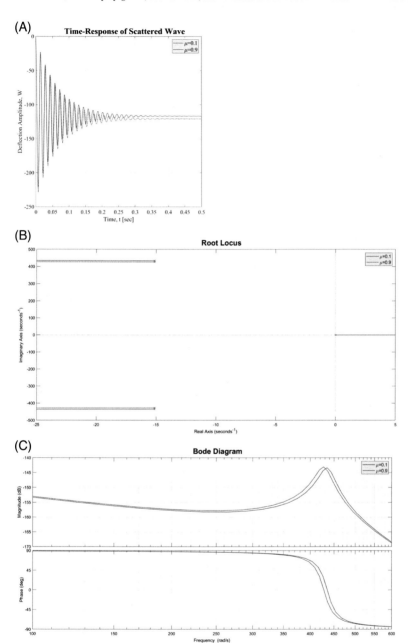

FIGURE 4.18 Effect of the volume fraction of the clusters on the wave dynamics and stability of the acoustical waves dispersed in MSH PNC beams rested on a three-parameter visco-Pasternak medium ($L/h = 100$, $\eta = 1$, $w_r = 0.1$, $p = 2$, $V_F = 0.2$, $K_W = 10$, $K_P = 1$, $C_d = 0.5$). In this diagram, the full agglomeration state is considered. Subplot (A) illustrates the time–response of the system; whereas, subplots (B) and (C) exhibit the stability of the system in the framework of root locus and bode criteria, respectively.

enhancement in the steady-state response of the system. This improvement can also be observed in the transient response, too. The physical reason of this trend is the positive effect of addition of the agglomeration parameter μ on the total stiffness of the continuous system. In other words, the system moves toward constructing a great cluster including all of the agglomerated nanofillers inside while a greater value is assigned to the volume fraction of the clusters. So, it is now clarified that why the deflection-time response of the system is shifted upward in the case of increasing the volume fraction of the clusters. Based on the Root Locus plotted in Fig. 4.18B, it can be concluded that a negative gain system is in hand whose transfer function is as same as what introduced in the previous paragraphs. Obviously, change of the agglomeration parameter μ cannot affect the real part of the pole's location; however, the imaginary part varies as the volume fraction of the clusters differs. It is observable that a rise in the value of the volume fraction of the clusters leads to an improve in the imaginary part of the pole's location. Because of the negative being of the system's gain, no intersection between the Root Locus and real axis can be found. If the system had a positive gain, such an intersection could be observed. In the former sentences, it was mentioned that the imaginary part of the pole varies as the volume fraction of the clusters changes. This change leads to greater θ for the case of $\mu = 0.9$ thanks to the greater damped frequency in this case. In addition, the Bode plot proves that the present system possesses a negative gain because the phase of the system has started to decrease from $\pi/2$ radian instead of $-\pi/2$. Also, existence of the system's complex pole can be certified based on the Bode diagram due to the fact that the phase of the system tends to the asymptotic value of $-\pi/2$ radian. To give more insight about the dynamics of this system, it can be pointed out that greater damped frequency of the system with $\mu = 0.9$ results in a lower peak time for this case due to the inverse relation between damped frequency and peak time. On the other hand, although variation of the agglomeration parameter μ cannot manipulate the damping ratio of the system as well, it can be stated that the greater peak of overshoot belongs to the system whose damping ratio is smaller, that is, corresponding with the case of $\mu = 0.9$. Due to the same reason, the settling time of the system with $\mu = 0.9$ is larger than the other one.

In another example, the role of the nanofillers' mass fraction on the determination of the dynamic characteristics of the system is illustrated in Fig. 4.19. Based on the first subplot in this figure, it can be simply found that an increase in the value of nanofillers' mass fraction results in a reduction in the deflection amplitude of the system because of the stiffer behavior of the system in the case of using a higher content of the reinforcing nanofiller in the composition of the MSH PNC system. Like all of the previous examples, it can be another time certified that this difference can be better observed while the system reaches to its steady-state response. It is noteworthy that

4.1 Wave propagation, free vibration, and nonlinear forced vibration of beams 257

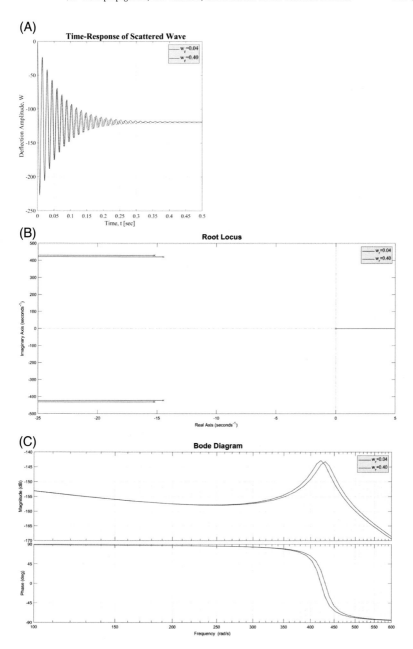

FIGURE 4.19 Effect of the mass fraction of the nanofillers on the wave dynamics and stability of the acoustical waves dispersed in MSH PNC beams rested on a three-parameter visco-Pasternak medium ($L/h = 100$, $\eta = 1$, $\mu = 0.4$, $p = 2$, $V_F = 0.2$, $K_W = 10$, $K_P = 1$, $C_d = 0.5$). In this diagram, the full agglomeration state is considered. Subplot (A) illustrates the time–response of the system; whereas, subplots (B) and (C) exhibit the stability of the system in the framework of root locus and bode criteria, respectively.

the positive influence of the mass fraction of the CNTs on the equivalent stiffness of the MSH PNC beam is not too clear due to the fact that the full agglomeration case is considered in the present case study. From control-based point of view, this system is a negative gain one. This issue can be well observed by taking a brief look at the Root Locus of the system shown in Fig. 4.19B. It can be found that both real and imaginary parts of the system's poles vary as the CNTs' mass fraction changes. This variation takes place in a way that PNC beam with lower content of reinforcing nanofillers is more stable than the system with higher amount of nanofillers. In fact, addition of the CNTs in the composition of the system leads to stiffer behavior of the continuous system and consequently moves the poles of the system toward right-hand side. Thus, the stability margin will be smaller than its original limit. Again, it is observed that the Root Locus and real axis do not intersect each other at all due to the negative value of the system's gain. It is worth mentioning that change of the CNTs' mass fraction causes variations in both damping ratio and damped frequency of the system. These variations appear in a way that the greater θ corresponds with MSH PNC beams with $w_r = 0.40$. According to the Bode diagram (refer to Fig. 4.19C), the negative value of the system's gain can be understood easily by taking a brief look at the starting and ending point of the phase curve. Such a condition (i.e., corresponding with start from $\pi/2$ radian and end at $-\pi/2$) cannot be observed in a type-1 system unless the gain is smaller than zero. On the other hand, bigger damped frequency of the system with $w_r = 0.04$ results in a smaller peak time for this case in comparison with the case of utilization of $w_r = 0.40$. Moreover, bigger damping ratio of the MSH PNC beams with $w_r = 0.04$ results in a small difference between the peak of overshoot in the studied cases so that the peak of overshoot for the mentioned system is smaller than the system possessing $w_r = 0.40$. With the same logic, it can be claimed that the settling time of MSH PNC beams with $w_r = 0.40$ is greater than that of MSH PNC system with $w_r = 0.04$.

On the other hand, impact of the wavy shape of the nanofillers on the dynamic responses of the scattered waves in a MSH PNC beam is highlighted in Fig. 4.20. According to the time–response of the propagated waves, one can understand that the deflection amplitude can be intensified by if curvy nanofillers are employed in the manufacturing procedure. This negative influence is because of the destroying effect of the wavy shape of the CNTs on the mechanism of reinforcement in the PNC. In fact, this wavy shape of the reinforcing elements does not allow the polymer matrix to be reinforced with the ideal potential of the nanofillers. So, the stiffness of the obtained nanomaterial will be smaller than the predicted ideal value. So, and with regard to the inverse relationship between stiffness and deflection, the continuous system experiences greater deflections. This issue can be easily observed in the steady-state response of the system.

4.1 Wave propagation, free vibration, and nonlinear forced vibration of beams

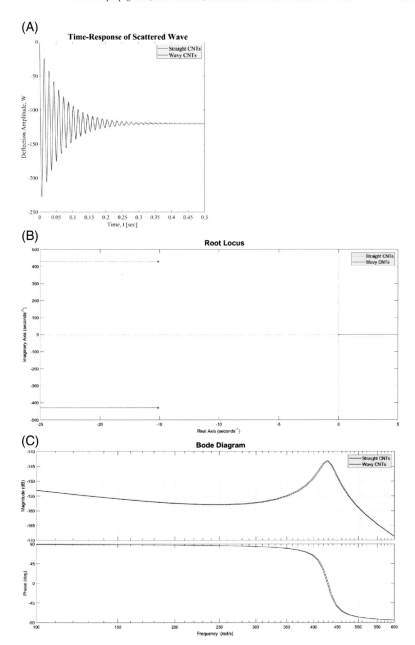

FIGURE 4.20 Effect of the wavy shape of the nanofillers on the wave dynamics and stability of the acoustical waves dispersed in MSH PNC beams rested on a three-parameter visco-Pasternak medium ($L/h = 100$, $V_r = 2\%$, $V_F = 0.2$, $K_W = 10$, $K_P = 1$, $C_d = 0.5$). In this diagram, the full agglomeration state is considered. Subplot (A) illustrates the time–response of the system; whereas, subplots (B) and (C) exhibit the stability of the system in the framework of root locus and bode criteria, respectively.

Based on the Root Locus illustrated in Fig. 4.20B, sensing the differences between the exact location of the systems' poles is difficult. However, based on the differences between the stiffnesses related to the cases of using straight and wavy CNTs, it can be perceived that the poles are not similar. The differences are in a way so that the imaginary part of the pole related to the case of using wavy CNTs is smaller than that corresponding to the case of using straight nanofillers in the fabrication. This is not the only outcome of waviness phenomenon. Due to the aforesaid difference, the damped frequency of the MSH PNC beam decreases as wavy CNTs are utilized in the fabrication process. Similar to previous case studies, the negative gain of the system results in existence of no intersection between the Root Locus and real axis. Recalling the difference between the imaginary parts of the poles corresponding to straight and wavy CNTs, it can be concluded that the angle θ will be reduced in the case of using wavy nanotubes. Based on the Bode diagram, the negative being of the system's gain can be reconfirmed notifying the fact that the phase of the system starts to reduce from $\pi/2$ radian instead of $-\pi/2$ radian. Based on the greater damped frequency of the MSH PNC systems including straight CNTs, such systems possess smaller peak time. Furthermore, based on the inverse relationship between damping ratio and peak of overshoot and by recalling the smaller damping ratio of the systems containing straight CNTs, such systems have higher peaks of overshoot. Also, it is worth mentioning that the settling time of the MSH PNC beams with straight CNTs is bigger than that of continuous systems containing wavy nanofillers.

In the final illustration in this subsection, the influence of the length-to-diameter ratio of the CNTs on the wave dispersion characteristics of beam-type elements fabricated from MSH PNCs is tracked. As shown in Fig. 4.21A, there exists a very small difference between the time–response of the propagated waves in PNC beams. This small difference originates from the positive influence of the thin nanofillers on the improvement of the nanomaterial's stiffness. Hence, in this condition, the deflection of the beam will be decreased because of the inverse relationship between stiffness and deflection. It must be pointed out that it is not logical to increase the length-to-diameter ratio of the CNTs limitless. This is due to the unpleasant probability of existence of wavy nanofillers in the nanomaterial if very thin and flexible CNTs are selected to synthesize the MSH PNC material. The characteristics equation and transfer function of this system are similar to that of a negative gain system with three poles among which two of them are complex conjugates. Based on the Root Locus plotted in Fig. 4.21B, it can be perceived that the real part of the conjugate poles does not vary whenever length-to-diameter ratio of the CNTs is changed from $\alpha = 20$ to $\alpha = 2000$. However, the imaginary part of the latter case is greater than that of MSH PNC systems reinforced via CNTs whose length-to-diameter ratio is $\alpha = 20$. Besides, availability of no intersection between Root Locus and real axis certifies the negative being of the system's gain.

4.1 Wave propagation, free vibration, and nonlinear forced vibration of beams 261

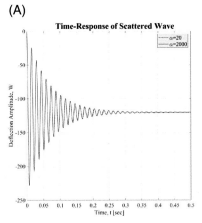
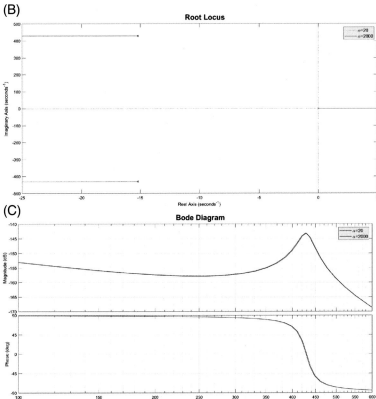

FIGURE 4.21 Effect of the nanofillers' length-to-diameter ratio on the wave dynamics and stability of the acoustical waves dispersed in MSH PNC beams rested on a three-parameter visco-Pasternak medium ($L/h = 100$, $W_r = 2\%$, $V_F = 0.2$, $K_W = 10$, $K_P = 1$, $C_d = 0.5$). In this diagram, the full agglomeration state is considered. Subplot (A) illustrates the time–response of the system; whereas, subplots (B) and (C) exhibit the stability of the system in the framework of root locus and bode criteria, respectively.

Motivated by the bigger imaginary part of the poles in the case of using $\alpha = 2000$, it can be demonstrated that the angle θ will be enlarged if thinner nanofillers are implemented in the fabrication procedure. Similarly, it can be concluded that the damping ratio will be decreased if CNTs with higher length-to-diameter ratio are hired. Looking at the Bode diagram of the system (refer to Fig. 4.21C), it can be realized that the oscillating system has a negative gain. The proof of this claim is the start of the decreasing trend of the system's phase angle from $\pi/2$ radian instead of $-\pi/2$ radian. Recalling previous sentences about the comparison of the imaginary parts of the poles in the analyzed cases, it can be simply found that the peak time of the system whose CNTs possess $\alpha = 2000$ is smaller than that of the system including $\alpha = 20$ thanks to the greater damped frequency of the first system. According to the inverse relationship between θ and damping ratio ξ, it can be stated that the smallest damping ratio among analyzed cases belongs to the MSH PNC beams with $\alpha = 2000$. Thus, it is obvious that the peak of overshoot for this condition will be greater than that of the remaining case. By recalling the order of the damping ratio for MSH PNC beams with different nanofillers' length-to-diameter ratio, it is obvious that big settling times correspond with thin nanofillers.

4.2 Wave propagation, free vibration, and nonlinear forced vibration of plates

Like what we observed in the previous subsection, this one undergoes with dynamic characteristics of MSH PNC plate-type elements with regard to the influences of agglomeration of the nanofillers, wavy shape of them, and their length-to-diameter ratio on the system's behavior. Free vibration, nonlinear forced oscillation, and wave propagation analyses will be carried out in this subsection. To these ends, the kinematic relations introduced in Sections 2.2.2.1.1, 2.2.2.1.2, and 2.2.2.2.3 of Chapter 2 will be implemented as well as homogenization schemes introduced in Sections 2.1.1.3, 2.1.1.5, and 2.1.1.6 of the same chapter.

4.2.1 Free vibration analysis of MSH PNC plates

Herein, the modal analysis of MSH PNC plates will be accomplished in the framework of an analytical approach. The natural frequency of the plate's free vibrations in the linear domain will be computed by the means of the eigenvalue problem introduced in Section 2.2.2.2.3 of Chapter 2. Due to the fact that the influences of shear deflection on the frequency of the plate are captured, thick-walled plates will be considered in the following case studies. In all of the future examples, the plate's thickness will be considered to be $h = 2$ mm. In order to simplify the procedure of interpreting the results, the following dimensionless forms of the

TABLE 4.3 Comparison of the dimensionless fundamental frequency of the CNTR PNC square plates, rested on Pasternak foundation, with SSSS BC ($a/h = 10$).

(K_w, K_p)	Source	$V_{CNT} = 0.11$	$V_{CNT} = 0.14$	$V_{CNT} = 0.17$
(0, 0)	Ref. [736]	0.1357	0.1438	0.1685
	Present	0.1319	0.1400	0.1638
(100, 0)	Ref. [736]	0.1390	0.1469	0.1712
	Present	0.1353	0.1432	0.1665
(100, 50)	Ref. [736]	0.1683	0.1747	0.1954
	Present	0.1654	0.1718	0.1915

foundation parameters and natural frequency will be utilized in the numerical examples:

$$K_W = \frac{k_W a^4}{E_m h^3}, \quad K_P = \frac{k_P a^2}{E_m h^3}, \quad \Omega = \omega h \sqrt{\frac{\rho_m}{E_m}} \quad (4.4)$$

Before starting to go through analyzing the results of present text, it is better to be sure from the accuracy of the implemented methodology. To this purpose, the natural frequencies of CNTR PNC square plates are regenerated with the aid of present model and the results are compared with those reported in Ref. [736]. Based on the data tabulated in Table 4.3, it can be inferred that the results of present model are in an excellent agreement with those reported in a previously published valid study. Hence, the numerical data generated by the present modeling can be trusted in the linear domain for thick-type PNC plates.

In the first case study, the impact of nanofillers' mass fraction on the variation of the dimensionless frequency of the MSH PNC plate against volume fraction of the CNTs inside the clusters is monitored in Fig. 4.22. According to this diagram, it can be observed that an increase in the mass fraction of the nanoparticles can be resulted in a reciprocal improvement in the frequency of the plate's free vibrations. This influence is because of the positive effect of addition of the CNTs to the composition of the nanomaterial on the equivalent stiffness of the continua. So, it is natural to see that the frequency curve of the system is shifted upward. In reverse, it is shown that the frequency of the plate's oscillations will be decreased as the volume fraction of the CNT inside the clusters grows. This negative effect is due to the destroying effect of the aggregation of the nanotubes in the clusters on the stiffness reinforcing mechanism in the MSH PNC material. In addition, in any arbitrary values of the agglomeration parameter η and mass fraction w_r, plates with fully clamped edges are able to support higher natural frequencies compared with those with all edges simply supported.

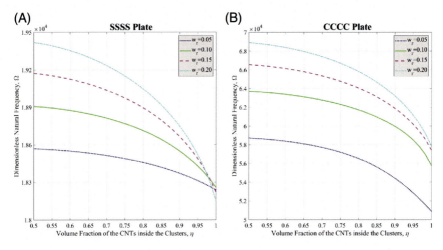

FIGURE 4.22 Variation of the dimensionless natural frequency of the MSH PNC square plates with (A) SSSS and (B) CCCC BCs against volume fraction of the CNTs inside the clusters for different mass fractions of the nanofillers ($a/h = 10$, $\mu = 0.4$, $p = 2$, $V_F = 0.2$, $K_W = 20$, $K_P = 5$).

The physical reason of this trend is the greater structural stiffness of the clamped BC in comparison with the simply supported one. It is worth regarding that in SSSS plates, it is possible to observe a reverse reinforcing mechanism in the cases close to full agglomeration. In such cases, increment of the CNTs' mass fraction reduces the natural frequency.

In the next illustration, that is, Fig. 4.23, the influence of gradient index in the volume fraction of the nanofillers on the dynamic behaviors of the MSH PNC plate is shown. Based on this example, it is observed that an increase in the gradient index decreases the natural frequency of the plate because it leads to a reduction in the modulus of the MSH PNC plate in any desired thickness. For illustrative proof of this claim, one can refer to Fig. 2.6 of Chapter 2. Based on the results perceived from this figure, it seems logical to choose high gradient indices in the theoretical investigations to avoid from engineering overestimations while designing MSH PNC devices. Again, it is revealed that the dimensionless frequency will be decreased gradually if higher values are assigned to the volume fraction of the CNTs inside the clusters. This negative effect is due to the destroying influence of the agglomeration phenomenon on the mechanism of stiffness improvement in the nanomaterials. Like previous case study, CCCC plates are able to promote greater natural frequencies in comparison with SSSS ones thanks to the greater structural stiffness of clamped support.

Moreover, investigation of effect of the volume fraction of the clusters on the free vibration characteristics of MSH PNC plates is fulfilled in the context of Fig. 4.24. In this figure, the variation of dimensionless frequency

4.2 Wave propagation, free vibration, and nonlinear forced vibration of plates

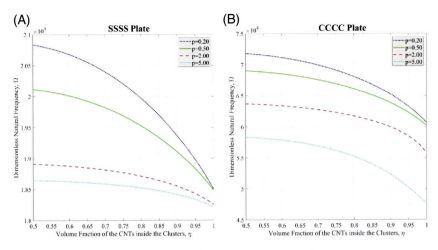

FIGURE 4.23 Variation of the dimensionless natural frequency of the MSH PNC square plates with (A) SSSS and (B) CCCC BCs against volume fraction of the CNTs inside the clusters for different gradient indices ($a/h = 10$, $\mu = 0.4$, $w_r = 0.1$, $V_F = 0.2$, $K_W = 20$, $K_P = 5$).

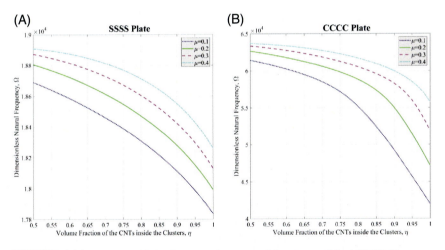

FIGURE 4.24 Variation of the dimensionless natural frequency of the MSH PNC square plates with (A) SSSS and (B) CCCC BCs against volume fraction of the CNTs inside the clusters for different volume fractions of the clusters ($a/h = 10$, $w_r = 0.1$, $p = 2$, $V_F = 0.2$, $K_W = 20$, $K_P = 5$).

versus agglomeration parameter η is plotted whenever the other parameter of agglomeration, that is, volume fraction of the clusters μ, is varied. It can be easily seen that greater natural frequencies can be supported by the continuous system if a bigger value is assigned to the volume fraction of the clusters. This trend is appeared because of the positive effect of

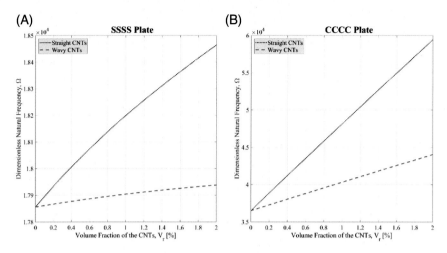

FIGURE 4.25 Variation of the dimensionless natural frequency of the MSH PNC square plates with (A) SSSS and (B) CCCC BCs against volume fraction of the CNTs for PNC structures consisted of straight and wavy CNTs ($a/h = 10$, $V_F = 0.2$, $K_W = 20$, $K_P = 5$).

increasing agglomeration parameter μ on the stiffness of the PNC system. In other words, if this parameter is aggrandized, the nanomaterial becomes closer to the state of containing a giant cluster inside itself so that all of the nanofillers are entangled in it. Thus, it can be stated that this condition is roughly equivalent to the case of having no agglomerated nanofiller in the composition of the MSH PNC. Just as same as former illustrations, CCCC plates possess greater frequencies compared with SSSS ones because of their bigger structural stiffness.

Changing the atmosphere of the discussion and moving toward investigation of the impact of curved shape of the CNTs on the dynamic response of the MSH PNC plates, Fig. 4.25 is presented to compare natural frequencies of MSH PNC plates reinforced via straight and wavy CNTs together. In this diagram, a qualitative viewpoint is hired to see the influence of the waviness phenomenon on the stiffness reinforcing mechanism in nanoengineered structures. According to this figure, it can be realized that wavy shape of the nanofillers can be resulted in a remarkable reduction in the natural frequency of the PNC plates. The physical reason of this phenomenon is the negative effect of waviness phenomenon on the mechanism of reinforcement in the microstructure of the MSH PNC material. Indeed, ideal theoretical approximations about stiffness of the PNC materials cannot be met in practice because of the curvy shape of the nanofillers. Besides, it is demonstrated that PNC structures with fully clamped BC possess bigger natural frequencies in comparison with those containing simply supported edges.

Fig. 4.26 is depicted to the study of the quantitative impact of the waviness coefficient on the dynamic behaviors of MSH PNC plates containing

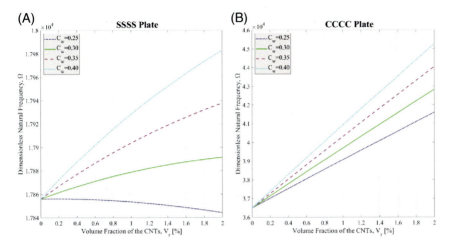

FIGURE 4.26 Variation of the dimensionless natural frequency of the MSH PNC square plates with (A) SSSS and (B) CCCC BCs against volume fraction of the CNTs for different waviness coefficients ($a/h = 10$, $V_F = 0.2$, $K_W = 20$, $K_P = 5$).

wavy nanofillers. It can be inferred that assignment of higher values to the waviness coefficient results an improve in the natural frequency of the PNC plate. The physical reason of this issue is that by using great waviness coefficient, the amplitude of the wavy shape of the CNTs will be assumed to be smaller. Hence, the CNTs will be presumed to be close to their ideal straight shape. According to these facts, it is natural to see an enhancement in the natural frequency of the plate if waviness coefficient is intensified. It is noteworthy that the negative influence of the waviness phenomenon can be even resulted in a severe reduction in the total stiffness of the nanomaterial so that the natural frequency becomes smaller by adding the CNTs' loading in the composition of the MSH PNC structure. This case can be observed in Fig. 4.26A once the waviness coefficient is assumed to be $C_w = 0.25$. Obviously, CCCC plates are capable of providing bigger dimensionless frequencies in comparison with those with all edges simply supported.

As the last case study in this subsection, the effect of thin or thick being of the nanofillers on the determination of the dynamic response of MSH PNC plates is tracked in Fig. 4.27. In this figure, the variation of the dimensionless frequency of MSH PNC plate against length-to-diameter ratio of the CNTs is plotted once different values are assigned to the weight fraction of the nanofillers. It is clear that greater natural frequencies can be enriched by adding the weight fraction of the CNTs. It must be kept in mind that issues like agglomeration and waviness are dismissed in this figure and the provided data are ideal ones. By taking a look at this diagram, one can understand that natural frequency of the plate can be increased if CNTs with great length-to-diameter ratio are utilized in the fabrication. However,

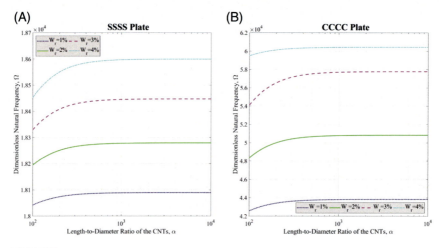

FIGURE 4.27 Variation of the dimensionless natural frequency of the MSH PNC square plates with (A) SSSS and (B) CCCC BCs against length-to-diameter ratio of the CNTs for different weight fractions of the nanofillers ($a/h = 10$, $V_F = 0.2$, $K_W = 20$, $K_P = 5$).

it must be considered that this positive effect contains an efficiency limit. In other words, it is not logical at all to implement CNTs possessing $\alpha > 1000$ because not only it does not have any remarkable positive influence on the stiffness of the nanomaterial, but also increases the probability of the wavy being of the reinforcing nanofillers because of their high flexibility.

4.2.2 Nonlinear forced vibration analysis of MSH PNC plates

In this part of present chapter, the harmonically stimulated nonlinear dynamic characteristics of MSH PNC plates embedded on a four-parameter nonlinear viscoelastic substrate will be monitored. Like other nonlinear forced oscillation analyses in this text book, the primary resonance of the PNC structure is going to be studied herein. The geometrical formulation of the problem is accomplished with the aid of the classical theory of thin-walled plates, as instructed in Section 2.2.2.1.1 of Chapter 2 in detail. Like previous subsection, impacts of existence of CNT agglomerates and curvy CNTs in the microstructure of the hybrid nanomaterial on the nonlinear dynamic response of the PNC thin-walled plates will be included. Also, the effect of the nanofillers' length-to-diameter ratio on the dynamic behaviors of the system is paid attention, too. It is worth regarding that in all of the case studies, the plate's thickness is fixed on $h = 2$ mm. Similar to Section 4.1.2 of present chapter, below dimensionless terms will be employed for the sake of simplicity:

$$K_L = \frac{k_L a^4}{E_m h^3}, \quad K_P = \frac{k_P a^2}{E_m h^3}, \quad K_{NL} = \frac{k_{NL} a^4}{E_m h}, \quad C_d = \frac{c_d a^2}{h^2 \sqrt{\rho_m E_m}}, \quad F = \frac{\bar{f} M a^2}{E_m h^2} \tag{4.5}$$

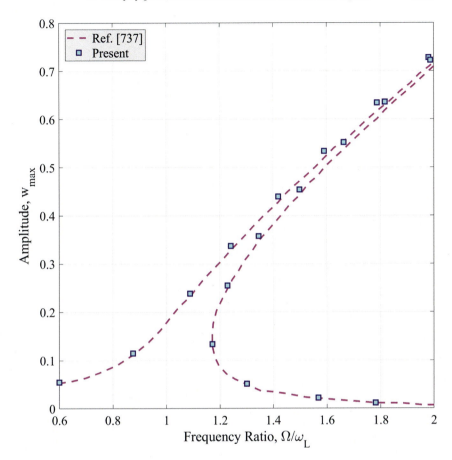

FIGURE 4.28 Comparison of the frequency–response curve of CNTR PNC square plates with all edges simply supported ($a/h = 10$, $V_{CNT} = 0.28$, $c = 0.01$, $f = 0.06$). CNTs are assumed to be distributed across the plate's thickness in uniform manner.

In the first step in this subsection, the validity of the provided material is double-checked by comparing the results of present work with those recommended in Ref. [737]. To this end, the frequency–response curves of PNC square plates reinforced via uniformly distributed CNTs were plotted in Fig. 4.28. According to this figure, it can be claimed that present model can estimate the nonlinear forced vibration responses of PNC plates with an acceptable precision. It is obvious that there exist minor differences between the frequency–response introduced in references and that of ours. However, this difference is due to the fact that in Ref. [737], HSDT is chosen by the authors; whereas, we used classical theory. This difference is the maximum possible difference because thick-walled plates with aspect ratio of $a/h = 10$ are studied in Ref. [737]. So, the results of our work which are

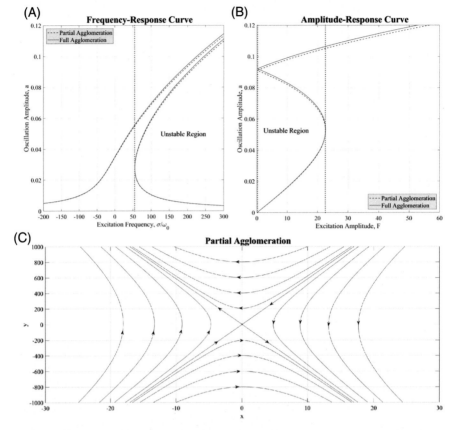

FIGURE 4.29 Effect of both partial and full agglomeration states on the nonlinear forced vibration characteristics of MSH PNC square plates ($a/h = 100$, $\mu = 0.2$, $p = 2$, $w_r = 0.1$, $V_F = 0.2$, $C_d = 100$, $K_L = 1000$, $K_P = 100$, $K_{NL} = 10$, and $F = 200$). In subplots (A) and (B), the frequency–response and amplitude–response curves of MSH PNC plates with both partial and full agglomeration states are plotted, respectively. Part (B) is plotted at point $\sigma/\omega_0 = 200$, that is, in the unstable region of the frequency–response diagram. Subplots (C) and (D) correspond with the phase portraits of the linearized system in the unstable region with partial and full agglomeration states, respectively. In the latter subplots, points $(a^*, \sigma/\omega_0) = (0.07532, 150.2)$ and $(a^*, \sigma/\omega_0) = (0.07628, 150.2)$ are chosen, respectively, to plot the trajectories.

all concerned with thin-walled plates containing aspect ratio of $a/h = 100$ are accurate and reliable for sure.

In Fig. 4.29, the influence of agglomeration degree on the nonlinear vibration characteristics of MSH PNC plates is investigated. In the first two subplots of this diagram, frequency–response and amplitude–response curves of the plates consisted of MSH PNC materials with both partial and full agglomeration conditions are drawn. It is clear that change of the

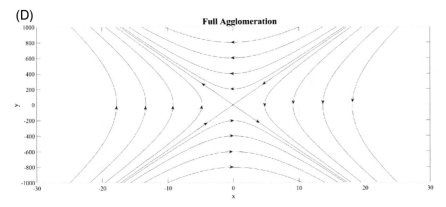

FIGURE 4.29 Continued

agglomeration pattern from partial to full will affect the nonlinear forced oscillation response of the PNC system negatively. In fact, the frequency–response curve of the system bends toward left-hand side if full agglomeration is considered. This trend is a softening trend which means that the destroying effect of the aggregation phenomenon on the stiffness of the hybrid nanomaterial will be intensified in the full agglomeration condition. The same trends can be observed in the amplitude–response curve of the system. Although agglomeration pattern affects the nonlinear dynamic characteristics of the MSH PNC plate, it does not have a remarkable effect on the unstable region of the oscillator. In both cases, the unstable region between two high- and low-energy limit cycles is similar. In order to analyze the mechanical behavior of the system in its unstable region, a geometric viewpoint is used in Fig. 4.29C and D. In these subplots, desired points on the frequency–response curve of the system are chose for both partial and full agglomeration cases. Afterward, linearization procedure is passed for the nonlinear oscillator to find the Jacobian matrix which can describe the linearized system corresponding to the nonlinear one. One can review the instructions required to linearize the system by referring to the instructions of Section 2.2.1.1.2 of Chapter 2. Based on the performed linearization for both partial and full agglomeration cases, it is revealed that the linearization can be trusted because it recommends hyperbolic fixed points for the linearized system. Referring to the Hartman–Grobman theorem in the nonlinear dynamics [723], it can be simply found that the conducted linearization can be topologically trusted without any concern. Based on the phase portraits, the fluctuating PNC plate possesses two asymptotic manifolds so that one of them is stable and another one is unstable. It is obvious that trajectories will move toward the unstable manifold as time exceeds. However, if a backward integration is implemented, the

272 4. Dynamic analysis of multiscale hybrid nanocomposite structures

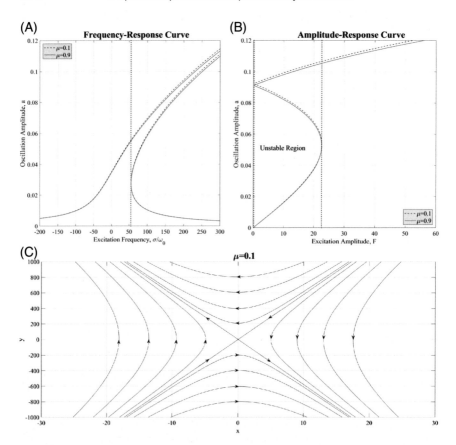

FIGURE 4.30 Effect of volume fraction of the clusters on the nonlinear forced vibration characteristics of MSH PNC square plates ($a/h = 100$, $\eta = 1$, $p = 2$, $w_r = 0.1$, $V_F = 0.2$, $C_d = 100$, $K_L = 1000$, $K_P = 100$, $K_{NL} = 10$, and $F = 200$). In subplots (A) and (B), the frequency–response and amplitude–response curves of MSH PNC plates with both partial and full agglomeration states are plotted, respectively. Part (B) is plotted at point $\sigma/\omega_0 = 200$, that is, in the unstable region of the frequency–response diagram. Subplots (C) and (D) correspond with the phase portraits of the linearized system in the unstable region while the volume fraction of the clusters is fixed on 0.1 and 0.9, respectively. In the latter subplots, points $(a^*, \sigma/\omega_0) = (0.07640, 150.2)$ and $(a^*, \sigma/\omega_0) = (0.07544, 150.2)$ are chosen, respectively, to plot the trajectories.

trajectories will be parallel to the stable manifold. It must be pointed out that phase portraits shown in Fig. 4.29C and D are not identical. Their similarity is because of the weak effect of the agglomeration type on the nonlinear dynamic behaviors of the PNC structure.

In next example, it is tried to show that how can the nonlinear vibration response of the oscillating PNC plate be affected if the volume fraction of the clusters is changed. Based on Fig. 4.30, it can be demonstrated that an increase in the value assigned to the agglomeration parameter μ leads to

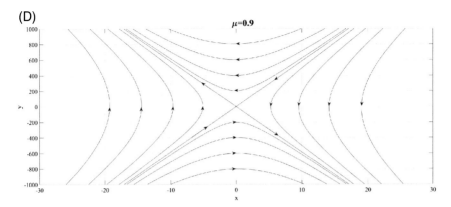

FIGURE 4.30 Continued

stiffer behavior of the plate. In other words, the MSH PNC plate will be a stiffer material if the volume fraction of the clusters is increased. The reason of this issue is that in this circumstance, the nanomaterial moves toward constructing a giant cluster containing all nanofillers together. Therefore, the destroying effect of the agglomeration phenomenon will be weakened in this way. With regard to these physical reasons and by taking a look at the frequency response curve plotted in Fig. 4.30A, it can be simply understood why the frequency–response curve of the plate is bent toward right-hand side. It can be stated that addition of the value of the volume fraction of the clusters seems like increasing the nonlinear stiffness of a Duffing oscillator in positive manner. All of the trends observed in the frequency–response curve can be certified by taking a brief look at the amplitude–response curve illustrated in Fig. 4.30B, too. Moreover, it can be claimed that the unstable region of the PNC system is enlarged if the limiting factor for the measurement is the oscillation amplitude. In addition to the quantitative studies accomplished based on Fig. 4.30A and B, qualitative investigation with the aid of linearization technique is presented in subplots Fig. 4.30C and D. According to the phase portraits plotted in these subplots, it can be seen that the system contains two stable and unstable manifolds. With regard to our physical understanding from such an oscillator and by considering the Hartman–Grobman theorem in the literature of nonlinear dynamics, the above statements can be trusted. However, it must be declared that the phase portraits of the nonlinear systems may be different in value because of the accomplished linearization.

As the final example concerned with both qualitative and quantitative investigations of the nonlinear dynamic characteristics of the MSH PNC plates, the influence ultrasmall and very big gradient indices on the nonlinear vibration response of the PNC plate is tracked in Fig. 4.31. It

274　　4. Dynamic analysis of multiscale hybrid nanocomposite structures

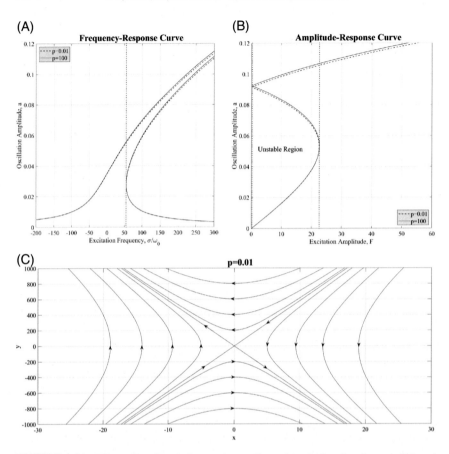

FIGURE 4.31 Effect of gradient index on the nonlinear forced vibration characteristics of MSH PNC square plates ($a/h = 100$, $\eta = 1$, $\mu = 0.4$, $w_r = 0.1$, $V_F = 0.2$, $C_d = 100$, $K_L = 1000$, $K_P = 100$, $K_{NL} = 10$, and $F = 200$). In subplots (A) and (B), the frequency–response and amplitude–response curves of MSH PNC plates for different gradient indices are plotted, respectively. Part (B) is plotted at point $\sigma/\omega_0 = 200$, that is, in the unstable region of the frequency–response diagram. Subplots (C) and (D) correspond with the phase portraits of the linearized system in the unstable region while the gradient index is fixed on 0.01 and 100, respectively. In the latter subplots, points $(a^*, \sigma/\omega_0) = (0.07568, 150.3)$ and $(a^*, \sigma/\omega_0) = (0.07652, 150.2)$ are chosen, respectively, to plot the trajectories.

can be easily inferred that the system behaves softer if greater values are assigned to the gradient index. This soft behavior can be clearly related to the bending of branches of system's limit cycles to the left-hand side in the case of using high gradient indices. This effect could be predicted before, too. Indeed, if one possesses Fig. 2.6 of Chapter 2 in mid, understanding of this issue is not difficult at all. In other words, implementation of high gradient index corresponds with the reduction of the total modulus of

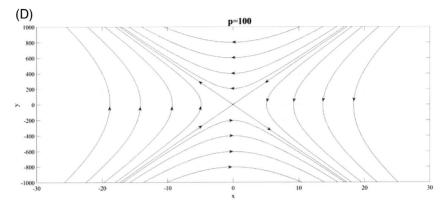

FIGURE 4.31 Continued

the nanomaterial in any desired thickness. Hence, the observed trend is logical. This issue can also be perceived by taking a look at the amplitude–response curve shown in Fig. 4.31B. Based on both of the amplitude–response and frequency–response curves, it can be realized that the area of unstable region will be decreased by a small value if the criterion is assumed to be the oscillation amplitude. On the other hand, the phase portraits illustrated in Fig. 4.31C and D demonstrate the same physical system which was previously seen in Figs. 4.29 and 4.30. Based on the obtained phase portrait, this system will tend to its unstable branch if the time is tended to infinity. It must be noticed that the results of these phase portraits are qualitatively accurate, but they may be different in magnitude if quantitative calculations are performed.

Furthermore, the influence of CNTs' mass fraction on the nonlinear forced vibration response of the nanomaterial plates is highlighted in the framework of Fig. 4.32. Based on this figure, both frequency–response and amplitude–response curves of the MSH PNC plate can be observed. It is obvious that PNC systems with a higher content of the reinforcing nanofillers promote stiffer behavior from themselves. This stiffer exhibition can be observed in the frequency–response curve as greater tendency of the frequency curve to the right-hand side if a bigger value is assigned to the mass fraction of the nanofillers. Similarly, amplitude–response curve shows this issue, too. This outcome is because of the extraordinary Young's modulus of the CNTs which leads to an improvement in the equivalent stiffness of the system. Also, it can be mentioned that the region of instability of the oscillating plate will be enlarged if the content of the nanofillers in the composition of the MSH PNC is increased. This consequence can be observed in any other case that the stiffness of the MSH PNC plate has risen.

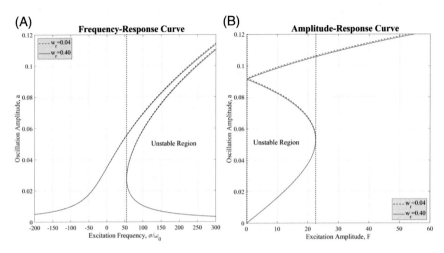

FIGURE 4.32 Effect of nanofillers' mass fraction on the nonlinear forced vibration characteristics of MSH PNC square plates ($a/h = 100$, $\eta = 1$, $\mu = 0.4$, $p = 2$, $V_F = 0.2$, $C_d = 100$, $K_L = 1000$, $K_P = 100$, $K_{NL} = 10$, and $F = 200$). In subplots (A) and (B), the frequency–response and amplitude–response curves of MSH PNC plates for different mass fractions of the CNTs are plotted, respectively. Part (B) is plotted at point $\sigma/\omega_0 = 200$, that is, in the unstable region of the frequency–response diagram.

Another issue that is included in this subsection is the investigation of the role of either straight or wavy being of the CNTs in the determination of the nonlinear dynamic characteristics of MSH PNC plates. This is the topic of Fig. 4.33. In this diagram, both frequency–response and amplitude–response curves of the MSH PNC plates are plotted for two cases. In the first case, the PNC material is assumed to be manufactured from ideal (straight) CNTs and in another one, wavy CNTs are assumed to be used in the fabrication of the nanomaterial. Although the curves for both of these cases are close to each other, the negative effect of the waviness phenomenon on the dynamic response of the PNC plate is observable. In high excitation frequencies, the frequency–response of the PNC plates consisted of wavy CNTs is located behind that of the PNC structures reinforced via ideal CNTs. This is in complete agreement with our former knowledge about the impact of the waviness phenomenon on the structural behaviors of the continuous systems.

At the end of this subsection, Fig. 4.34 is depicted to study the influence of the CNTs' length-to-diameter ratio on the nonlinear oscillation characteristics of MSH PNC plates by considering the 3D distribution of the nanofillers in the media. According to this figure, it can be perceived that an increase in the length-to-diameter ratio of the nanotubes leads to stiffer behavior of the PNC structure. The reason of this issue is the positive influence of thin being of the reinforcing elements on the reinforcement efficiency in the nanomaterials. Hence, the frequency–response curve of

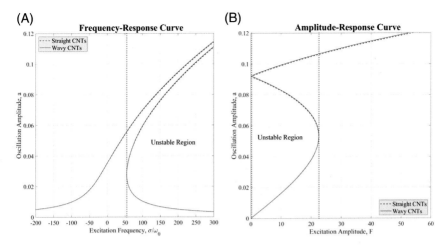

FIGURE 4.33 Qualitative effect of waviness phenomenon on the nonlinear forced vibration characteristics of MSH PNC square plates ($a/h = 100$, $V_r = 0.02$, $V_F = 0.2$, $C_d = 100$, $K_L = 1000$, $K_P = 100$, $K_{NL} = 10$, and $F = 200$). In subplots (A) and (B), the frequency–response and amplitude–response curves of MSH PNC plates for PNC plates consisted of straight and wavy CNTs, respectively, are plotted. Part (B) is plotted at point $\sigma/\omega_0 = 200$, that is, in the unstable region of the frequency–response diagram.

the MSH PNC plates containing thinner CNTs has greater tendency to the right-hand side which indicates on the stiffer behavior of such a system. Based on the frequency–response and amplitude–response curves, it can be claimed that the unstable region of the PNC plate will be enlarged in the case of using length-to-diameter ratio of $\alpha = 200$.

4.2.3 Wave propagation analysis of MSH PNC plates

This subsection of present chapter is dedicated to the investigation of the influences of CNT agglomerates and waviness phenomenon on the dynamic behaviors of elastic waves dispersed in MSH PNC plates. In addition, it is shown that how can the dispersion curves of the plate be influenced if the length-to-diameter ratio of the CNTs is changed. The studied structure is considered to be a thin-walled one whose thickness remains $h = 2$ mm in all of the case studies. Due to the fact that in the thin-walled plates the effect of shear deformation on the mechanical response of the continuous systems can be ignored, kinematic relations of the plate are extended based on the Kirchhoff–Love hypothesis, as introduced in Section 2.2.2.1.2 of Chapter 2. Besides, the stability of the PNC plate in any case will be double-checked in this subsection by the means of the fundamentals of control engineering in the linear domain. To this purpose, Bode and Root Locus diagrams will be plotted for each of the depicted case studies. In all of the case studies of present part, the following dimensionless terms are

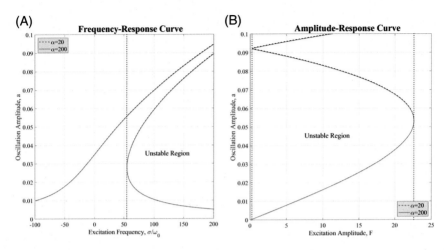

FIGURE 4.34 Effect of nanofillers' length-to-diameter ratio on the nonlinear forced vibration characteristics of MSH PNC square plates ($a/h = 100$, $W_r = 0.02$, $V_F = 0.2$, $C_d = 100$, $K_L = 1000$, $K_P = 100$, $K_{NL} = 10$, and $F = 200$). In subplots (A) and (B), the frequency–response and amplitude–response curves of MSH PNC plates for PNC plates consisted of CNTs with $\alpha = 20$ and $\alpha = 200$, respectively, are plotted. Part (B) is plotted at point $\sigma/\omega_0 = 200$, that is, in the unstable region of the frequency–response diagram.

hired to simplify the calculations:

$$K_W = \frac{k_W a^4}{E_m h^3}, \quad K_P = \frac{k_P a^2}{E_m h^3}, \quad C_d = \frac{c_d a^2}{h^2 \sqrt{\rho_m E_m}}, \quad W = \frac{w(t, a/2, b/2) E_m h}{f_0 a^2} \quad (4.6)$$

In the first illustration in this part, the validity of the presented modeling is surveyed by comparing the results of this work with those reported in Ref. [633]. This comparison can be observed in Fig. 4.35. It is obvious that the results achieved from present model differ from those reported by the authors in their former study. However, this difference is due to the fact that in Ref. [633], HSDT is implemented to analyze thick plates, whereas, present data are produced for thin-walled plates. The differences are small enough to call the presented methodology a valid one. Thus, the data provided in the following examples can be trusted.

In Fig. 4.36, the influence of the gradient index in the volume fraction of the nanofillers (refer to Eq. (2.42) for detailed data) on the wave propagation characteristics of MSH PNC plates is tracked. According to the first subplot in this illustration, it can be seen that the deflection amplitude is enlarged once a bigger value is assigned to the gradient index. Indeed, it can be inferred that the plate becomes softer if the gradient index has risen. This trend could be found by us before, thanks to our knowledge about the effect of gradient index on the total stiffness of the hybrid nanomaterial (see

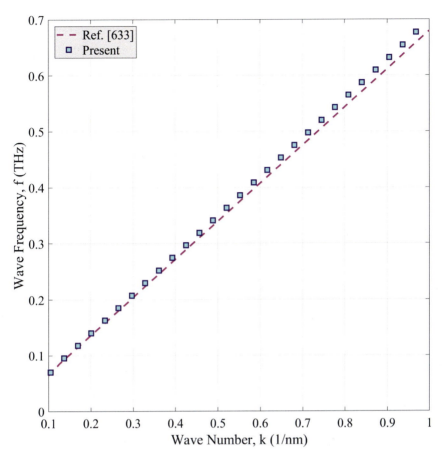

FIGURE 4.35 Comparison of the frequency–wave number curve of inhomogeneous plates ($\mu = \eta = 0$, $V = \Omega = 0$).

Fig. 2.6 of Chapter 2). This softening influence can be better observed once the steady-state response of the problem is paid attention. It is interesting to point out that adding the gradient index can be resulted in a retardation in the dynamic response of the plate. This phenomenon can be clearly seen by looking at the mid-range times in Fig. 4.36A. On the other hand, the stability diagrams plotted in Fig. 4.36B and C indicate the fact that the present system possesses a negative gain. Based on the Root Locus of the system, it can be inferred that variation of the gradient index results in changes in both real and imaginary parts of the system's poles. This change acts in a way so that the MSH PNC plate will be in a more stable state if a big value is assigned to the gradient index. Furthermore, there exists no intersection between the real axis and Root Locus of the present system because of the negative being of the system's gain. Recalling the

280   4. Dynamic analysis of multiscale hybrid nanocomposite structures

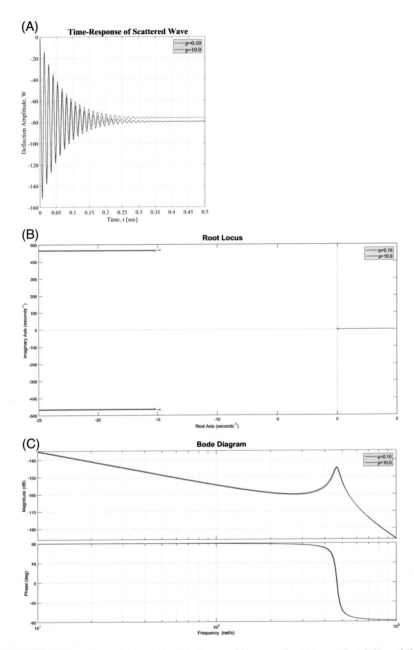

FIGURE 4.36 Effect of the gradient index on the wave dynamics and stability of the acoustical waves dispersed in MSH PNC square plates rested on a three-parameter visco-Pasternak medium ($a/h = 100$, $\eta = 1$, $\mu = 0.4$, $w_r = 0.1$, $V_F = 0.2$, $K_W = 10$, $K_P = 1$, $C_d = 0.5$). In this diagram, the full agglomeration state is considered. Subplot (A) illustrates the time–response of the system; whereas, subplots (B) and (C) exhibit the stability of the system in the framework of root locus and bode criteria, respectively.

definitions introduced in Section 4.1.3 of this chapter and by glancing the Root Locus of the present system, it can be perceived that the damping coefficient of the system is enlarged whenever the gradient index is increased. So, it can be demonstrated that the angle θ corresponding with the case in which $p = 10.0$ is smaller than that related to the case of $p = 0.10$. Based on the transfer function of this system, it is evident that the damping ratio of this system varies in the range of $0 < \xi < 1$. Due to the smaller θ angle of the $p = 10.0$ case compared with $p = 0.10$ one, the first case has a bigger damping ratio. In addition, the Bode diagram reveals that a pole at the origin exists in the system. Based on the type of the system, that is, type-1, existence of a pole at the origin cannot be justified unless the system's gain is a negative value. Also, the final value of the system's phase, that is, $-\pi/2$ radian, indicates on the existence of complex poles in the left-hand side of imaginary axis which can be simply seen in Fig. 4.36B. The Root Locus shows that the damped frequency is bigger if a small value is assigned to the gradient index; therefore, the peak time of the MSH PNC plate will be added if the gradient index is increased gradually. Moreover, very close damping ratio of the PNC system in both $p = 0.10$ and $p = 10.0$ cases shows that peak of overshoot cannot be dramatically influenced by changing the gradient index. Based on the smaller damping ratio of the oscillating system in $p = 0.10$ compared with the case of $p = 10.0$ and with regard to the definition of the settling time, it can be concluded that the settling time will be decreased if a greater gradient index is chosen in the theoretical calculations.

In another example, effect of agglomeration pattern on the dynamic properties of the elastic waves scattered in the PNC structure will be studied in detail. To this purpose, two partial and full agglomeration cases are selected and their dispersion characteristics in the time domain and their stability status in the frequency domain are monitored in Fig. 4.37. Based on the transient analysis shown in the first part of the mentioned figure, it can be declared that a remarkable downward shift in the deflection amplitude of the MSH PNC plate is made whenever all of the nanofillers are assumed to be entangled in the clusters. This phenomenon can be related to the worst negative impact of the full agglomeration condition on the stiffness of the nanomaterial compared with partial agglomeration. Therefore, it is natural to see such an enlargement in the amplitude due to the inverse relationship between deflection and stiffness. Fig. 4.37B denotes that the MSH PNC plate possesses a negative gain. It is clear that agglomeration pattern has affected the position of the imaginary part of the system's complex pole so that the imaginary part of the system with fully aggregated nanofillers is smaller than that of the system containing partially agglomerated CNTs in its microstructure. However, the real part of the system's pole is not influenced by changing the pattern of agglomeration. Therefore, it can be inferred that the damped frequency of the

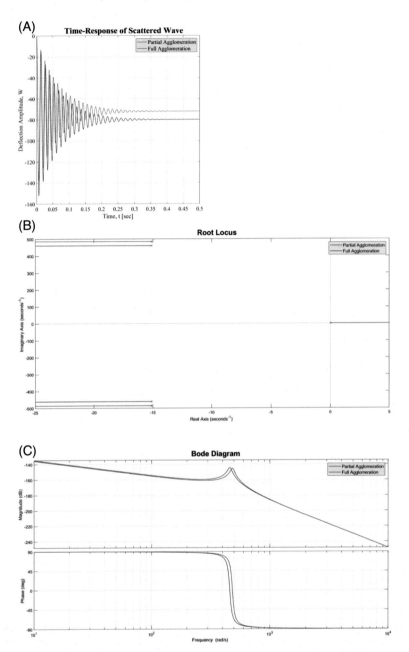

FIGURE 4.37 Effect of the agglomeration pattern on the wave dynamics and stability of the acoustical waves dispersed in MSH PNC square plates rested on a three-parameter visco-Pasternak medium ($a/h = 100$, $\mu = 0.2$, $w_r = 0.1$, $p = 2$, $V_F = 0.2$, $K_W = 10$, $K_P = 1$, $C_d = 0.5$). In this diagram, the full agglomeration state is considered. Subplot (A) illustrates the time–response of the system; whereas, subplots (B) and (C) exhibit the stability of the system in the framework of root locus and bode criteria, respectively.

PNC plate in the partial agglomeration condition is bigger than that of the same system if full agglomeration is considered. Like previous case study, the Root Locus does not intersect the real axis because of the negative value of the system's gain. According to the Root Locus, it can be easily concluded that the θ angle of the PNC system with partially agglomerated nanofillers is bigger than that of the same system whose reinforcement is accomplished with fully agglomerated CNTs. Based on the inverse relation between θ and ξ, PNC structures with partially agglomerated nanofillers have smaller damping ratio. Based on the Bode plot shown in Fig. 4.37C, the type-1 system has a pole at the origin. By considering the beginning value of the phase from $\pi/2$ radian, the negative value of the system's gain can be easily certified. Back to the bigger damped frequency in the partial agglomeration case, it can be realized that the peak time for the mentioned case is smaller than that of the case corresponding with full agglomeration. Similarly, based on the order of the damping ratio and damped frequency of the system in the studied cases, it can be inferred that the peak of overshoot for the case of partial agglomeration is bigger in comparison with the case of full agglomeration. With the same logic, the settling time of the plate reinforced via partially agglomerated CNTs is greater than that corresponding to PNC systems reinforced with fully entangled nanofillers.

Moreover, the effect of agglomeration parameter μ on the wave propagation response of the MSH PNC plate is investigated in the framework of Fig. 4.38. This figure reveals that if a bigger value is considered for the volume fraction of the clusters, the amplitude of the deflection will be decreased. The physical reason of this reduction is the positive role of the agglomeration parameter μ on the reinforcing mechanism in such hybrid nanomaterials. So, based on the inverse relation between deflection amplitude and stiffness, the plate's deflection is decreased. From another point of view, increment of the volume fraction of the clusters results in an accelerated motion in the media. Based on the Root Locus and with regard to the system's transfer function in this example, one can state that the present system possesses a negative gain. It is obvious that only the imaginary part of the system's complex pole will be changed if a substitution is made in the value of the agglomeration parameter μ. This change is in a way so that the imaginary part of the pole will be enlarged by adding the value of volume fraction of the clusters. Once again, it is illustrated that the Root Locus of the system does not have an intersection with the real axis because of the negative gain of the system. According to the greater damped frequency in the case of $\mu = 0.9$, it is clear that the angle θ will be bigger in this condition compared with the case of $\mu = 0.1$. So, based on this issue and with respect to the inverse relationship between θ and ξ, it can be claimed that the damping ratio is decreased once $\mu = 0.9$ is hired instead of $\mu = 0.1$. Looking at the Bode diagram plotted in

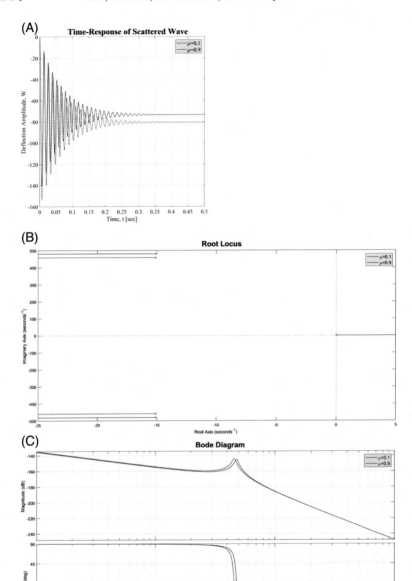

FIGURE 4.38 Effect of the volume fraction of the clusters on the wave dynamics and stability of the acoustical waves dispersed in MSH PNC square plates rested on a three-parameter visco-Pasternak medium ($a/h = 100$, $\eta = 1$, $w_r = 0.1$, $p = 2$, $V_F = 0.2$, $K_W = 10$, $K_P = 1$, $C_d = 0.5$). In this diagram, the full agglomeration state is considered. Subplot (A) illustrates the time–response of the system; whereas, subplots (B) and (C) exhibit the stability of the system in the framework of root locus and bode criteria, respectively.

Fig. 4.38C, it can be perceived that the phase of this type-1 system is begun from $\pi/2$ radian instead of $-\pi/2$. Therefore, the system cannot possess a positive gain. Also, the existence of a pole at the origin can be realized based on this figure. Furthermore, the greater damped frequency in the case of $\mu = 0.9$ reveals that the peak time of the system will be decreased if a bigger value is assigned to the agglomeration parameter μ. Also, it can be stated that because of the smaller damping ratio in the case of $\mu = 0.9$, the peak of overshoot will be greater in this condition. With the same logic and by recalling the definition of the settling time, it can be understood that the enhancement of the agglomeration parameter μ leads to an increase in the settling time.

The major goal of Fig. 4.39 is to survey the influence of CNTs' mass fraction on the propagation behaviors of elastic waves dispersed in MSH PNC plates. From this figure, it can be realized that the amplitude of the deflection will be reduced if a bigger value is assigned to the mass fraction of the nanofillers. This effect is attributed to the positive impact of addition of the CNTs' mass fraction on the equivalent stiffness of the nanomaterial that is used to manufacture the continua. So, the plate behaves in stiffer manner and by keeping the inverse relation between deflection and stiffness, it will be logical to see such a trend. Moreover, the Root Locus depicted in Fig. 4.39B states that the transfer function corresponding to the underobservation system contains a negative gain. It is clear that both real and imaginary parts of the complex pole of the system are changed when the mass fraction is varied. This change denotes that the pole of the stiffer MSH PNC plate, that is, corresponding with the state of $w_r = 0.40$, is closer to the origin. No intersection between Root Locus and real axis is observed that is because of the system's negative gain. The difference between the position of the system's poles in the study cases indicates on the bigger θ of the PNC plate with $w_r = 0.40$ in comparison with the system whose nanofillers' mass fraction is assumed to be $w_r = 0.04$. In consequence, the damping ratio of the stiffer system will be smaller, that is, $\xi_{w_r=0.04} > \xi_{w_r=0.40}$. According to Fig. 4.39C, it can be observed that the phase of the system starts to decrease from $\pi/2$ radian instead of $-\pi/2$. By considering this issue and with regard to the type-1 being of the system, the negative value of the system's gain will be certified. Also, the existence of a pole at the origin that was formerly observed in the system's Root Locus can be inferred from the Bode diagram, too. The reason of this claim is that the ultimate value of the system's phase is $-\pi/2$. Moreover, bigger damped frequency of the system whose nanofiller's mass fraction is $w_r = 0.04$ indicates on the fact that the stiffer system ($w_r = 0.40$) possesses a greater peak time. Also, even though the difference between the peak of overshoot in the studied cases is very small, this parameter is increased if the stiffer system is considered in mind. With the same reason, the stiffer system ($w_r = 0.40$) possesses a bigger settling time rather than the softer one ($w_r = 0.04$).

286 4. Dynamic analysis of multiscale hybrid nanocomposite structures

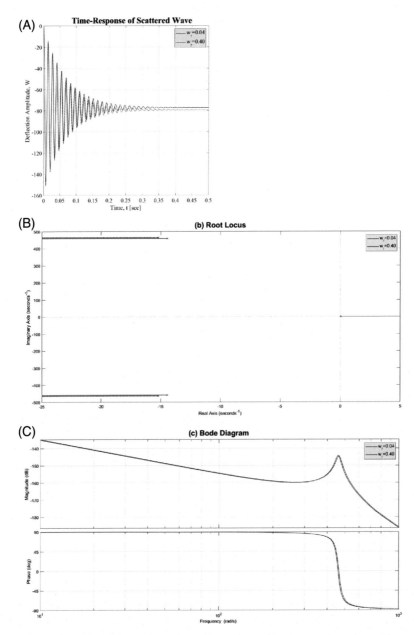

FIGURE 4.39 Effect of the mass fraction of the nanofillers on the wave dynamics and stability of the acoustical waves dispersed in MSH PNC square plates rested on a three-parameter visco-Pasternak medium ($L/h = 100$, $\eta = 1$, $\mu = 0.4$, $p = 2$, $V_F = 0.2$, $K_W = 10$, $K_P = 1$, $C_d = 0.5$). In this diagram, the full agglomeration state is considered. Subplot (A) illustrates the time–response of the system; whereas, subplots (B) and (C) exhibit the stability of the system in the framework of root locus and bode criteria, respectively.

In all of the previous case studies, the concentration was on the investigation of the impacts of various parameters of agglomeration on the dynamic characteristics of MSH PNC plates attacked by elastic waves. In Fig. 4.40 however, the focus will be on the measurement of the impact of the curvy shape of the CNTs on the wave dispersion behaviors of the aforesaid system. In the first part of this illustration, it can be simply found that the implementation of wavy nanofillers for the goal of reinforcement in the fabrication of MSH PNC plates results in an increase in the amplitude of the plate's deflection. This is exactly the outcome of the negative impact of the waviness phenomenon on the stiffness of the nanomaterial. Based on this destroying effect and by recalling the inverse relation between deflection and stiffness, it seems logical to see such a trend in the mentioned figure. By looking at the stability diagrams, it can be another time observed that a negative gain system is studied in this example. It is obvious that the Root Locus does not intersect the real axis thanks to the negative value of the system's gain. It can be perceived that if nonideal wavy nanofillers are implemented in the modeling, the position of the complex poles of the system will be closer to the origin and therefore, the stability of the system will be reduced. Based on the calculations, waviness phenomenon results in a decrease in the angle θ and thus, the damping ratio will be intensified due to this phenomenon. As sated before, the present system is a type-1 system whose phase is started to reduce from $\pi/2$ which means that the gain of this system cannot be positive. Also, the final value of the system's phase angle indicates on the availability of a zero pole at the origin which was previously observed in Fig. 4.40B. According to the fact that PNC plates consisted of ideal nanotubes possess higher damped frequency, their peak time will be smaller than those manufactured with wavy CNTs. On the other hand, bigger damping ratio in the case of using wavy nanofiller in the fabrication leads to smaller peak of overshoot for the system in this case. Based on the similar reason, if the constituent material is synthesized via straight nanofillers, the settling time will be bigger due to the inverse relationship between settling time and damping ratio.

Finally, the impact of length-to-diameter ratio of the CNTs on both time-dependent and stability characteristics of MSH PNC plates was monitored in Fig. 4.41. Based on this figure, the absolute value of the deflection is decreased once thinner CNTs ($\alpha = 2000$) are assumed to be used in the modeling. This outcome is because of the positive effect of the thin being of the nanofillers on the total stiffness of the PNC structure. Hence, it is not strange to see such a trend due to the inverse relationship between deflection and stiffness. The stability diagrams denote that present system has a negative gain like former ones. It is observable that the imaginary part of the system's poles will be increased in magnitude if $\alpha = 20$ is replaced with $\alpha = 2000$. As another benchmark, it is worth mentioning that the negative gain of the system results in existence of no intersection

288　　4. Dynamic analysis of multiscale hybrid nanocomposite structures

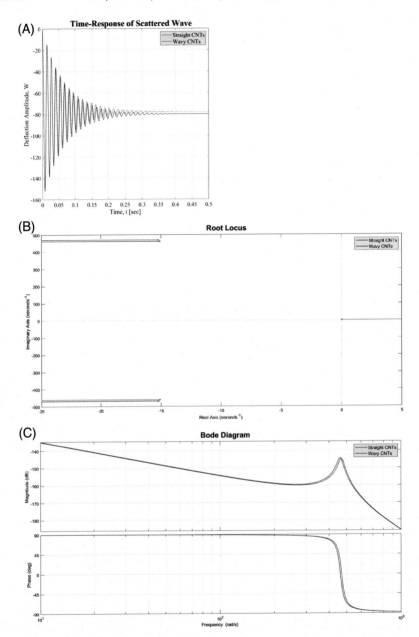

FIGURE 4.40 Effect of the wavy shape of the nanofillers on the wave dynamics and stability of the acoustical waves dispersed in MSH PNC square plates rested on a three-parameter visco-Pasternak medium ($a/h = 100$, $V_r = 2\%$, $V_F = 0.2$, $K_W = 10$, $K_P = 1$, $C_d = 0.5$). In this diagram, the full agglomeration state is considered. Subplot (A) illustrates the time-response of the system; whereas, subplots (B) and (C) exhibit the stability of the system in the framework of root locus and bode criteria, respectively.

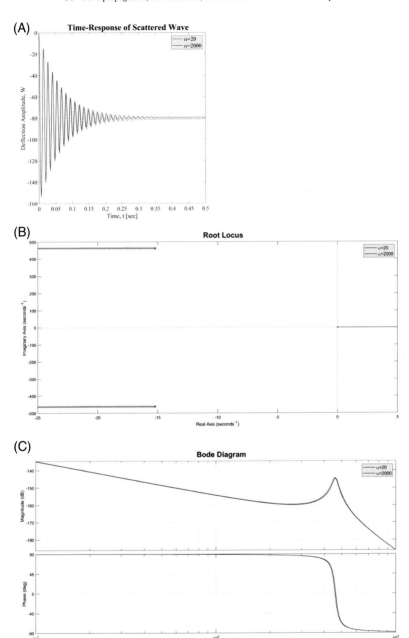

FIGURE 4.41 Effect of the nanofillers' length-to-diameter ratio on the wave dynamics and stability of the acoustical waves dispersed in MSH PNC square plates rested on a three-parameter visco-Pasternak medium ($a/h = 100$, $W_r = 2\%$, $V_F = 0.2$, $K_W = 10$, $K_P = 1$, $C_d = 0.5$). In this diagram, the full agglomeration state is considered. Subplot (A) illustrates the time-response of the system; whereas, subplots (B) and (C) exhibit the stability of the system in the framework of root locus and bode criteria, respectively.

between real axis and Root Locus of the system. Based on the difference in the imaginary part of the complex poles of the system while length-to-diameter ratio of the nanofillers is varied, it is clear that if thinner CNTs are considered in the modeling, the damped frequency will be enlarged and it leads to greater θ angle in consequence. Also, it can be demonstrated that due to the inverse relation between θ and ξ, the damping ratio will be decreased if a bigger value is assigned to the CNTs' length-to-diameter ratio in the modeling. According to the fact that the studied system is of type-1 and the phase of the system has begun its reducing trend from $\pi/2$ radian, the negative value of the gain can be perceived from the Bode diagram. In addition, the existence of a pole at the origin can be realized by looking at the Bode diagram because of the fact that the final value of the system's phase is $-\pi/2$. Moreover, it can be simply found that increment of the CNTs' length-to-diameter ratio results in a decrease in the peak time due to the inverse relation between damped frequency and peak time. Also, addition of the length-to-diameter ratio of the nanotubes induces a rise in the peak of overshoot of the system because of its reducing impact on the damping ratio of the system. Based on the same reason, the settling will be increased if the length-to-diameter ratio of the CNTs is intensified.

4.3 Wave propagation, free vibration, and nonlinear forced vibration of shells

In this section of present chapter, the dynamic behaviors of MSH PNC structures of shell type will be investigated in detail. In the future subsections, free vibration, nonlinear forced vibration, and wave dispersion analyses of MSH PNC shells will be conducted. It is worth regarding that in the mentioned analyses, the effects of nanofiller agglomerates, wavy shape of the CNTs, and their length-to-diameter ratio on the dynamic response of the PNC structures will be included. It must be considered that structure's modeling methods used in this section are like those instructed in Sections 2.2.3.1, 2.2.3.2.2, and 2.2.3.2.3 of Chapter 2. In what follows, free vibration characteristics of MSH PNC conical shells will be studied at first. Thereafter, nonlinear forced oscillation analysis of MSH PNC cylinders will be presented. As the final examination of this book, dispersion curves of elastic waves scattered in MSH PNC cylinders will be investigated in detail.

4.3.1 Free vibration analysis of MSH PNC shells

Within the present subsection, several case studies will be depicted in order to probe the variation of the natural frequency of MSH PNC conical

TABLE 4.4 Comparison of the dimensionless frequencies of CNTR PNC conical shells with clamped edges once the semivertex cone angle is changed ($h/R_1 = 0.05$, $L/R_1 = 2.5$, $V_{CNT} = 0.28$).

	α = 15°		α = 30°	
Source	λ_{01}	λ_{02}	λ_{01}	λ_{02}
Ref. [738]	0.878	0.642	0.822	0.559
Ref. [739]	0.907	0.657	0.824	0.606
Present	0.863	0.637	0.819	0.596

shells while the material properties of the constituent nanomaterial are varied. According to the former explanations, the classical theory of conical shells is utilized here and therefore, we will be about to analyze thin-walled PNC shells. In all of the future case studies, a shell of thickness $h = 2$ mm will be studied length of the shell will be fixed so that $L/h = 100$. It must be considered that in the numerical examples of this part of present chapter, more than 80 grid points are gathered to extract the natural frequency of the PNC conical shell. Because of the fact that only 20 grid points are sufficient for convergence of the answer in a generalized differential quadrature-assisted problem in the linear domain, such a selection guarantees the convergence of the response. In addition, the following dimensionless terms will be utilized for the sake of simplicity:

$$K_W = \frac{k_W L^4}{E_m h^3}, \quad K_P = \frac{k_P L^2}{E_m h^3}, \quad \Omega = \omega h \sqrt{\frac{\rho_m}{E_m}} \quad (4.7)$$

For the goal of being sure from the accuracy of the model presented in this text, a comparison between the dimensionless frequencies of truncated conical shells consisted of PNC materials reinforced via CNTs is organized. Based on the results tabulated in Table 4.4, it can be inferred that the results of present work and those reported in Refs. [738] and [739] are in an excellent agreement. Thus, it can be claimed that the following results are valid and can be trusted without concern.

The main objective of the first case study is to show the influence of volume fraction of the CNTs inside the clusters on the variation of dimensionless frequency of the conical shell. Based on Fig. 4.42, it can be observed that the frequency curve of the MSH PNC conical shell is continuously decreased as the volume fraction of the CNTs inside the clusters is intensified. This trend can be physically justified by recalling the negative influence of the agglomeration parameter η on the equivalent stiffness of the constituent nanomaterial. Therefore, the natural frequency is reduced because of the direct relation between frequency and stiffness. Also, it must be mentioned that the value of natural frequency in any

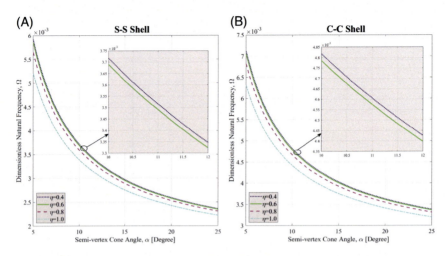

FIGURE 4.42 Variation of the dimensionless natural frequency of the MSH PNC conical shells with (A) S–S and (B) C–C BCs against semivertex cone angle for various values of volume fraction of the CNTs inside the clusters ($\mu = 0.4$, $p = 2$, $w_r = 0.1$, $V_F = 0.2$, $K_W = 10$, $K_P = 1$).

desired value of η is in its maximum whenever C–C BC is considered in the theoretical calculations. The main reason of this issue is the greater structural stiffness of the clamped condition in comparison with simply supported one.

Furthermore, the effect of volume fraction of the clusters on the dynamic characteristics of MSH PNC conical sells is probed in the framework of Fig. 4.43. Based on this diagram, it can be perceived that the continuous system will be enabled to support bigger frequencies in any arbitrary semivertex cone angle if a greater value is assigned to the volume fraction of the clusters in the modeling. This positive influence of agglomeration parameter μ on the dimensionless frequency is due to the same influence of this term on the total stiffness of the PNC material. It is obvious that the bigger is the volume fraction of the clusters, the nanomaterial will be closer to the construction of a giant cluster containing all of the nanofillers. Thus, the negative influence of CNT agglomerates on the stiffness of the MSH PNC will be decreased. Besides, it is another time shown that C–C MSH PNC conical shells possess greater natural frequencies compared with S–S ones. This is exactly because of higher structural stiffness of clamped support rather than another one.

Fig. 4.44 is depicted in order to highlight the impact of CNTs' mass fraction on the variation of the natural frequency of the MSH PNC conical shells. In this diagram, the variation of dimensionless frequency versus semivertex cone angle is drawn for different values of nanofillers' mass fraction. According to the generated data, it is clear that the frequency curve of the PNC shell is shifted upward while a bigger value is assigned to

4.3 Wave propagation, free vibration, and nonlinear forced vibration of shells

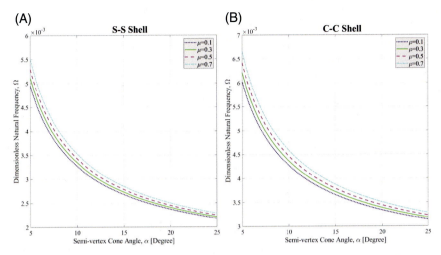

FIGURE 4.43 Variation of the dimensionless natural frequency of the MSH PNC conical shells with (A) S–S and (B) C–C BCs against semivertex cone angle for various values of volume fraction of the clusters ($\eta = 1$, $p = 2$, $w_r = 0.1$, $V_F = 0.2$, $K_W = 10$, $K_P = 1$).

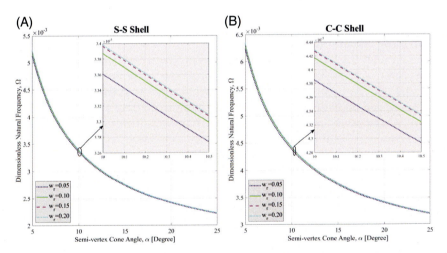

FIGURE 4.44 Variation of the dimensionless natural frequency of the MSH PNC conical shells with (A) S–S and (B) C–C BCs against semivertex cone angle for various values of nanofillers' mass fraction ($\eta = 1$, $\mu = 0.4$, $p = 2$, $V_F = 0.2$, $K_W = 10$, $K_P = 1$).

the nanofillers' mass fraction. The main reason of this trend is the positive influence of the higher content of the nanofillers on the enhancement of the total stiffness of the MSH PNC material. Thus, it is in agreement with our predictions that the frequency is aggrandized by adding the nanofillers' mass fraction. Another time, it is clearly shown that the MSH PNC shells

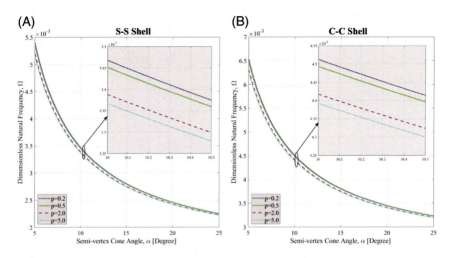

FIGURE 4.45 Variation of the dimensionless natural frequency of the MSH PNC conical shells with (A) S–S and (B) C–C BCs against semivertex cone angle for various gradient indices ($\eta = 1$, $\mu = 0.4$, $w_r = 0.1$, $V_F = 0.2$, $K_W = 10$, $K_P = 1$).

with C–C supports are able to promote bigger frequencies in comparison with those containing S–S edges.

In another investigation, the impact of gradient index on the dynamic behaviors of the MSH PNC shell is surveyed in detail. As shown in Fig. 4.45, it can be inferred that the dimensionless frequency of the PNC structure can be affected in a decreasing manner if a rise is induced in the gradient index. By recalling Fig. 2.6 of Chapter 2, it can be remembered that an increase in the gradient index corresponds with a reduction in the total stiffness of the MSH PNC material. Therefore, it is natural to see that the frequency of the nanoengineered system is decreased in the case of assigning a big value to the gradient index. Like former illustrations, the impact of changing the BC at the ends of the shell on the dimensionless frequency of the shell is shown in this diagram, too.

Leaving the discussion about the effects of parameters involved in the modeling of agglomeration phenomenon on the dynamic characteristics of the system, the next two diagrams are depicted to probe the frequency variations of the MSH PNC shells under action of waviness phenomenon. Based on Fig. 4.46, two PNC shells are considered and their dimensionless frequency is plotted for different semivertex cone angles. According to this study, it is proven that if wavy nanofillers are implemented in the fabrication of the PNCs, the structure cannot support big natural frequencies. This trend is appeared because of the negative influence of the waviness phenomenon on the reinforcement mechanism in such nanomaterials. So, if the direct relation between stiffness and natural frequency is reviewed,

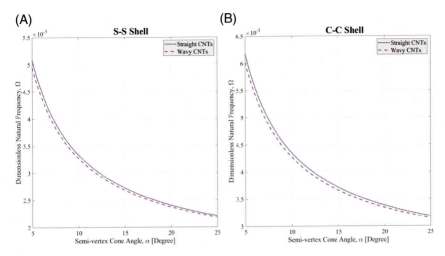

FIGURE 4.46 Qualitative effect of the waviness phenomenon on the free vibration response of the MSH PNC conical shells with (A) S–S and (B) C–C BCs ($V_r = 1\%$, $V_F = 0.2$, $K_W = 10$, $K_P = 1$).

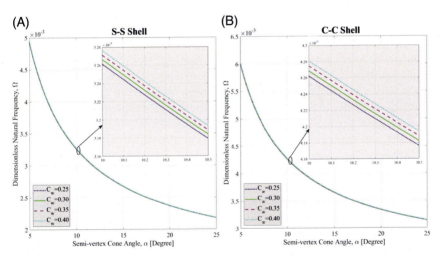

FIGURE 4.47 Variation of the dimensionless natural frequency of MSH PNC conical shells with (A) S–S and (B) C–C BCs against semivertex cone angle for various values of waviness coefficient ($V_r = 1\%$, $V_F = 0.2$, $K_W = 10$, $K_P = 1$).

it can be realized that the observed trend is not strange at all. In the next example shown in Fig. 4.47, it is illustrated that how can the vibration frequency be affected if the waviness coefficient is changed. This figure reveals that increment of the waviness coefficient results in a reciprocal enhancement in the dimensionless frequency of the PNC shell. This trend

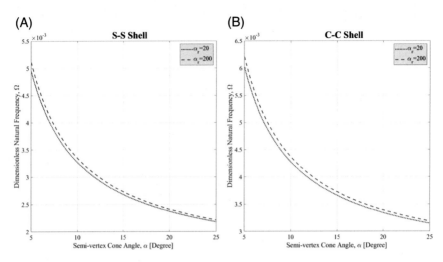

FIGURE 4.48 Variation of the dimensionless natural frequency of MSH PNC conical shells with (A) S–S and (B) C–C BCs against semivertex cone angle for various length-to-diameter ratios of the CNTs ($W_r = 2\%$, $V_F = 0.2$, $K_W = 10$, $K_P = 1$).

can be simply justified by remembering the definition of the waviness coefficient. According to the definition, the bigger is this coefficient, the smaller will be the amplitude of the wave existing in the chord of the CNT. Thus, it can be declared that big waviness coefficients correspond with CNTs close to the ideal state. Based on these explanations, it is natural to observe the reported trend. In both of the mentioned figures, the effects of both simply supported and clamped BCs on the dimensionless frequency of the MSH PNC conical shells are taken into consideration, too.

As the final case study in this subsection, Fig. 4.48 is plotted in order to show that the dimensionless frequency of the MSH PNC shells can be manipulated if the length-to-diameter ratio of the reinforcing nanofillers is varied. To avoid from the readers' sense of confusion, the length-to-diameter ratio of the CNTs is shown with sign α_r in this figure. This decision is made to generate a distinction between CNTs' length-to-diameter ratio and semivertex cone angle of the conical shell. Obviously, it can be demonstrated that an increase in the length-to-diameter ratio of the nanofillers leads to an increase in the natural frequency of the shell's free vibrations. The observed trend seems logical because of the fact that thinner nanofillers are able to reinforce PNCs in a more efficient manner. However, it must be kept in mind that the length-to-diameter ratio of the CNTs cannot be increased limitless. This is due to the fact that the mentioned rise can be resulted in increment of the possibility of existence of curvy nanofillers in the composition of the PNC conical shell.

4.3.2 Nonlinear forced vibration analysis of MSH PNC shells

Within this subsection, it will be tried to survey the nonlinear forced vibration behaviors of MSH PNC cylindrical shells. In order to do so, the kinematic relations of the shell's first-order shear deformation theory will be employed. Detailed data about the procedure of derivation of the governing equation can be found by referring to Section 2.2.3.2.3 of Chapter 2. The influences of agglomeration and waviness phenomena as well as that of the nanofillers' length-to-diameter ratio on the vibrational responses of the PNC system will be captured here. It is noteworthy that only quantitative analyses will be carried out in this part and geometric analysis of the phase portrait of the system will be dismissed because it was deeply investigated in Sections 4.1.2 and 4.2.2 of present chapter in detail. Therefore, it is avoided to present such analyses for the sake of brevity. In the future case studies, the following nondimensional form of the parameters will be utilized:

$$K_L = \frac{k_L L^4}{E_m h^3}, \quad K_P = \frac{k_P L^2}{E_m h^3}, \quad K_{NL} = \frac{k_{NL} L^4}{E_m h}, \quad C_d = \frac{c_d L^2}{h^2 \sqrt{\rho_m E_m}}, \quad F = \frac{\bar{f} M L^2}{E_m h^2} \tag{4.8}$$

In addition, it must be pointed out that in all of the following case studies, simply supported thin-walled shells are considered and therefore, the length and radius of the shell are assumed to be $L/h = 100$ and $L/R = 5$. For the goal of being sure from accuracy of the present modeling, a comparison between the results of this text and those reported in Ref. [1] is organized in Fig. 4.49. Based on this illustration, it can be well observed that the results of this work are in complete agreement with those reported by the authors in their former book [1]. Thus, the case studies of present work can be trusted.

In Fig. 4.50, the impact of the agglomeration pattern on the nonlinear dynamic response of the MSH PNC shell is probed. In this diagram, two cases of full and partial agglomeration are considered and it is shown that the nonlinear vibration response of the PNC shell can be dramatically affected by this issue. According to Fig. 4.50A, the frequency–response curve of the system with partially agglomerated nanofillers possesses a bigger tendency to the right-hand side. This is because of the greater stiffness of the PNC material in this condition compared with the case of considering hybrid nanomaterials reinforced with fully aggregated CNTs. This trend is appeared due to the worst negative impact of the full agglomeration state on the equivalent stiffness of the constituent nanomaterial. In addition to the system's frequency–response curve, the variation of the amplitude–response of the system, that is, shown in Fig. 4.50B reveals the same outcome. According to this diagram, the stiffer branch belongs to the case of partial agglomeration. Also, it can be easily found that the region of

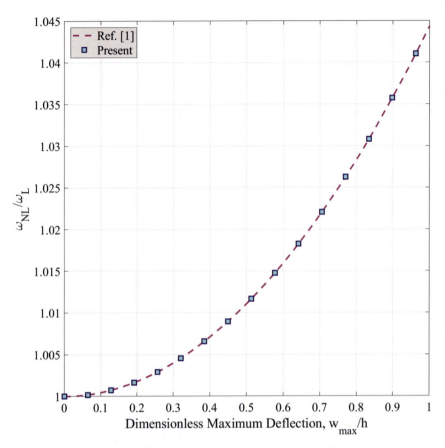

FIGURE 4.49 Comparison of the nonlinear vibration response of porous PNC shells reinforced via oxide of graphene with those reported in the literature ($m = 1, n = 3, L/h = 100, L/R = 20, e_1 = 0.1, W_{GO} = 1\%$). In this plot, GOR-A PNC structure with porosity type-I is considered.

instability of the system will be increased if all of the reinforcing nanofillers are entangled in the clusters, that is, corresponding with full agglomeration case.

In the next case study, the effect of another agglomeration parameter, that is, μ, on both frequency–response and amplitude–response curves of the MSH PNC plate is studied. Fig. 4.51 states that a rise in the volume fraction of the clusters makes the system stiffer and this stiffer nature leads to an enhancement in the nonlinear response of the system as a consequence. This is due to the direct relation between stiffness and dynamic response of the oscillating system. This physical issue can be observed in the amplitude–response of the system, too. It is clear that the branch whose stiffness is bigger is corresponding with the case of

4.3 Wave propagation, free vibration, and nonlinear forced vibration of shells

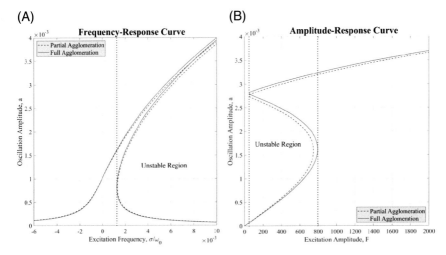

FIGURE 4.50 Effect of both partial and full agglomeration states on the nonlinear forced vibration characteristics of MSH PNC shells ($L/h = 100$, $\mu = 0.2$, $p = 2$, $w_r = 0.1$, $V_F = 0.2$, $C_d = 10$, $K_L = 100$, $K_P = 10$, $K_{NL} = 10^7$, and $F = 100$). In subplots (A) and (B), the frequency–response and amplitude–response curves of MSH PNC shells with both partial and full agglomeration states are plotted, respectively. Part (B) is plotted at point $\sigma/\omega_0 = 0.005$, that is, in the unstable region of the frequency–response diagram.

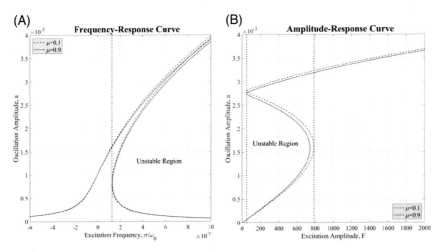

FIGURE 4.51 Effect of the volume fraction of the clusters on the nonlinear forced vibration characteristics of MSH PNC shells ($L/h = 100$, $\eta = 1$, $p = 2$, $w_r = 0.1$, $V_F = 0.2$, $C_d = 10$, $K_L = 100$, $K_P = 10$, $K_{NL} = 10^7$, and $F = 100$). In subplots (A) and (B), the frequency–response and amplitude–response curves of MSH PNC shells with both partial and full agglomeration states are plotted, respectively. Part (B) is plotted at point $\sigma/\omega_0 = 0.005$, that is, in the unstable region of the frequency–response diaELSST093-04-gram.

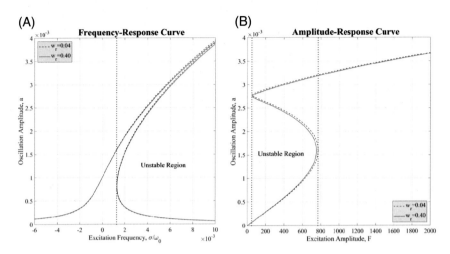

FIGURE 4.52 Effect of the nanofillers' mass fraction on the nonlinear forced vibration characteristics of MSH PNC shells ($L/h = 100$, $\eta = 1$, $\mu = 0.4$, $p = 2$, $V_F = 0.2$, $C_d = 10$, $K_L = 100$, $K_P = 10$, $K_{NL} = 10^7$, and $F = 100$). In subplots (A) and (B), the frequency–response and amplitude–response curves of MSH PNC shells with both partial and full agglomeration states are plotted, respectively. Part (B) is plotted at point $\sigma/\omega_0 = 0.005$, that is, in the unstable region of the frequency–response diagram.

assigning the value of $\mu = 0.9$ to the volume fraction of the clusters. It is interesting to mention that this positive impact leads to a reduction in the instability area of the system.

In the previous figures, influences of agglomeration parameters η and μ on the vibrational characteristics of MSH PNC shells subjected to harmonic excitation were investigated. In the next two diagrams, the effects of parameters involved in the determination of the volume fraction of the aggregated CNTs on the nonlinear oscillations of the PNC shells will be monitored. According to Fig. 4.52, it is obvious that the nonlinear dynamic response of the plate will be increased if a bigger content of the CNTs is implemented in the composition of the hybrid nanomaterial. This issue can be easily found from the mentioned figure by looking at the higher tendency of the frequency–response curve of the system to the right-hand side. Thus, the continuous system behaves in stiffer manner which is attributed to the stiffness enhancement of the PNC material. It can be seen that this stiffer behavior is shown in the amplitude–response curve of the system, too. It is worth mentioning that the unstable region of the oscillator that is surrounded between two low- and high-energy limit cycles will be affected if the nanofillers' mass fractions is varied. On the other hand, the impact of gradient index on the dynamic response of the continuous system is illustrated in Fig. 4.53. Based on this figure, it can be stated that the PNC shell will be a softer oscillator if a bigger value is assigned to the gradient index in the micromechanical modeling of the nanomaterial. The

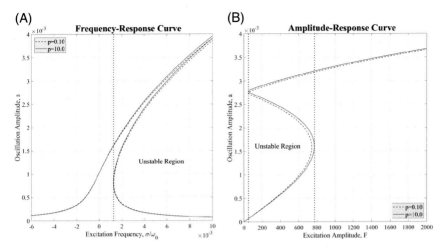

FIGURE 4.53 Effect of the gradient index on the nonlinear forced vibration characteristics of MSH PNC shells ($L/h = 100$, $\eta = 1$, $\mu = 0.4$, $w_r = 0.1$, $V_F = 0.2$, $C_d = 10$, $K_L = 100$, $K_P = 10$, $K_{NL} = 10^7$, and $F = 100$). In subplots (A) and (B), the frequency–response and amplitude–response curves of MSH PNC shells with both partial and full agglomeration states are plotted, respectively. Part (B) is plotted at point $\sigma/\omega_0 = 0.005$, that is, in the unstable region of the frequency–response diagram.

physical reason of this trend is the decreasing influence of the gradient index on the effective stiffness of the hybrid nanomaterial. This issue can be better perceived by taking a brief look at the Fig. 2.6 of Chapter 2.

Moreover, to give insight about the variation of the nonlinear dynamic response of the MSH PNC shell under action of the waviness phenomenon to the readers, Fig. 4.54 is depicted. In this figure, both frequency–response and amplitude–response curves of PNC shells consisted of either straight or wavy CNTs are drawn. It is shown that the system's response will be bent to the right-hand side with a higher intensity if ideal CNTs are utilized. Hence, it can be concluded that wavy shape of the nanofillers results in a softening change in the response of the system if a harmonic stimulation is applied on the MSH PNC shell. This trend is because of the destroying influence of wavy shape of the CNTs on the stiffness reinforcement mechanism in the constituent nanomaterial. It must be mentioned that all of the abovementioned trends can be certified if the amplitude–response shown in Fig. 4.54B is tracked. According to this subplot, it can be resulted that the unstable region of the oscillator will be enlarged in area if wavy nanofillers are employed in the process of manufacturing.

At the end of this subsection, the impact of changing the length-to-diameter ratio of the CNTs on the nonlinear forced vibration response of the MSH PNC shells can be well observed in Fig. 4.55. This illustration indicates on the positive impact of the CNTs' length-to-diameter ratio on the

302 4. Dynamic analysis of multiscale hybrid nanocomposite structures

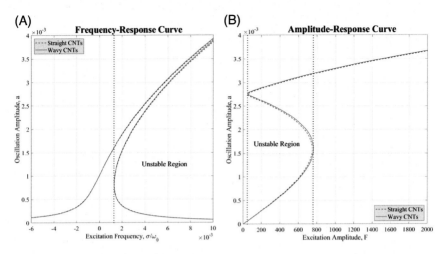

FIGURE 4.54 Effect of the wavy shape of the CNTs on the nonlinear forced vibration characteristics of MSH PNC shells ($L/h = 100$, $V_r = 0.02$, $V_F = 0.2$, $C_d = 10$, $K_L = 100$, $K_P = 10$, $K_{NL} = 10^7$, and $F = 100$). In subplots (A) and (B), the frequency–response and amplitude–response curves of MSH PNC shells with both partial and full agglomeration states are plotted, respectively. Part (B) is plotted at point $\sigma/\omega_0 = 0.005$, that is, in the unstable region of the frequency–response diagram.

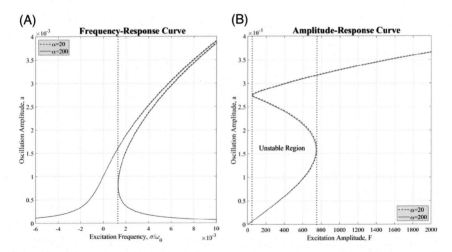

FIGURE 4.55 Effect of the CNTs' length-to-diameter ratio on the nonlinear forced vibration characteristics of MSH PNC shells ($L/h = 100$, $W_r = 0.02$, $V_F = 0.2$, $C_d = 10$, $K_L = 100$, $K_P = 10$, $K_{NL} = 10^7$, and $F = 100$). In subplots (A) and (B), the frequency–response and amplitude–response curves of MSH PNC shells with both partial and full agglomeration states are plotted, respectively. Part (B) is plotted at point $\sigma/\omega_0 = 0.005$, that is, in the unstable region of the frequency–response diagram.

dynamic response of the system. This issue can be simply found by looking at both frequency–response and amplitude–response curves. The main reason of the observed trend is the positive effect of the thin nanofillers on the mechanism of stiffness improvement in the hybrid nanomaterials. Therefore, the observed trend seems logical due to the direct relation between stiffness and vibrational response of the system. It is noteworthy to mention that limitless increase of the CNTs' length-to-diameter ratio cannot be recommended. The reason of this issue is that if ultrathin CNTs are selected in the fabrication of the MSH PNC material, the probability of existence of curved nanofillers in the microstructure of the nanomaterial will be risen.

4.3.3 Wave propagation analysis of MSH PNC shells

This part is the final illustrative part of present text. In this subsection, it is supposed to survey the dispersion behaviors of elastic waves traveling inside MSH PNC cylinders rested on a two-parameter elastic seat. In this regard, the kinematic modeling previously introduced in Section 2.2.3.2.2 of Chapter 2 will be employed. In the following results, thin-walled shells whose length-to-thickness and length-to-diameter ratios are, respectively, $L/h = 100$ and $L/R = 5$ are studied. It must be mentioned that the following data are generated based on the dimensionless variables similar to those defined in Eq. (4.7). As an opening, Fig. 4.56 is presented to show the accuracy of the present study. In this diagram, the variation of wave frequency of CNTs versus wave number is illustrated and compared with those previously reported in Ref. [740]. It can be clearly seen that although different theories are implemented in present work and Ref. [740], the results of our work are in a remarkable agreement with those reported in the mentioned reference. Hence, the future data can be trusted and hired if a similar problem is going to be solved.

In Fig. 4.57, the effect of volume fraction of the CNTs inside the clusters on the wave frequency in MSH PNC shells is monitored. It is shown that the influence of agglomeration parameter η on the frequency of the propagated waves in small and mid-range wave numbers cannot be seen clearly. However, this issue can be simply observed in high wave numbers. The reason of this issue is the direct impact of the longitudinal wave number on the total stiffness of the PNC shell. It is shown that the dimensionless frequency of the dispersed waves will be decreased if a bigger value is assigned to the volume fraction of the CNTs inside the clusters. This outcome is because of the reducing effect of CNT agglomerates on the stiffness of the hybrid nanomaterial.

Moreover, the impact of another parameter of agglomeration, that is, volume fraction of the clusters, on the dimensionless wave frequency is highlighted in Fig. 4.58. Based on this figure, it is obvious that greater values are estimated for the frequency of the scattered waves in the case of

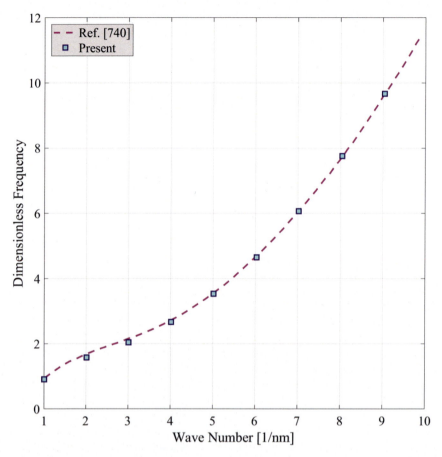

FIGURE 4.56 Comparison of frequency–wave number curve of shells with those available in the literature ($e_0 a = 0$).

using bigger agglomeration parameter μ in the theoretical modeling. This increasing trend is due to the positive impact of increment of the volume fraction of the clusters on the total stiffness of the PNC material. Thus, it is natural to see such a hardening phenomenon because of the direct relation between frequency of the system and stiffness of the hybrid nanomaterial. It is interesting to point out that the influence of agglomeration parameter μ on the dispersion curve of the system cannot be easily seen if small values of longitudinal wave number are considered. This is because of the dependency of the structure's stiffness matrix on the wave numbers in both longitudinal and circumferential directions.

In addition to the agglomeration parameters η and μ, changes in the volume fraction of the reinforcing nanofillers can induce variations in the wave frequency of MSH PNC shells. To survey such an issue, Figs. 4.59 and

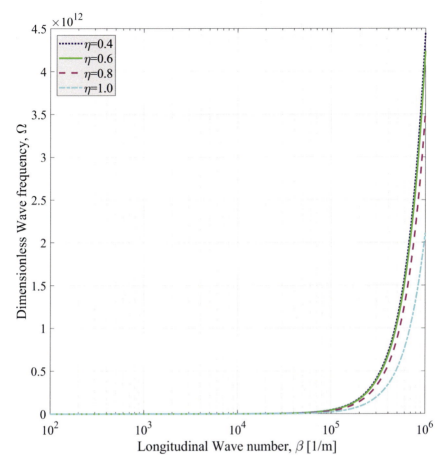

FIGURE 4.57 Variation of the dimensionless wave frequency of MSH PNC shells against longitudinal wave number for various amounts of volume fraction of the CNTs inside the clusters ($\mu = 0.4$, $w_r = 0.1$, $p = 2$, $V_F = 0.2$, $K_W = 20$, $K_P = 5$).

4.60 are plotted to highlight the effects of CNTs' mass fraction and gradient index on the wave frequency of the propagated waves, respectively. Based on Fig. 4.59, the dimensionless frequency of the dispersed waves will be enlarged if the CNTs' mass fraction is increased. However, the wave frequency is reduced if the gradient index is aggrandized as shown in Fig. 4.60. The mentioned increasing and decreasing trends are made in the dynamic response of the PNC shell because of the same effects that is generated in the equivalent stiffness of the constituent nanomaterial, respectively. Like former illustrations, it can be realized that the longitudinal wave number can amplify or weaken the aforementioned influences because of the direct effect of the wave number on the determination of the

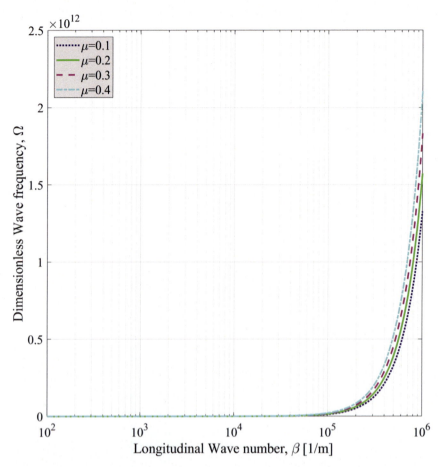

FIGURE 4.58 Variation of the dimensionless wave frequency of MSH PNC shells against longitudinal wave number for various amounts of volume fraction of the clusters ($\eta = 1$, $w_r = 0.1$, $p = 2$, $V_F = 0.2$, $K_W = 20$, $K_P = 5$).

stiffness of the continuous system. This trend leads to more powerful impact of nanofillers' mass fraction and gradient index on the wave frequency in big longitudinal wave numbers rather than small ones.

In another example, the influence of straight or wavy shape of the CNTs on the wave frequency of the propagated waves is studied. According to Fig. 4.61, the frequency of the dispersed waves in any desired longitudinal wave number will be reduced if wavy nanofillers are implemented in the manufacturing process. The physical reason of this phenomenon is the decreasing impact of the wavy shape of the CNTs on the effective Young's modulus of the MSH PNC material. Thus, this decreasing effect results in a softer PNC shell which cannot support big frequencies. In this

4.3 Wave propagation, free vibration, and nonlinear forced vibration of shells

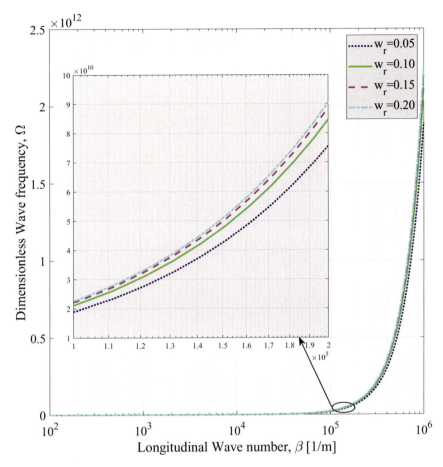

FIGURE 4.59 Variation of the dimensionless wave frequency of MSH PNC shells against longitudinal wave number for various amounts of nanofillers' mass fraction ($\eta = 1$, $\mu = 0.4$, $p = 2$, $V_F = 0.2$, $K_W = 20$, $K_P = 5$).

diagram, a comparative viewpoint is hired to discuss about the impact of nonideal shape of the CNTs on the dispersion curve of the MSH PNC shells. In the next illustration, however, the variation of wave frequency versus longitudinal wave number is drawn while waviness coefficient is varied. According to Fig. 4.62, it can be perceived that the frequency of the propagated waves will be aggrandized once the waviness coefficient is added. The reason of this issue is that reinforcing CNTs will be assumed to be ideal ones if a higher value is assigned to the waviness coefficient. So, the stiffness of the nanomaterial will be increased in this regard. Thus, if the direct relation between stiffness and frequency is considered, it can be figured out that the observed trend is logical.

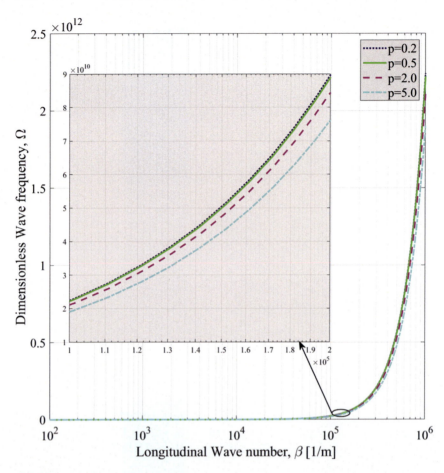

FIGURE 4.60 Variation of the dimensionless wave frequency of MSH PNC shells against longitudinal wave number for various gradient indices ($\eta = 1$, $\mu = 0.4$, $w_r = 0.1$, $V_F = 0.2$, $K_W = 20$, $K_P = 5$).

As the final case study, the impact of the nanofillers' length-to-diameter ratio on the variation of the wave frequency against longitudinal wave number is illustrated in Fig. 4.63. Based on this diagram, it can be simply found that the frequency of the elastic waves can be intensified if thin-type nanofillers are employed in the composition of the MSH PNC material. The major reason of this trend is that CNTs with big length-to-diameter ratios are able to improve the stiffness of the PNCs better than those possessing small length-to-diameter ratios. Therefore, based on the relationship existing between frequency and stiffness of the continuous system, the observed trend is natural. However, it is noteworthy that utilization of very thin CNTs leads to a rise in the probability of existence of wavy nanofillers

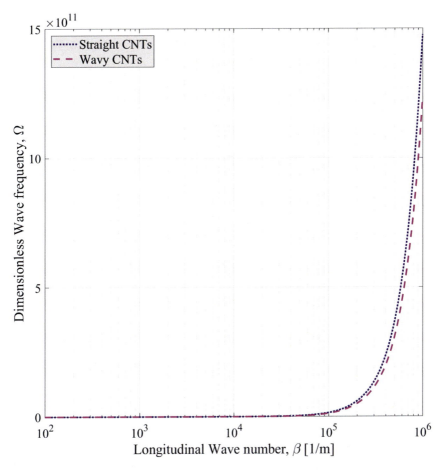

FIGURE 4.61 Qualitative effect of waviness phenomenon on the variation of the dimensionless wave frequency of MSH PNC shells against longitudinal wave number ($V_r = 1\%$, $V_F = 0.2$, $K_W = 20$, $K_P = 5$).

in the microstructure of the MSH PNC that results in a reduction in the dynamic response of the system in consequence.

4.4 Concluding remarks

In this chapter of present text, the main focus was on the investigation of the impacts of a wide range of parameters involved in the micromechanical homogenization of MSH PNCs on the dynamic responses of continuous systems consisted of such nanomaterials. In order to fulfill the above studies, continuum modeling of beams, plates, and shells with the aid of either

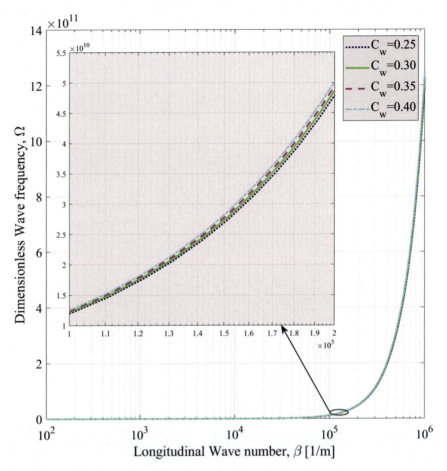

FIG.4.62 Variation of the dimensionless wave frequency of MSH PNC shells against longitudinal wave number for various waviness coefficients ($V_r = 1\%$, $V_F = 0.2$, $K_W = 20$, $K_P = 5$).

classical or higher-order hypotheses was carried out. It was tried to show how the vibration frequency, wave propagation response, and nonlinear dynamic behaviors of the MSH PNC structures can be influenced if practical issues like aggregation of nanofillers in the polymer matrix, nonideal shape of CNTs in the nanomaterial's microstructure, and thin/thick being of the nanofillers are captured in the homogenization scheme. Herein, and in the context of a couple of brief paragraphs, the provided material will be concluded.

According to this chapter, it must be deeply keep in mind of anyone who is supposed to design structural elements that dynamic reaction of the MSH PNC systems to environmental stimulations is totally different from

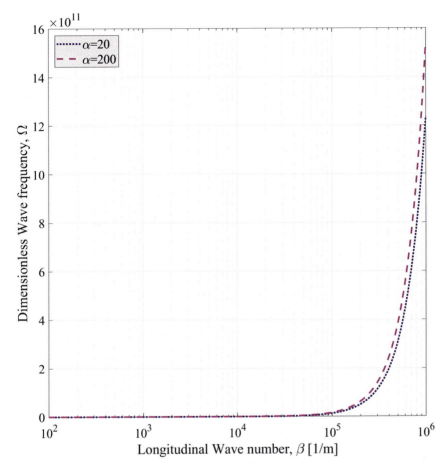

FIGURE 4.63 Variation of the dimensionless wave frequency of MSH PNC shells against longitudinal wave number for various nanofillers' length-to-diameter ratio ($W_r = 0.02$, $V_F = 0.2$, $K_W = 20$, $K_P = 5$).

what recommended by simple theoretical estimations. To be specifically involved in this sentence, it can be stated that existence of CNT agglomerates in the microstructure of the hybrid nanomaterials can be resulted in occurrence of the resonance phenomenon in frequencies smaller than those obtained from calculations with a large number of simplifying assumptions. Thus, it is recommended to consider the worst case introduced in this text to avoid from such undesired consequences in practice. The mentioned example can be stated in a more general form. For instance, if it is aimed to design MSH PNC structures that are assumed to be fluctuating under action of a harmonic loading, it must be noticed that the designed oscillator may promote soft behavior from itself because of the existence of bundles of nanofillers in the material's microstructure.

In the previous paragraph, the concentration was on the highlighting the destroying impact of agglomeration phenomenon on the system's vibrating characteristics in practical problems. One should pay attention to the fact that the same unwanted destructions can happen if it is assumed that all of the nanofillers are straight-shaped ones. In this condition, an overestimation from the nanomaterial's stiffness will be considered in the theoretical approximations. However, in practice, the designed structure behaves softer and all of the mentioned consequences will be observed. Moreover, it is noteworthy that the designers are not recommended to synthesize too long CNTs to manufacture ultrastiff MSH PNCs. The supporting reason of this recommendation is that by doing so, the probability of existence of nonideal curved nanofillers in the nanomaterial's microstructure will be increased, that is, not desirable at all. On the other hand, the condition can be even worse if the uncertainties are combined with the mentioned issues. To cover the latter issue, the results of present chapter must be modified by the designer with the aid of the stochastic modeling.

As the final concluding point in this chapter, the designers of MSH PNC structures that are going be implemented as a wave media must pay attention to the influences of aggregation and waviness phenomena due to two reasons. The first reason is similar to what discussed in the last two paragraphs. The latter one, however, is a little different. In such cases, the designers need to be sure from the stability status of the designed system if the pattern of the agglomeration, that is, partial or full agglomeration, in the microstructure of the hybrid nanomaterial is changed. To this purpose, an acceptable knowledge of control engineering is required.

References

[1] F. Ebrahimi, A. Dabbagh, Mechanics of Nanocomposites: Homogenization and Analysis, first ed., CRC Press, Boca Raton, FL, 2020. https://doi.org/10.1201/9780429316791.

[2] J.N. Reddy, Mechanics of Laminated Composite Plates and Shells: Theory and Analysis, CRC Press, Boca Raton, FL, 2003. https://doi.org/10.1201/b12409.

[3] A.K. Kaw, Mechanics of Composite Materials, second ed., CRC Press, Boca Raton, FL, 2005. https://doi.org/10.1201/9781420058291.

[4] D.H. Robbins, J.N. Reddy, Analysis of piezoelectrically actuated beams using a layerwise displacement theory, Comput. Struct. 41 (2) (1991) 265–279. https://doi.org/10.1016/0045-7949(91)90430-T.

[5] P. Beardmore, C.F. Johnson, The potential for composites in structural automotive applications, Compos. Sci. Technol. 26 (4) (1986) 251–281. https://doi.org/10.1016/0266-3538(86)90002-3.

[6] W. Hufenbach, et al., Polypropylene/glass fibre 3D-textile reinforced composites for automotive applications, Mater. Des. 32 (3) (2011) 1468–1476. https://doi.org/10.1016/j.matdes.2010.08.049.

[7] J. Holbery, D. Houston, Natural-fiber-reinforced polymer composites in automotive applications, JOM 58 (11) (2006) 80–86. https://doi.org/10.1007/s11837-006-0234-2.

[8] G. Koronis, A. Silva, M. Fontul, Green composites: a review of adequate materials for automotive applications, Compos. Part B: Eng. 44 (1) (2013) 120–127. https://doi.org/10.1016/j.compositesb.2012.07.004.

[9] P. Rohatgi, Cast aluminum-matrix composites for automotive applications, JOM 43 (4) (1991) 10–15. https://doi.org/10.1007/BF03220538.

[10] A.P. Mouritz, et al., Review of advanced composite structures for naval ships and submarines, Compos. Struct. 53 (1) (2001) 21–42. https://doi.org/10.1016/S0263-8223(00)00175-6.

[11] D. Mathijsen, Now is the time to make the change from metal to composites in naval shipbuilding, Reinf. Plast. 60 (5) (2016) 289–293. https://doi.org/10.1016/j.repl.2016.08.003.

[12] O.H. Hassoon, M. Tarfaoui, A. El Moumen, Progressive damage modeling in laminate composites under slamming impact water for naval applications, Compos. Struct. 167 (2017) 178–190. https://doi.org/10.1016/j.compstruct.2017.02.004.

[13] F. Le Lay, J. Gutierrez, Improvement of the fire behaviour of composite materials for naval application, Polym. Degrad. Stab. 64 (3) (1999) 397–401. https://doi.org/10.1016/S0141-3910(98)00140-2.

[14] U. Sorathia, C.M. Rollhauser, W.A. Hughes, Improved fire safety of composites for naval applications, Fire Mater. 16 (3) (1992) 119–125. https://doi.org/10.1002/fam.810160303.

[15] U. Sorathia, J. Ness, M. Blum, Fire safety of composites in the US Navy, Compos. Part A 30 (5) (1999) 707–713. https://doi.org/10.1016/S1359-835X(98)00112-2.

[16] S.P. Rawal, Metal-matrix composites for space applications, JOM 53 (4) (2001) 14–17. https://doi.org/10.1007/s11837-001-0139-z.

[17] A. Toldy, B. Szolnoki, G. Marosi, Flame retardancy of fibre-reinforced epoxy resin composites for aerospace applications, Polym. Degrad. Stab. 96 (3) (2011) 371–376. https://doi.org/10.1016/j.polymdegradstab.2010.03.021.

[18] Y. Liu, et al., Shape memory polymers and their composites in aerospace applications: a review, Smart Mater. Struct. 23 (2) (2014) 023001. https://doi.org/10.1088/0964-1726/23/2/023001.

[19] J. Leng, et al., Effect of the γ-radiation on the properties of epoxy-based shape memory polymers, J. Intell. Mater. Syst. Struct. 25 (10) (2014) 1256–1263. https://doi.org/10.1177/1045389x13504474.

[20] S. Arzberger, et al., Elastic memory composite technology for thin lightweight space- and ground-based deployable mirrors, in: Optical Science and Technology, SPIE's 48th Annual Meeting, 5179, SPIE, 2003.

[21] X. Lan, et al., Fiber reinforced shape-memory polymer composite and its application in a deployable hinge, Smart Mater. Struct. 18 (2) (2009) 024002. https://doi.org/10.1088/0964-1726/18/2/024002.

[22] P. Keller, et al., Development of a deployable boom for microsatellites using elastic memory composite material, in: 45th AIAA/ASME/ASCE/AHS/ASC Structures, Structural Dynamics & Materials Conference, Palm Springs, California, 2004.

[23] P. Keller, et al., Development of elastic memory composite stiffeners for a flexible precision reflector, in: 47th AIAA/ASME/ASCE/AHS/ASC Structures, Structural Dynamics, and Materials Conference, Newport, Rhode Island, 2006.

[24] S. Chen, et al., Experiment and analysis of morphing skin embedded with shape memory polymer composite tube, J. Intell. Mater. Syst. Struct. 25 (16) (2014) 2052–2059. https://doi.org/10.1177/1045389X13517307.

[25] J. Hinkle, J. Lin, D. Kling, Design and materials study on secondary structures in deployable planetary and space habitats, in: 52nd AIAA/ASME/ASCE/AHS/ASC Structures, Structural Dynamics and Materials Conference, 2011.

[26] P. Baldus, M. Jansen, D. Sporn, Ceramic fibers for matrix composites in high-temperature engine applications, Science 285 (5428) (1999) 699–703. https://doi.org/10.1126/science.285.5428.699.

[27] R. Riedel, et al., A silicoboron carbonitride ceramic stable to 2,000°C, Nature 382 (6594) (1996) 796–798. https://doi.org/10.1038/382796a0.

[28] H. Chen, et al., Thermal conductivity of polymer-based composites: fundamentals and applications, Prog. Polym. Sci. 59 (2016) 41–85. https://doi.org/10.1016/j.progpolymsci.2016.03.001.

[29] E.K. Akdogan, M. Allahverdi, A. Safari, Piezoelectric composites for sensor and actuator applications, IEEE Trans. Ultrason. Ferroelectr. Freq. Control 52 (5) (2005) 746–775. https://doi.org/10.1109/TUFFC.2005.1503962.

[30] K.A. Klicker, J.V. Biggers, R.E. Newnham, Composites of PZT and epoxy for hydrostatic transducer applications, J. Am. Ceram. Soc. 64 (1) (1981) 5–9. https://doi.org/10.1111/j.1151-2916.1981.tb09549.x.

[31] H.L.W. Chan, J. Unsworth, Simple model for piezoelectric ceramic/polymer 1-3 composites used in ultrasonic transducer applications, IEEE Trans. Ultrason. Ferroelectr. Freq. Control 36 (4) (1989) 434–441. https://doi.org/10.1109/58.31780.

[32] T. Ritter, et al., Single crystal PZN/PT-polymer composites for ultrasound transducer applications, IEEE Trans. Ultrason. Ferroelectr. Freq. Control 47 (4) (2000) 792–800. https://doi.org/10.1109/58.852060.

[33] K.C. Cheng, et al., Single crystal PMN-0.33PT/epoxy 1-3 composites for ultrasonic transducer applications, IEEE Trans. Ultrason. Ferroelectr. Freq. Control 50 (9) (2003) 1177–1183. https://doi.org/10.1109/TUFFC.2003.1235328.

[34] L. Van Den Einde, L. Zhao, F. Seible, Use of FRP composites in civil structural applications, Constr. Build. Mater. 17 (6) (2003) 389–403. https://doi.org/10.1016/S0950-0618(03)00040-0.

[35] J. Weeks, et al., The US-TCCMAR full-scale five-story masonry research building test: part V—repair and retest, University of California, San Diego (1994).

[36] F. Colangelo, et al., Epoxy/glass fibres composites for civil applications: comparison between thermal and microwave crosslinking routes, Compos. Part B: Eng. 126 (2017) 100–107. https://doi.org/10.1016/j.compositesb.2017.06.003.

[37] S.S. Pendhari, T. Kant, Y.M. Desai, Application of polymer composites in civil construction: a general review, Compos. Struct. 84 (2) (2008) 114–124. https://doi.org/10.1016/j.compstruct.2007.06.007.

[38] L. Yan, B. Kasal, L. Huang, A review of recent research on the use of cellulosic fibres, their fibre fabric reinforced cementitious, geo-polymer and polymer composites in civil engineering, Compos. Part B: Eng. 92 (2016) 94–132. https://doi.org/10.1016/j.compositesb.2016.02.002.

[39] H. Fang, et al., Connections and structural applications of fibre reinforced polymer composites for civil infrastructure in aggressive environments, Compos. Part B: Eng. 164 (2019) 129–143. https://doi.org/10.1016/j.compositesb.2018.11.047.

[40] P. Colombo, et al., Joining of SiC/SiCf ceramic matrix composites for fusion reactor blanket applications, J. Nucl. Mater. 278 (2) (2000) 127–135. https://doi.org/10.1016/S0022-3115(99)00268-8.

[41] Y. Katoh, et al., Progress in SiC-based ceramic composites for fusion applications, Fusion Sci. Technol. 44 (1) (2003) 155–162. https://doi.org/10.13182/FST03-A326.

[42] L.L. Snead, et al., High thermal conductivity of graphite fiber silicon carbide composites for fusion reactor application, J. Nucl. Mater. 307-311 (2002) 1200–1204. https://doi.org/10.1016/S0022-3115(02)01115-7.

[43] Y. Katoh, et al., Property tailorability for advanced CVI silicon carbide composites for fusion, Fusion Eng. Des. 81 (8) (2006) 937–944. https://doi.org/10.1016/j.fusengdes.2005.08.045.

[44] Y. Katoh, et al., Current status and critical issues for development of SiC composites for fusion applications, J. Nucl. Mater. 367-370 (2007) 659–671. https://doi.org/10.1016/j.jnucmat.2007.03.032.

[45] T. Nozawa, et al., Recent advances and issues in development of silicon carbide composites for fusion applications, J. Nucl. Mater. 386-388 (2009) 622–627. https://doi.org/10.1016/j.jnucmat.2008.12.305.

[46] Y. Katoh, et al., Thermophysical and mechanical properties of near-stoichiometric fiber CVI SiC/SiC composites after neutron irradiation at elevated temperatures, J. Nucl. Mater. 403 (1) (2010) 48–61. https://doi.org/10.1016/j.jnucmat.2010.06.002.

[47] L.L. Snead, et al., Silicon carbide composites as fusion power reactor structural materials, J. Nucl. Mater. 417 (1) (2011) 330–339. https://doi.org/10.1016/j.jnucmat.2011.03.005.

[48] Y. Katoh, et al., Radiation effects in SiC for nuclear structural applications, Curr. Opin. Solid State Mater. Sci. 16 (3) (2012) 143–152. https://doi.org/10.1016/j.cossms.2012.03.005.

[49] Y. Katoh, et al., Continuous SiC fiber, CVI SiC matrix composites for nuclear applications: properties and irradiation effects, J. Nucl. Mater. 448 (1) (2014) 448–476. https://doi.org/10.1016/j.jnucmat.2013.06.040.

[50] Y. Katoh, et al., Radiation-tolerant joining technologies for silicon carbide ceramics and composites, J. Nucl. Mater. 448 (1) (2014) 497–511. https://doi.org/10.1016/j.jnucmat.2013.10.002.

[51] A.A. Campbell, et al., Method for analyzing passive silicon carbide thermometry with a continuous dilatometer to determine irradiation temperature, Nucl. Instrum. Methods Phys. Res. Sect. B 370 (2016) 49–58. https://doi.org/10.1016/j.nimb.2016.01.005.

[52] Y. Katoh, L.L. Snead, Silicon carbide and its composites for nuclear applications – historical overview, J. Nucl. Mater. 526 (2019) 151849. https://doi.org/10.1016/j.jnucmat.2019.151849.

[53] M. Shahinpoor, K.J. Kim, Ionic polymer–metal composites: IV. Industrial and medical applications, Smart Mater. Struct. 14 (1) (2004) 197. https://doi.org/10.1088/0964-1726/14/1/020.

[54] H.-y. Cheung, et al., Natural fibre-reinforced composites for bioengineering and environmental engineering applications, Compos. Part B: Eng. 40 (7) (2009) 655–663. https://doi.org/10.1016/j.compositesb.2009.04.014.

[55] M.J. Mirzaali, et al., Crumpling-based soft metamaterials: the effects of sheet pore size and porosity, Sci. Rep. 7 (1) (2017) 13028. https://doi.org/10.1038/s41598-017-12821-6.

[56] M.J. Mirzaali, et al., Length-scale dependency of biomimetic hard-soft composites, Sci. Rep. 8 (1) (2018) 12052. https://doi.org/10.1038/s41598-018-30012-9.

[57] M.J. Mirzaali, et al., Shape-matching soft mechanical metamaterials, Sci. Rep. 8 (1) (2018) 965. https://doi.org/10.1038/s41598-018-19381-3.

[58] M.J. Mirzaali, H. Pahlavani, A.A. Zadpoor, Auxeticity and stiffness of random networks: lessons for the rational design of 3D printed mechanical metamaterials, Appl. Phys. Lett. 115 (2) (2019) 021901. https://doi.org/10.1063/1.5096590.

[59] S. Janbaz, et al., Strain rate-dependent mechanical metamaterials, Sci. Adv. 6 (25) (2020) eaba0616. https://doi.org/10.1126/sciadv.aba0616.

[60] S. Iijima, Helical microtubules of graphitic carbon, Nature 354 (6348) (1991) 56–58. https://doi.org/10.1038/354056a0.

[61] S. Iijima, T. Ichihashi, Single-shell carbon nanotubes of 1-nm diameter, Nature 363 (6430) (1993) 603–605. https://doi.org/10.1038/363603a0.

[62] R. Rafiee, Carbon Nanotube-reinforced Polymers: From Nanoscale to Macroscale, first ed., Elsevier, Oxford, UK, 2018. https://doi.org/10.1016/C2016-0-00371-6.

[63] M.R. Ayatollahi, S. Shadlou, M.M. Shokrieh, Fracture toughness of epoxy/multi-walled carbon nanotube nano-composites under bending and shear loading conditions, Mater. Des. 32 (4) (2011) 2115–2124. https://doi.org/10.1016/j.matdes.2010.11.034.

[64] A. Zeinedini, M.M. Shokrieh, A. Ebrahimi, The effect of agglomeration on the fracture toughness of CNTs-reinforced nanocomposites, Theor. Appl. Fract. Mech. 94 (2018) 84–94. https://doi.org/10.1016/j.tafmec.2018.01.009.

[65] M.S.P. Shaffer, A.H. Windle, Fabrication and characterization of carbon nanotube/poly(vinyl alcohol) composites, Adv. Mater. 11 (11) (1999) 937–941. https://doi.org/10.1002/(sici)1521-4095(199908)11:11<937::aid-adma937>3.0.co;2-9.

[66] F.T. Fisher, R.D. Bradshaw, L.C. Brinson, Effects of nanotube waviness on the modulus of nanotube-reinforced polymers, Appl. Phys. Lett. 80 (24) (2002) 4647–4649. https://doi.org/10.1063/1.1487900.

[67] B. Lim, et al., The effects of interfacial bonding on mechanical properties of single-walled carbon nanotube reinforced copper matrix nanocomposites, Nanotechnology 17 (23) (2006) 5759–5764. https://doi.org/10.1088/0957-4484/17/23/008.

[68] H. Tan, et al., The effect of van der Waals-based interface cohesive law on carbon nanotube-reinforced composite materials, Compos. Sci. Technol. 67 (14) (2007) 2941–2946. https://doi.org/10.1016/j.compscitech.2007.05.016.

[69] W.B. Lu, et al., A cohesive law for interfaces between multi-wall carbon nanotubes and polymers due to the van der Waals interactions, Comput. Meth. Appl. Mech. Eng. 197 (41) (2008) 3261–3267. https://doi.org/10.1016/j.cma.2007.12.008.

[70] A.J. Crosby, J.Y. Lee, Polymer nanocomposites: the "nano" effect on mechanical properties, Polym. Rev. 47 (2) (2007) 217–229. https://doi.org/10.1080/15583720701271278.

[71] G. Pandey, E.T. Thostenson, Carbon nanotube-based multifunctional polymer nanocomposites, Polym. Rev. 52 (3) (2012) 355–416. https://doi.org/10.1080/15583724.2012.703747.

[72] L.S. Schadler, S.C. Giannaris, P.M. Ajayan, Load transfer in carbon nanotube epoxy composites, Appl. Phys. Lett. 73 (26) (1998) 3842–3844. https://doi.org/10.1063/1.122911.

[73] P.M. Ajayan, et al., Single-walled carbon nanotube–polymer composites: strength and weakness, Adv. Mater. 12 (10) (2000) 750–753. https://doi.org/10.1002/(SICI)g1521-4095(200005)12:10%3C750::AID-ADMA750%3E3.0.CO;2-6.

[74] K.-t. Lau, S.-q. Shi, Failure mechanisms of carbon nanotube/epoxy composites pretreated in different temperature environments, Carbon 40 (15) (2002) 2965–2968. https://doi.org/10.1016/S0008-6223(02)00245-2.

[75] S.J.V. Frankland, et al., Molecular simulation of the influence of chemical cross-links on the shear strength of carbon nanotube−polymer interfaces, J. Phys. Chem. B 106 (12) (2002) 3046–3048. https://doi.org/10.1021/jp015591+.

[76] M. Wong, et al., Physical interactions at carbon nanotube-polymer interface, Polymer 44 (25) (2003) 7757–7764. https://doi.org/10.1016/j.polymer.2003.10.011.

[77] F. Ebrahimi, A. Dabbagh, A brief review on the influences of nanotubes' entanglement and waviness on the mechanical behaviors of CNTR polymer nanocomposites, J. Comput. Appl. Mech. 51 (1) (2020) 247–252. https://doi.org/10.22059/jcamech.2020.304476.517.

[78] N.G. Sahoo, et al., Polymer nanocomposites based on functionalized carbon nanotubes, Prog. Polym. Sci. 35 (7) (2010) 837–867. https://doi.org/10.1016/gj.progpolymsci.2010.03.002.

[79] V.J. Cruz-Delgado, et al., Nanocomposites based on plasma-polymerized carbon nanotubes and Nylon-6, Polym. J. 44 (9) (2012) 952–958. https://doi.org/10.1038/pj.2012.49.

[80] S. Jamali, M.C. Paiva, J.A. Covas, Dispersion and re-agglomeration phenomena during melt mixing of polypropylene with multi-wall carbon nanotubes, Polym. Test. 32 (4) (2013) 701–707. https://doi.org/10.1016/j.polymertesting.2013.03.005.

[81] Y. Han, et al., Bio-inspired aggregation control of carbon nanotubes for ultra-strong composites, Sci. Rep. 5 (1) (2015) 11533. https://doi.org/10.1038/srep11533.

[82] B.K. Shrestha, et al., Globular shaped polypyrrole doped well-dispersed functionalized multiwall carbon nanotubes/nafion composite for enzymatic glucose biosensor application, Sci. Rep. 7 (1) (2017) 16191. https://doi.org/10.1038/s41598-017-16541-9.

[83] M. Zhang, J. Li, Carbon nanotube in different shapes, Mater. Today 12 (6) (2009) 12–18. https://doi.org/10.1016/S1369-7021(09)70176-2.

[84] M.M. Shokrieh, R. Rafiee, Development of a full range multi-scale model to obtain elastic properties of CNT/polymer composites, Iran. Polym. J. 21 (6) (2012) 397–402. https://doi.org/10.1007/s13726-012-0043-0.

[85] J.B. Bai, A. Allaoui, Effect of the length and the aggregate size of MWNTs on the improvement efficiency of the mechanical and electrical properties of nanocomposites—experimental investigation, Compos. Part A 34 (8) (2003) 689–694. https://doi.org/10.1016/S1359-835X(03)00140-4.

[86] M.R. Ayatollahi, et al., Effect of multi-walled carbon nanotube aspect ratio on mechanical and electrical properties of epoxy-based nanocomposites, Polym. Test. 30 (5) (2011) 548–556. https://doi.org/10.1016/j.polymertesting.2011.04.008.

[87] R. Gulotty, et al., Effects of functionalization on thermal properties of single-wall and multi-wall carbon nanotube–polymer nanocomposites, ACS Nano 7 (6) (2013) 5114–5121. https://doi.org/10.1021/nn400726g.

[88] C.-F. Kuan, et al., Flame retardance and thermal stability of carbon nanotube epoxy composite prepared from sol–gel method, J. Phys. Chem. Solids 71 (4) (2010) 539–543. https://doi.org/10.1016/j.jpcs.2009.12.031.

[89] H. Jung, et al., Enhancement of thermo-mechanical stability for nanocomposites containing plasma treated carbon nanotubes with an experimental study and molecular dynamics simulations, Sci. Rep. 10 (1) (2020) 405. https://doi.org/10.1038/s41598-019-56976-w.

[90] J. Suhr, et al., Viscoelasticity in carbon nanotube composites, Nat. Mater. 4 (2) (2005) 134–137. https://doi.org/10.1038/nmat1293.

[91] P.M. Ajayan, J. Suhr, N. Koratkar, Utilizing interfaces in carbon nanotube reinforced polymer composites for structural damping, J. Mater. Sci. 41 (23) (2006) 7824–7829. https://doi.org/10.1007/s10853-006-0693-4.

[92] D.S. Bethune, et al., Cobalt-catalysed growth of carbon nanotubes with single-atomic-layer walls, Nature 363 (6430) (1993) 605–607. https://doi.org/10.1038/363605a0.

[93] C. Journet, et al., Large-scale production of single-walled carbon nanotubes by the electric-arc technique, Nature 388 (6644) (1997) 756–758. https://doi.org/10.1038/41972.

[94] T. Guo, et al., Catalytic growth of single-walled manotubes by laser vaporization, Chem. Phys. Lett. 243 (1) (1995) 49–54. https://doi.org/10.1016/0009-2614(95)00825-O.

[95] A.G. Rinzler, et al., Large-scale purification of single-wall carbon nanotubes: process, product, and characterization, Appl. Phys. A 67 (1) (1998) 29–37. https://doi.org/10.1007/s003390050734.

[96] P. Nikolaev, et al., Gas-phase catalytic growth of single-walled carbon nanotubes from carbon monoxide, Chem. Phys. Lett. 313 (1) (1999) 91–97. https://doi.org/10.1016/S0009-2614(99)01029-5.

[97] Z.F. Ren, et al., Synthesis of large arrays of well-aligned carbon nanotubes on glass, Science 282 (5391) (1998) 1105–1107. https://doi.org/10.1126/science.282.5391.1105.

[98] J.L. Hutchison, et al., Double-walled carbon nanotubes fabricated by a hydrogen arc discharge method, Carbon 39 (5) (2001) 761–770. https://doi.org/10.1016/S0008-6223(00)00187-1.

[99] H. Li, et al., Direct synthesis of high purity single-walled carbon nanotube fibers by arc discharge, J. Phys. Chem. B 108 (15) (2004) 4573–4575. https://doi.org/10.1021/jp036563p.

[100] K. Imasaka, et al., Production of carbon nanoonions and nanotubes using an intermittent arc discharge in water, Thin Solid Films 506-507 (2006) 250–254. https://doi.org/10.1016/j.tsf.2005.08.024.

[101] S. Yaragalla, et al. (Eds.), Carbon-Based Nanofillers and Their Rubber Nanocomposites: Carbon Nano-Objects. 2019, Elsevier: Cambridge, MA, 402. https://doi.org/10.1016/C2016-0-03648-3.

[102] S. Kumar, et al., Carbon nanotubes: a novel material for multifaceted applications in human healthcare, Chem. Soc. Rev. 46 (1) (2017) 158–196. https://doi.org/10.1039/C6CS00517A.

[103] H. Dai, et al., Nanotubes as nanoprobes in scanning probe microscopy, Nature 384 (6605) (1996) 147–150. https://doi.org/10.1038/384147a0.

[104] Z.F. Ren, et al., Growth of a single freestanding multiwall carbon nanotube on each nanonickel dot, Appl. Phys. Lett. 75 (8) (1999) 1086–1088. https://doi.org/10.1063/1.124605.

[105] P.-X. Hou, C. Liu, H.-M. Cheng, Purification of carbon nanotubes, Carbon 46 (15) (2008) 2003–2025. https://doi.org/10.1016/j.carbon.2008.09.009.

[106] F.V. Ferreira, et al., Correlation of surface treatment, dispersion and mechanical properties of HDPE/CNT nanocomposites, Appl. Surf. Sci. 389 (2016) 921–929. https://doi.org/10.1016/j.apsusc.2016.07.164.

[107] P. Avouris, Z. Chen, V. Perebeinos, Carbon-based electronics, Nat. Nanotechnol. 2 (10) (2007) 605–615. https://doi.org/10.1038/nnano.2007.300.

[108] P.G. Collins, M.S. Arnold, P. Avouris, Engineering carbon nanotubes and nanotube circuits using electrical breakdown, Science 292 (5517) (2001) 706–709. https://doi.org/10.1126/science.1058782.

[109] S. Yaragalla (Ed.), Carbon-Based Nanofillers and Their Rubber Nanocomposites, Elsevier, Oxford, UK, 2019. https://doi.org/10.1016/C2016-0-03648-3.

[110] M. Yudasaka, et al., Mechanism of the effect of NiCo, Ni and Co catalysts on the yield of single-wall carbon nanotubes formed by pulsed Nd:YAG laser ablation, J. Phys. Chem. B 103 (30) (1999) 6224–6229. https://doi.org/10.1021/jp9908451.

[111] H. Kataura, et al., Diameter control of single-walled carbon nanotubes, Carbon 38 (11) (2000) 1691–1697. https://doi.org/10.1016/S0008-6223(00)00090-7.

[112] M.M. Tavakoli, et al., Synergistic roll-to-roll transfer and doping of CVD-graphene using parylene for ambient-stable and ultra-lightweight photovoltaics, Adv. Funct. Mater. 30 (31) (2020) 2001924. https://doi.org/10.1002/adfm.202001924.

[113] Z. Spitalsky, et al., Carbon nanotube–polymer composites: chemistry, processing, mechanical and electrical properties, Prog. Polym. Sci. 35 (3) (2010) 357–401. https://doi.org/10.1016/j.progpolymsci.2009.09.003.

[114] P.M. Ajayan, et al., Aligned carbon nanotube arrays formed by cutting a polymer resin—nanotube composite, Science 265 (5176) (1994) 1212–1214. https://doi.org/10.1126/science.265.5176.1212.

[115] T.K. Gupta, et al., Improved nanoindentation and microwave shielding properties of modified MWCNT reinforced polyurethane composites, J. Mater. Chem. A 1 (32) (2013) 9138–9149. https://doi.org/10.1039/C3TA11611E.

[116] L. Jin, C. Bower, O. Zhou, Alignment of carbon nanotubes in a polymer matrix by mechanical stretching, Appl. Phys. Lett. 73 (9) (1998) 1197–1199. https://doi.org/10.1063/1.122125.

[117] H.Z. Geng, et al., Fabrication and properties of composites of poly(ethylene oxide) and functionalized carbon nanotubes, Adv. Mater. 14 (19) (2002) 1387–1390. https://doi.org/10.1002/1521-4095(20021002)14:19%3C1387::AID-ADMA1387%3E3.0.CO;2-Q.

[118] F. Du, J.E. Fischer, K.I. Winey, Coagulation method for preparing single-walled carbon nanotube/poly(methyl methacrylate) composites and their modulus, electrical conductivity, and thermal stability, J. Polym. Sci. Part B Polym. Phys. 41 (24) (2003) 3333–3338. https://doi.org/10.1002/polb.10701.

[119] C.Y. Li, et al., Nanohybrid shish-kebabs: periodically functionalized carbon nanotubes, Adv. Mater. 17 (9) (2005) 1198–1202. https://doi.org/10.1002/adma.200401977.

[120] R. Haggenmueller, J.E. Fischer, K.I. Winey, Single wall carbon nanotube/polyethylene nanocomposites: nucleating and templating polyethylene crystallites, Macromolecules 39 (8) (2006) 2964–2971. https://doi.org/10.1021/ma0527698.

[121] J.-H. Lee, S.K. Kim, N.H. Kim, Effects of the addition of multi-walled carbon nanotubes on the positive temperature coefficient characteristics of carbon-black-filled high-density polyethylene nanocomposites, Scr. Mater. 55 (12) (2006) 1119–1122. https://doi.org/10.1016/j.scriptamat.2006.08.051.

[122] K. Kalaitzidou, H. Fukushima, L.T. Drzal, A new compounding method for exfoliated graphite–polypropylene nanocomposites with enhanced flexural properties and lower percolation threshold, Compos. Sci. Technol. 67 (10) (2007) 2045–2051. https://doi.org/10.1016/j.compscitech.2006.11.014.

[123] A.S. Babal, et al., Synergistic effect on static and dynamic mechanical properties of carbon fiber-multiwalled carbon nanotube hybrid polycarbonate composites, RSC Adv. 6 (72) (2016) 67954–67967. https://doi.org/10.1039/C6RA08487G.

[124] J. Jyoti, et al., Detailed dynamic rheological studies of multiwall carbon nanotube-reinforced acrylonitrile butadiene styrene composite, J. Mater. Sci. 51 (5) (2016) 2643–2652. https://doi.org/10.1007/s10853-015-9578-8.

[125] R. Haggenmueller, et al., Aligned single-wall carbon nanotubes in composites by melt processing methods, Chem. Phys. Lett. 330 (3) (2000) 219–225. https://doi.org/10.1016/S0009-2614(00)01013-7.

[126] H. Kim, Y. Miura, C.W. Macosko, Graphene/polyurethane nanocomposites for improved gas barrier and electrical conductivity, Chem. Mater. 22 (11) (2010) 3441–3450. https://doi.org/10.1021/cm100477v.

[127] Z. Jin, et al., Dynamic mechanical behavior of melt-processed multi-walled carbon nanotube/poly(methyl methacrylate) composites, Chem. Phys. Lett. 337 (1) (2001) 43–47. https://doi.org/10.1016/S0009-2614(01)00186-5.

[128] P. Pötschke, T.D. Fornes, D.R. Paul, Rheological behavior of multiwalled carbon nanotube/polycarbonate composites, Polymer 43 (11) (2002) 3247–3255. https://doi.org/10.1016/S0032-3861(02)00151-9.

[129] Z. Jia, et al., Study on poly(methyl methacrylate)/carbon nanotube composites, Mater. Sci. Eng.: A 271 (1) (1999) 395–400. https://doi.org/10.1016/S0921-5093(99)00263-4.

[130] C. Velasco-Santos, et al., Improvement of thermal and mechanical properties of carbon nanotube composites through chemical functionalization, Chem. Mater. 15 (23) (2003) 4470–4475. https://doi.org/10.1021/cm034243c.

[131] J. Fan, et al., Synthesis, characterizations, and physical properties of carbon nanotubes coated by conducting polypyrrole, J. Appl. Polym. Sci. 74 (11) (1999) 2605–2610. https://doi.org/10.1002/(sici)1097-4628(19991209)74:11<2605::aid-app6>3.0.co;2-r.

[132] B.Z. Tang, H. Xu, Preparation, alignment, and optical properties of soluble poly(phenylacetylene)-wrapped carbon nanotubes, Macromolecules 32 (8) (1999) 2569–2576. https://doi.org/10.1021/ma981825k.

[133] C. Downs, et al., Efficient polymerization of aniline at carbon nanotube electrodes, Adv. Mater. 11 (12) (1999) 1028–1031. https://doi.org/10.1002/(sici)1521-4095(199908)11:12<1028::aid-adma1028>3.0.co;2-n.

[134] G.Z. Chen, et al., Carbon nanotube and polypyrrole composites: coating and doping, Adv. Mater. 12 (7) (2000) 522–526. https://doi.org/10.1002/s(sici)1521-4095(200004)12:7<522::aid-adma522>3.0.co;2-s.

[135] T. Kimura, et al., Polymer composites of carbon nanotubes aligned by a magnetic field, Adv. Mater. 14 (19) (2002) 1380–1383. https://doi.org/10.1002/1521-4095(20021002)14:19<1380::aid-adma1380>3.0.co;2-v.

[136] E.S. Choi, et al., Enhancement of thermal and electrical properties of carbon nanotube polymer composites by magnetic field processing, J. Appl. Phys. 94 (9) (2003) 6034–6039. https://doi.org/10.1063/1.1616638.

[137] C.A. Martin, et al., Electric field-induced aligned multi-wall carbon nanotube networks in epoxy composites, Polymer 46 (3) (2005) 877–886. https://doi.org/10.1016/j.polymer.2004.11.081.

[138] C.H. Liu, et al., Thermal conductivity improvement of silicone elastomer with carbon nanotube loading, Appl. Phys. Lett. 84 (21) (2004) 4248–4250. https://doi.org/10.1063/1.1756680.

[139] F.H. Gojny, et al., Carbon nanotube-reinforced epoxy-composites: enhanced stiffness and fracture toughness at low nanotube content, Compos. Sci. Technol. 64 (15) (2004) 2363–2371. https://doi.org/10.1016/j.compscitech.2004.04.002.

[140] A.A. Mamedov, et al., Molecular design of strong single-wall carbon nanotube/polyelectrolyte multilayer composites, Nat. Mater. 1 (3) (2002) 190–194. https://doi.org/10.1038/nmat747.

[141] S. Srivastava, N.A. Kotov, Composite layer-by-layer (LBL) assembly with inorganic nanoparticles and nanowires, Acc. Chem. Res. 41 (12) (2008) 1831–1841. https://doi.org/10.1021/ar8001377.

[142] M.T. Byrne, Y.K. Gun'ko, Recent advances in research on carbon nanotube–polymer composites, Adv. Mater. 22 (15) (2010) 1672–1688. https://doi.org/10.1002/adma.200901545.

[143] J.N. Coleman, et al., Improving the mechanical properties of single-walled carbon nanotube sheets by intercalation of polymeric adhesives, Appl. Phys. Lett. 82 (11) (2003) 1682–1684. https://doi.org/10.1063/1.1559421.

[144] M.F. Arif, S. Kumar, T. Shah, Tunable morphology and its influence on electrical, thermal and mechanical properties of carbon nanostructure-buckypaper, Mater. Des. 101 (2016) 236–244. https://doi.org/10.1016/j.matdes.2016.03.122.

[145] Z. Wang, et al., Processing and property investigation of single-walled carbon nanotube (SWNT) buckypaper/epoxy resin matrix nanocomposites, Compos. Part A 35 (10) (2004) 1225–1232. https://doi.org/10.1016/j.compositesa.2003.09.029.

[146] F.M. Blighe, et al., Observation of percolation-like scaling – far from the percolation threshold – in high volume fraction, high conductivity polymer-nanotube composite films, Adv. Mater. 19 (24) (2007) 4443–4447. https://doi.org/10.1002/adma.200602912.

[147] S. Teotia, et al., Multifunctional, robust, light-weight, free-standing MWCNT/phenolic composite paper as anodes for lithium ion batteries and EMI shielding material, RSC Adv. 4 (63) (2014) 33168–33174. https://doi.org/10.1039/C4RA04183F.

[148] P. Larkin, Infrared and RamanSpectroscopy: Principles and Spectral Interpretation, first ed., Elsevier, Waltham, MA, 2011. https://doi.org/10.1016/C2010-0-68479-3.

[149] L. Valentini, et al., Morphological characterization of single-walled carbon nanotubes-PP composites, Compos. Sci. Technol. 63 (8) (2003) 1149–1153. https://doi.org/10.1016/S0266-3538(03)00036-8.

[150] C.C. Kao, R.J. Young, A Raman spectroscopic investigation of heating effects and the deformation behaviour of epoxy/SWNT composites, Compos. Sci. Technol. 64 (15) (2004) 2291–2295. https://doi.org/10.1016/j.compscitech.2004.01.019.

[151] T.E. Chang, et al., Conductivity and mechanical properties of well-dispersed single-wall carbon nanotube/polystyrene composite, Polymer 47 (22) (2006) 7740–7746. https://doi.org/10.1016/j.polymer.2006.09.013.

[152] S. Wang, et al., Load-transfer in functionalized carbon nanotubes/polymer composites, Chem. Phys. Lett. 457 (4) (2008) 371–375. https://doi.org/10.1016/j.cplett.2008.04.037.

[153] B. Vigolo, et al., Multiscale characterization of single-walled carbon nanotube/polymer composites by coupling Raman and Brillouin spectroscopy, J. Phys. Chem. C 113 (41) (2009) 17648–17654. https://doi.org/10.1021/jp903960f.

[154] A. de la Vega, et al., Combined Raman and dielectric spectroscopy on the curing behaviour and stress build up of carbon nanotube–epoxy composites, Compos. Sci. Technol. 69 (10) (2009) 1540–1546. https://doi.org/10.1016/j.compscitech.2008.09.015.

[155] N. Lachman, et al., Raman response of carbon nanotube/PVA fibers under strain, J. Phys. Chem. C 113 (12) (2009) 4751–4754. https://doi.org/10.1021/jp900355k.

[156] D. Roy, et al., Measurement of interfacial shear strength in single wall carbon nanotubes reinforced composite using Raman spectroscopy, J. Appl. Phys. 107 (4) (2010) 043501. https://doi.org/10.1063/1.3295907.

[157] V.G. Hadjiev, et al., Raman microscopy of residual strains in carbon nanotube/epoxy composites, Carbon 48 (6) (2010) 1750–1756. https://doi.org/10.1016/j.carbon.2010.01.018.

[158] A. de la Vega, et al., Simultaneous global and local strain sensing in SWCNT–epoxy composites by Raman and impedance spectroscopy, Compos. Sci. Technol. 71 (2) (2011) 160–166. https://doi.org/10.1016/j.compscitech.2010.11.004.

[159] P. Nikolaev, et al., Raman probing of adhesion loss in carbon nanotube-reinforced composite, Compos. Part A 42 (11) (2011) 1681–1686. https://doi.org/10.1016/j.compositesa.2011.07.022.

[160] L. Bokobza, J.-L. Bruneel, M. Couzi, Raman spectroscopic investigation of carbon-based materials and their composites. Comparison between carbon nanotubes and carbon black, Chem. Phys. Lett. 590 (2013) 153–159. https://doi.org/10.1016/j.cplett.2013.10.071.

[161] S. Laurenzi, et al., Experimental study of impact resistance in multi-walled carbon nanotube reinforced epoxy, Compos. Struct. 99 (2013) 62–68. https://doi.org/10.1016/j.compstruct.2012.12.002.

[162] A.A. Azeez, et al., Application of cryomilling to enhance material properties of carbon nanotube reinforced chitosan nanocomposites, Compos. Part B: Eng. 50 (2013) 127–134. https://doi.org/10.1016/j.compositesb.2013.01.010.

[163] J. Jyoti, et al., Superior mechanical and electrical properties of multiwall carbon nanotube reinforced acrylonitrile butadiene styrene high performance composites, Compos. Part B: Eng. 83 (2015) 58–65. https://doi.org/10.1016/j.compositesb.2015.08.055.

[164] O. Büyüköztürk, et al., Structural solution using molecular dynamics: fundamentals and a case study of epoxy-silica interface, Int. J. Solids Struct. 48 (14) (2011) 2131–2140. https://doi.org/10.1016/j.ijsolstr.2011.03.018.

[165] J.-L. Tsai, S.-H. Tzeng, Y.-T. Chiu, Characterizing elastic properties of carbon nanotubes/polyimide nanocomposites using multi-scale simulation, Compos. Part B: Eng. 41 (1) (2010) 106–115. https://doi.org/10.1016/j.compositesb.2009.06.003.

[166] Y. Li, et al., Pull-out simulations on interfacial properties of carbon nanotube-reinforced polymer nanocomposites, Comput. Mater. Sci. 50 (6) (2011) 1854–1860. https://doi.org/10.1016/j.commatsci.2011.01.029.

[167] M. Mahboob, M. Zahabul Islam, Molecular dynamics simulations of defective CNT-polyethylene composite systems, Comput. Mater. Sci. 79 (2013) 223–229. https://doi.org/10.1016/j.commatsci.2013.05.042.

[168] S. Yang, et al., Nonlinear multiscale modeling approach to characterize elastoplastic behavior of CNT/polymer nanocomposites considering the interphase and interfacial imperfection, Int. J. Plast. 41 (2013) 124–146. https://doi.org/10.1016/j.ijplas.2012.09.010.

[169] B. Arash, Q. Wang, V.K. Varadan, Mechanical properties of carbon nanotube/polymer composites, Sci. Rep. 4 (1) (2014) 6479. https://doi.org/10.1038/srep06479.

[170] Q. Jiang, et al., Molecular dynamics simulations of the effect of the volume fraction on unidirectional polyimide–carbon nanotube nanocomposites, Carbon 67 (2014) 440–448. https://doi.org/10.1016/j.carbon.2013.10.016.

[171] S. Rouhi, Y. Alizadeh, R. Ansari, Molecular dynamics simulations of the single-walled carbon nanotubes/poly (phenylacetylene) nanocomposites, Superlattices Microstruct. 72 (2014) 204–218. https://doi.org/10.1016/j.spmi.2013.10.046.

[172] J.-H. Sul, B.G. Prusty, D.W. Kelly, Application of molecular dynamics to evaluate the design performance of low aspect ratio carbon nanotubes in fibre reinforced polymer resin, Compos. Part A 65 (2014) 64–72. https://doi.org/10.1016/j.compositesa.2014.03.004.

[173] R. Rafiee, M. Mahdavi, Characterizing nanotube–polymer interaction using molecular dynamics simulation, Comput. Mater. Sci. 112 (2016) 356–363. https://doi.org/10.1016/j.commatsci.2015.10.041.

[174] A.R. Alian, S.A. Meguid, Molecular dynamics simulations of the effect of waviness and agglomeration of CNTs on interface strength of thermoset nanocomposites, PCCP 19 (6) (2017) 4426–4434, doi:10.1039/C6CP07464B.

[175] R. Chawla, S. Sharma, Molecular dynamics simulation of carbon nanotube pull-out from polyethylene matrix, Compos. Sci. Technol. 144 (2017) 169–177. https://doi.org/10.1016/j.compscitech.2017.03.029.

[176] J. Zhang, X. Peng, Superior interfacial mechanical properties of boron nitride-carbon nanotube reinforced nanocomposites: a molecular dynamics study, Mater. Chem. Phys. 198 (2017) 250–257. https://doi.org/10.1016/j.matchemphys.2017.05.064.

[177] A.R. Alian, S.A. Meguid, Large-scale atomistic simulations of CNT-reinforced thermoplastic polymers, Compos. Struct. 191 (2018) 221–230. https://doi.org/10.1016/j.compstruct.2018.02.056.

[178] H.K. Choi, et al., Interfacial effects of nitrogen-doped carbon nanotubes on mechanical and thermal properties of nanocomposites: a molecular dynamics study, Compos. Part B: Eng. 167 (2019) 615–620. https://doi.org/10.1016/j.compositesb.2019.03.036.

[179] S. Doagou-Rad, et al., Multiscale molecular dynamics-FE modeling of polymeric nanocomposites reinforced with carbon nanotubes and graphene, Compos. Struct. 217 (2019) 27–36. https://doi.org/10.1016/j.compstruct.2019.03.017.

[180] W. Jian, D. Lau, Creep performance of CNT-based nanocomposites: a parametric study, Carbon 153 (2019) 745–756. https://doi.org/10.1016/j.carbon.2019.07.069.

[181] J. Song, H. Lei, G. Zhao, Improved mechanical and tribological properties of polytetrafluoroethylene reinforced by carbon nanotubes: a molecular dynamics study, Comput. Mater. Sci. 168 (2019) 131–136. https://doi.org/10.1016/j.commatsci.2019.05.058.

[182] M. Hadipeykani, F. Aghadavoudi, D. Toghraie, A molecular dynamics simulation of the glass transition temperature and volumetric thermal expansion coefficient of thermoset polymer based epoxy nanocomposite reinforced by CNT: a statistical study, Physica A 546 (2020) 123995. https://doi.org/10.1016/j.physa.2019.123995.

[183] M.F. Kai, L.W. Zhang, K.M. Liew, Carbon nanotube-geopolymer nanocomposites: a molecular dynamics study of the influence of interfacial chemical bonding upon the structural and mechanical properties, Carbon 161 (2020) 772–783. https://doi.org/10.1016/j.carbon.2020.02.014.

[184] S. Norouzi, A. Kianfar, M.M. Seyyed Fakhrabadi, Multiscale simulation study of anisotropic nanomechanical properties of graphene spirals and their polymer nanocomposites, Mech. Mater. 145 (2020) 103376. https://doi.org/10.1016/j.mechmat.2020.103376.

[185] S. Norouzi, M.M. Seyyed Fakhrabadi, Anisotropic nature of thermal conductivity in graphene spirals revealed by molecular dynamics simulations, J. Phys. Chem. Solids 137 (2020) 109228. https://doi.org/10.1016/j.jpcs.2019.109228.

[186] R. Rafiee, et al., Challenges of the modeling methods for investigating the interaction between the CNT and the surrounding polymer, Adv. Mater. Sci. Eng. 2013 (2013) 1–10. https://doi.org/10.1155/2013/183026.

[187] B. Arash, H.S. Park, T. Rabczuk, Mechanical properties of carbon nanotube reinforced polymer nanocomposites: a coarse-grained model, Compos. Part B: Eng. 80 (2015) 92–100. https://doi.org/10.1016/j.compositesb.2015.05.038.

[188] B. Arash, H.S. Park, T. Rabczuk, Tensile fracture behavior of short carbon nanotube reinforced polymer composites: a coarse-grained model, Compos. Struct. 134 (2015) 981–988. https://doi.org/10.1016/j.compstruct.2015.09.001.

[189] B. Arash, H.S. Park, T. Rabczuk, Coarse-grained model of the J-integral of carbon nanotube reinforced polymer composites, Carbon 96 (2016) 1084–1092. https://doi.org/10.1016/j.carbon.2015.10.058.

[190] A.A. Mousavi, et al., A coarse-grained model for the elastic properties of cross linked short carbon nanotube/polymer composites, Compos. Part B: Eng. 95 (2016) 404–411. https://doi.org/10.1016/j.compositesb.2016.03.044.

[191] K. Duan, et al., Importance of interface in the coarse-grained model of CNT/epoxy nanocomposites, Nanomaterials 9 (10) (2019) 1479. https://doi.org/10.3390/nano9101479.

[192] K. Duan, et al., Machine-learning assisted coarse-grained model for epoxies over wide ranges of temperatures and cross-linking degrees, Mater. Des. 183 (2019). https://doi.org/10.1016/j.matdes.2019.108130.

[193] K. Ma, et al., Improving ductility of bimodal carbon nanotube/2009Al composites by optimizing coarse grain microstructure via hot extrusion, Compos. Part A 140 (2021). https://doi.org/10.1016/j.compositesa.2020.106198.

[194] J. Moon, S. Yang, M. Cho, Interfacial strengthening between graphene and polymer through Stone-Thrower-Wales defects: ab initio and molecular dynamics simulations, Carbon 118 (2017) 66–77. https://doi.org/10.1016/j.carbon.2017.03.021.

[195] A.H. Mashhadzadeh, A. Fereidoon, M. Ghorbanzadeh Ahangari, Surface modification of carbon nanotubes using 3-aminopropyltriethoxysilane to improve mechanical properties of nanocomposite based polymer matrix: experimental and

density functional theory study, Appl. Surf. Sci. 420 (2017) 167–179. https://doi.org/10.1016/j.apsusc.2017.05.148.

[196] S.M. Rahimian-Koloor, S.M. Hashemianzadeh, M.M. Shokrieh, Effect of CNT structural defects on the mechanical properties of CNT/epoxy nanocomposite, Physica B 540 (2018) 16–25. https://doi.org/10.1016/j.physb.2018.04.012.

[197] R. Hill, The elastic behaviour of a crystalline aggregate, Proc. Phys. Soc. London Sect. A 65 (5) (1952) 349–354. https://doi.org/10.1088/0370-1298/65/5/307.

[198] J.D. Eshelby, R.E. Peierls, The determination of the elastic field of an ellipsoidal inclusion, and related problems, Proc. R. Soc. Lond. Ser. A Math. Phys. Sci. 241 (1226) (1957) 376–396. https://doi.org/10.1098/rspa.1957.0133.

[199] T.I. Zohdi, P. Wriggers, An introduction to computational micromechanics, in: F. Pfeiffer, P. Wriggers (Eds.), Lecture Notes in Applied and Computational Mechanics, first ed., Springer-Verlag, Berlin, Heidelberg, 2005, pp. 1–183. https://doi.org/10.1007/978-3-540-32360-0.

[200] T. Mori, K. Tanaka, Average stress in matrix and average elastic energy of materials with misfitting inclusions, Acta Metall. 21 (5) (1973) 571–574. https://doi.org/10.1016/0001-6160(73)90064-3.

[201] F.T. Fisher, R.D. Bradshaw, L.C. Brinson, Fiber waviness in nanotube-reinforced polymer composites—I: modulus predictions using effective nanotube properties, Compos. Sci. Technol. 63 (11) (2003) 1689–1703. https://doi.org/10.1016/S0266-3538(03)00069-1.

[202] D.-L. Shi, et al., The effect of nanotube waviness and agglomeration on the elastic property of carbon nanotube-reinforced composites, J. Eng. Mater. Technol. 126 (3) (2004) 250–257. https://doi.org/10.1115/1.1751182.

[203] A. Haque, A. Ramasetty, Theoretical study of stress transfer in carbon nanotube reinforced polymer matrix composites, Compos. Struct. 71 (1) (2005) 68–77. https://doi.org/10.1016/j.compstruct.2004.09.029.

[204] A. Bagchi, S. Nomura, On the effective thermal conductivity of carbon nanotube reinforced polymer composites, Compos. Sci. Technol. 66 (11) (2006) 1703–1712. https://doi.org/10.1016/j.compscitech.2005.11.003.

[205] K. Li, X.L. Gao, A.K. Roy, Micromechanical modeling of viscoelastic properties of carbon nanotube-reinforced polymer composites, Mech. Adv. Mater. Struct. 13 (4) (2006) 317–328. https://doi.org/10.1080/15376490600583931.

[206] G.D. Seidel, D.C. Lagoudas, Micromechanical analysis of the effective elastic properties of carbon nanotube reinforced composites, Mech. Mater. 38 (8) (2006) 884–907. https://doi.org/10.1016/j.mechmat.2005.06.029.

[207] K. Li, S. Saigal, Micromechanical modeling of stress transfer in carbon nanotube reinforced polymer composites, Mater. Sci. Eng.: A 457 (1) (2007) 44–57. https://doi.org/10.1016/j.msea.2006.12.018.

[208] G.D. Seidel, D.C. Lagoudas, A micromechanics model for the thermal conductivity of nanotube-polymer nanocomposites, J. Appl. Mech. 75 (4) (2008) 1–9 https://doi.org/10.1115/1.2871265041025.

[209] P. Barai, G.J. Weng, A theory of plasticity for carbon nanotube reinforced composites, Int. J. Plast. 27 (4) (2011) 539–559. https://doi.org/10.1016/j.ijplas.2010.08.006.

[210] J. Silva, et al., The influence of matrix mediated hopping conductivity, filler concentration, aspect ratio and orientation on the electrical response of carbon nanotube/polymer nanocomposites, Compos. Sci. Technol. 71 (5) (2011) 643–646. https://doi.org/10.1016/j.compscitech.2011.01.005.

[211] C. Feng, L. Jiang, Micromechanics modeling of the electrical conductivity of carbon nanotube (CNT)–polymer nanocomposites, Compos. Part A 47 (2013) 143–149. https://doi.org/10.1016/j.compositesa.2012.12.008.

[212] J.E. Jam, et al., Characterizing elastic properties of carbon nanotube-based composites by using an equivalent fiber, Polym. Compos. 34 (2) (2013) 241–251, doi:10.1002/pc.22401.
[213] J. Nafar Dastgerdi, G. Marquis, M. Salimi, The effect of nanotubes waviness on mechanical properties of CNT/SMP composites, Compos. Sci. Technol. 86 (2013) 164–169. https://doi.org/10.1016/j.compscitech.2013.07.012.
[214] Y. Zhang, et al., An analytical solution on interface debonding for large diameter carbon nanotube-reinforced composite with functionally graded variation interphase, Compos. Struct. 104 (2013) 261–269. https://doi.org/10.1016/j.compstruct.2013.04.029.
[215] J. Nafar Dastgerdi, G. Marquis, M. Salimi, Micromechanical modeling of nanocomposites considering debonding and waviness of reinforcements, Compos. Struct. 110 (2014) 1–6. https://doi.org/10.1016/j.compstruct.2013.11.017.
[216] C.S. Jarali, S.F. Patil, S.C. Pilli, Hygro-thermo-electric properties of carbon nanotube epoxy nanocomposites with agglomeration effects, Mech. Adv. Mater. Struct. 22 (6) (2015) 428–439. https://doi.org/10.1080/15376494.2013.769654.
[217] R. Ansari, M.K. Hassanzadeh Aghdam, Micromechanics-based viscoelastic analysis of carbon nanotube-reinforced composites subjected to uniaxial and biaxial loading, Compos. Part B: Eng. 90 (2016) 512–522. https://doi.org/10.1016/j.compositesb.2015.10.048.
[218] R. Ansari, M.K. Hassanzadeh-Aghdam, Micromechanical investigation of creep-recovery behavior of carbon nanotube-reinforced polymer nanocomposites, Int. J. Mech. Sci. 115-116 (2016) 45–55. https://doi.org/10.1016/j.ijmecsci.2016.06.005.
[219] J. Pan, L. Bian, Influence of agglomeration parameters on carbon nanotube composites, Acta Mech. 228 (6) (2017) 2207–2217, doi:10.1007/s00707-017-1820-9.
[220] H. Souri, et al., A theoretical study on the piezoresistive response of carbon nanotubes embedded in polymer nanocomposites in an elastic region, Carbon 120 (2017) 427–437. https://doi.org/10.1016/j.carbon.2017.05.059.
[221] Y. Zare, K.Y. Rhee, The mechanical behavior of CNT reinforced nanocomposites assuming imperfect interfacial bonding between matrix and nanoparticles and percolation of interphase regions, Compos. Sci. Technol. 144 (2017) 18–25. https://doi.org/10.1016/j.compscitech.2017.03.012.
[222] M.K. Hassanzadeh-Aghdam, M.J. Mahmoodi, Micromechanical modeling of thermal conducting behavior of general carbon nanotube-polymer nanocomposites, Mater. Sci. Eng.: B 229 (2018) 173–183. https://doi.org/10.1016/j.mseb.2017.12.039.
[223] J.-M. Zhu, Y. Zare, K.Y. Rhee, Analysis of the roles of interphase, waviness and agglomeration of CNT in the electrical conductivity and tensile modulus of polymer/CNT nanocomposites by theoretical approaches, Colloids Surf. A 539 (2018) 29–36. https://doi.org/10.1016/j.colsurfa.2017.12.001.
[224] M.K. Hassanzadeh-Aghdam, R. Ansari, Thermal conductivity of shape memory polymer nanocomposites containing carbon nanotubes: a micromechanical approach, Compos. Part B: Eng. 162 (2019) 167–177. https://doi.org/10.1016/j.compositesb.2018.11.003.
[225] M.K. Hassanzadeh-Aghdam, M.J. Mahmoodi, R. Ansari, Creep performance of CNT polymer nanocomposites – an emphasis on viscoelastic interphase and CNT agglomeration, Compos. Part B: Eng. 168 (2019) 274–281. https://doi.org/10.1016/j.compositesb.2018.12.093.
[226] F. Zhu, C. Park, G.J. Yun, An extended Mori-Tanaka micromechanics model for wavy CNT nanocomposites with interface damage, Mech. Adv. Mater. Struct. 28 (3) (2021) 297–305. https://doi.org/10.1080/15376494.2018.1562135.
[227] Y. Zare, K.Y. Rhee, Advancement of a model for electrical conductivity of polymer nanocomposites reinforced with carbon nanotubes by a known model for thermal conductivity, Eng. Comput. (2020). https://doi.org/10.1007/s00366-020-01220-7.

[228] C. Li, T.-W. Chou, Failure of carbon nanotube/polymer composites and the effect of nanotube waviness, Compos. Part A 40 (10) (2009) 1580–1586. https://doi.org/10.1016/j.compositesa.2009.07.002.

[229] G.I. Giannopoulos, S.K. Georgantzinos, N.K. Anifantis, A semi-continuum finite element approach to evaluate the Young's modulus of single-walled carbon nanotube reinforced composites, Compos. Part B: Eng. 41 (8) (2010) 594–601. https://doi.org/10.1016/j.compositesb.2010.09.023.

[230] A. Needleman, et al., Effect of an interphase region on debonding of a CNT reinforced polymer composite, Compos. Sci. Technol. 70 (15) (2010) 2207–2215. https://doi.org/10.1016/j.compscitech.2010.09.002.

[231] M.M. Shokrieh, R. Rafiee, Investigation of nanotube length effect on the reinforcement efficiency in carbon nanotube based composites, Compos. Struct. 92 (10) (2010) 2415–2420. https://doi.org/10.1016/j.compstruct.2010.02.018.

[232] M.M. Shokrieh, R. Rafiee, Prediction of mechanical properties of an embedded carbon nanotube in polymer matrix based on developing an equivalent long fiber, Mech. Res. Commun. 37 (2) (2010) 235–240. https://doi.org/10.1016/j.mechrescom.2009.12.002.

[233] R. Rafiee, A. Fereidoon, M. Heidarhaei, Influence of non-bonded interphase on crack driving force in carbon nanotube reinforced polymer, Comput. Mater. Sci. 56 (2012) 25–28. https://doi.org/10.1016/j.commatsci.2011.12.025.

[234] M.A. Bhuiyan, et al., Understanding the effect of CNT characteristics on the tensile modulus of CNT reinforced polypropylene using finite element analysis, Comput. Mater. Sci. 79 (2013) 368–376. https://doi.org/10.1016/j.commatsci.2013.06.046.

[235] K.I. Tserpes, A. Chanteli, Parametric numerical evaluation of the effective elastic properties of carbon nanotube-reinforced polymers, Compos. Struct. 99 (2013) 366–374. https://doi.org/10.1016/j.compstruct.2012.12.004.

[236] P. Kumar, J. Srinivas, Numerical evaluation of effective elastic properties of CNT-reinforced polymers for interphase effects, Comput. Mater. Sci. 88 (2014) 139–144. https://doi.org/10.1016/j.commatsci.2014.03.002.

[237] S. Paunikar, S. Kumar, Effect of CNT waviness on the effective mechanical properties of long and short CNT reinforced composites, Comput. Mater. Sci. 95 (2014) 21–28. https://doi.org/10.1016/j.commatsci.2014.06.034.

[238] A. Chanteli, K.I. Tserpes, Finite element modeling of carbon nanotube agglomerates in polymers, Compos. Struct. 132 (2015) 1141–1148. https://doi.org/10.1016/j.compstruct.2015.07.033.

[239] Y. Jia, Z. Chen, W. Yan, A numerical study on carbon nanotube pullout to understand its bridging effect in carbon nanotube reinforced composites, Compos. Part B: Eng. 81 (2015) 64–71. https://doi.org/10.1016/j.compositesb.2015.07.003.

[240] K. Alasvand Zarasvand, H. Golestanian, Determination of nonlinear behavior of multi-walled carbon nanotube reinforced polymer: experimental, numerical, and micromechanical, Mater. Des. 109 (2016) 314–323. https://doi.org/10.1016/j.matdes.2016.07.071.

[241] D. Banerjee, T. Nguyen, T.-J. Chuang, Mechanical properties of single-walled carbon nanotube reinforced polymer composites with varied interphase's modulus and thickness: a finite element analysis study, Comput. Mater. Sci. 114 (2016) 209–218. https://doi.org/10.1016/j.commatsci.2015.12.026.

[242] N. Khani, M. Yildiz, B. Koc, Elastic properties of coiled carbon nanotube reinforced nanocomposite: a finite element study, Mater. Des. 109 (2016) 123–132. https://doi.org/10.1016/j.matdes.2016.06.126.

[243] A. Pontefisso, L. Mishnaevsky, Nanomorphology of graphene and CNT reinforced polymer and its effect on damage: micromechanical numerical study, Compos. Part B: Eng. 96 (2016) 338–349. https://doi.org/10.1016/j.compositesb.2016.04.006.

[244] M. Ahmadi, R. Ansari, S. Rouhi, Finite element investigation of temperature dependence of elastic properties of carbon nanotube reinforced polypropylene, Eur. Phys. J. Appl. Phys. 80 (3) (2017) 30401. https://doi.org/10.1051/epjap/2017170169.

[245] M. Karimi, A. Montazeri, R. Ghajar, On the elasto-plastic behavior of CNT-polymer nanocomposites, Compos. Struct. 160 (2017) 782–791. https://doi.org/10.1016/j.compstruct.2016.10.053.

[246] M. Karimi, R. Ghajar, A. Montazeri, A novel interface-treated micromechanics approach for accurate and efficient modeling of CNT/polymer composites, Compos. Struct. 201 (2018) 528–539. https://doi.org/10.1016/j.compstruct.2018.05.140.

[247] Q. Liu, S.V. Lomov, L. Gorbatikh, Spatial distribution and orientation of nanotubes for suppression of stress concentrations optimized using genetic algorithm and finite element analysis, Mater. Des. 158 (2018) 136–146. https://doi.org/10.1016/j.matdes.2018.08.019.

[248] X. Chen, A.R. Alian, S.A. Meguid, Modeling of CNT-reinforced nanocomposite with complex morphologies using modified embedded finite element technique, Compos. Struct. 227 (2019) 111329. https://doi.org/10.1016/j.compstruct.2019.111329.

[249] M.A. Maghsoudlou, et al., Effect of interphase, curvature and agglomeration of SWCNTs on mechanical properties of polymer-based nanocomposites: experimental and numerical investigations, Compos. Part B: Eng. 175 (2019) 107119. https://doi.org/10.1016/j.compositesb.2019.107119.

[250] A. Negi, et al., Analysis of CNT reinforced polymer nanocomposite plate in the presence of discontinuities using XFEM, Theor. Appl. Fract. Mech. 103 (2019) 102292. https://doi.org/10.1016/j.tafmec.2019.102292.

[251] X. Lu, et al., Numerical modeling and experimental characterization of the AC conductivity and dielectric properties of CNT/polymer nanocomposites, Compos. Sci. Technol. 194 (2020) 108150. https://doi.org/10.1016/j.compscitech.2020.108150.

[252] X.L. Gao, K. Li, A shear-lag model for carbon nanotube-reinforced polymer composites, Int. J. Solids Struct. 42 (5) (2005) 1649–1667. https://doi.org/10.1016/j.ijsolstr.2004.08.020.

[253] C. Li, T.-W. Chou, Multiscale modeling of compressive behavior of carbon nanotube/polymer composites, Compos. Sci. Technol. 66 (14) (2006) 2409–2414. https://doi.org/10.1016/j.compscitech.2006.01.013.

[254] K.I. Tserpes, et al., Multi-scale modeling of tensile behavior of carbon nanotube-reinforced composites, Theor. Appl. Fract. Mech. 49 (1) (2008) 51–60. https://doi.org/10.1016/j.tafmec.2007.10.004.

[255] A. Montazeri, R. Naghdabadi, Study the effect of viscoelastic matrix model on the stability of CNT/polymer composites by multiscale modeling, Polym. Compos. 30 (11) (2009) 1545–1551. https://doi.org/10.1002/pc.20797.

[256] M.M. Shokrieh, R. Rafiee, Stochastic multi-scale modeling of CNT/polymer composites, Comput. Mater. Sci. 50 (2) (2010) 437–446. https://doi.org/10.1016/j.commatsci.2010.08.036.

[257] M.M. Shokrieh, R. Rafiee, On the tensile behavior of an embedded carbon nanotube in polymer matrix with non-bonded interphase region, Compos. Struct. 92 (3) (2010) 647–652. https://doi.org/10.1016/j.compstruct.2009.09.033.

[258] M.R. Ayatollahi, S. Shadlou, M.M. Shokrieh, Multiscale modeling for mechanical properties of carbon nanotube reinforced nanocomposites subjected to different types of loading, Compos. Struct. 93 (9) (2011) 2250–2259. https://doi.org/10.1016/j.compstruct.2011.03.013.

[259] A. Montazeri, H. Rafii-Tabar, Multiscale modeling of graphene- and nanotube-based reinforced polymer nanocomposites, Phys. Lett. A 375 (45) (2011) 4034–4040. https://doi.org/10.1016/j.physleta.2011.08.073.

[260] G.D. Seidel, A.S. Puydupin-Jamin, Analysis of clustering, interphase region, and orientation effects on the electrical conductivity of carbon nanotube–polymer nanocomposites via computational micromechanics, Mech. Mater. 43 (12) (2011) 755–774. https://doi.org/10.1016/j.mechmat.2011.08.010.

[261] S. Kirtania, D. Chakraborty, Multi-scale modeling of carbon nanotube reinforced composites with a fiber break, Mater. Des. 35 (2012) 498–504. https://doi.org/10.1016/j.matdes.2011.09.041.

[262] J.M. Wernik, B.J. Cornwell-Mott, S.A. Meguid, Determination of the interfacial properties of carbon nanotube reinforced polymer composites using atomistic-based continuum model, Int. J. Solids Struct. 49 (13) (2012) 1852–1863. https://doi.org/10.1016/j.ijsolstr.2012.03.024.

[263] S. Yang, et al., Multiscale modeling of size-dependent elastic properties of carbon nanotube/polymer nanocomposites with interfacial imperfections, Polymer 53 (2) (2012) 623–633. https://doi.org/10.1016/j.polymer.2011.11.052.

[264] R. Rafiee, Influence of carbon nanotube waviness on the stiffness reduction of CNT/polymer composites, Compos. Struct. 97 (2013) 304–309. https://doi.org/10.1016/j.compstruct.2012.10.028.

[265] S. Gong, Z.H. Zhu, S.A. Meguid, Carbon nanotube agglomeration effect on piezoresistivity of polymer nanocomposites, Polymer 55 (21) (2014) 5488–5499. https://doi.org/10.1016/j.polymer.2014.08.054.

[266] M. Quaresimin, M. Salviato, M. Zappalorto, A multi-scale and multi-mechanism approach for the fracture toughness assessment of polymer nanocomposites, Compos. Sci. Technol. 91 (2014) 16–21. https://doi.org/10.1016/j.compscitech.2013.11.015.

[267] R. Rafiee, V. Firouzbakht, Multi-scale modeling of carbon nanotube reinforced polymers using irregular tessellation technique, Mech. Mater. 78 (2014) 74–84. https://doi.org/10.1016/j.mechmat.2014.07.021.

[268] K.N. Spanos, S.K. Georgantzinos, N.K. Anifantis, Investigation of stress transfer in carbon nanotube reinforced composites using a multi-scale finite element approach, Compos. Part B: Eng. 63 (2014) 85–93. https://doi.org/10.1016/j.compositesb.2014.03.020.

[269] S. Ebrahimi, H. Rafii-Tabar, Influence of hydrogen functionalization on mechanical properties of graphene and CNT reinforced in chitosan biological polymer: multi-scale computational modelling, Comput. Mater. Sci. 101 (2015) 189–193. https://doi.org/10.1016/j.commatsci.2015.01.036.

[270] V.S. Romanov, et al., Stress magnification due to carbon nanotube agglomeration in composites, Compos. Struct. 133 (2015) 246–256. https://doi.org/10.1016/j.compstruct.2015.07.069.

[271] A.R. Alian, S. El-Borgi, S.A. Meguid, Multiscale modeling of the effect of waviness and agglomeration of CNTs on the elastic properties of nanocomposites, Comput. Mater. Sci. 117 (2016) 195–204. https://doi.org/10.1016/j.commatsci.2016.01.029.

[272] A.K. Gupta, S.P. Harsha, Analysis of mechanical properties of carbon nanotube reinforced polymer composites using multi-scale finite element modeling approach, Compos. Part B: Eng. 95 (2016) 172–178. https://doi.org/10.1016/j.compositesb.2016.04.005.

[273] M.M. Shokrieh, R. Ghajar, A.R. Shajari, The effect of time-dependent slightly weakened interface on the viscoelastic properties of CNT/polymer nanocomposites, Compos. Struct. 146 (2016) 122–131. https://doi.org/10.1016/j.compstruct.2016.03.022.

[274] M.M. Shokrieh, A. Zeinedini, Effect of CNTs debonding on mode I fracture toughness of polymeric nanocomposites, Mater. Des. 101 (2016) 56–65. https://doi.org/10.1016/j.matdes.2016.03.134.

[275] Y. Li, G.D. Seidel, Multiscale modeling of the interface effects in CNT-epoxy nanocomposites, Comput. Mater. Sci. 153 (2018) 363–381. https://doi.org/10.1016/j.commatsci.2018.07.015.

[276] A.R. Shajari, R. Ghajar, M.M. Shokrieh, Multiscale modeling of the viscoelastic properties of CNT/polymer nanocomposites, using complex and time-dependent homogenizations, Comput. Mater. Sci. 142 (2018) 395–409. https://doi.org/10.1016/j.commatsci.2017.10.006.

[277] E. García-Macías, et al., Multiscale modeling of the elastic moduli of CNT-reinforced polymers and fitting of efficiency parameters for the use of the extended rule-of-mixtures, Compos. Part B: Eng. 159 (2019) 114–131. https://doi.org/10.1016/j.compositesb.2018.09.057.

[278] J.A. Palacios, R. Ganesan, Dynamic response of carbon-nanotube-reinforced-polymer materials based on multiscale finite element analysis, Compos. Part B: Eng. 166 (2019) 497–508. https://doi.org/10.1016/j.compositesb.2019.02.039.

[279] M. Javid, H. Biglari, A multi-scale finite element approach to mechanical performance of polyurethane/CNT nanocomposite foam, Mater. Today Commun. 24 (2020) 101081. https://doi.org/10.1016/j.mtcomm.2020.101081.

[280] M.H. Namdari Pour, G. Payganeh, M. Tajdari, Mechanical behavior of carbon nanotube reinforced polymethylmethacrylate foam: a multi-scale finite element method approach, Eur. J. Mech. A Solids 83 (2020) 104019. https://doi.org/10.1016/j.euromechsol.2020.104019.

[281] R. Rafiee, M. Sharaei, Investigating the influence of bonded and non-bonded interactions on the interfacial bonding between carbon nanotube and polymer, Compos. Struct. 238 (2020) 111996. https://doi.org/10.1016/j.compstruct.2020.111996.

[282] R. Rafiee, M. Sahraei, Characterizing delamination toughness of laminated composites containing carbon nanotubes: experimental study and stochastic multiscale modeling, Compos. Sci. Technol. 201 (2021) 108487. https://doi.org/10.1016/j.compscitech.2020.108487.

[283] L.-L. Ke, J. Yang, S. Kitipornchai, Nonlinear free vibration of functionally graded carbon nanotube-reinforced composite beams, Compos. Struct. 92 (3) (2010) 676–683. https://doi.org/10.1016/j.compstruct.2009.09.024.

[284] M.H. Yas, N. Samadi, Free vibrations and buckling analysis of carbon nanotube-reinforced composite Timoshenko beams on elastic foundation, Int. J. Press. Vessels Pip. 98 (2012) 119–128. https://doi.org/10.1016/j.ijpvp.2012.07.012.

[285] P. Zhu, Z.X. Lei, K.M. Liew, Static and free vibration analyses of carbon nanotube-reinforced composite plates using finite element method with first order shear deformation plate theory, Compos. Struct. 94 (4) (2012) 1450–1460. https://doi.org/10.1016/j.compstruct.2011.11.010.

[286] L.-L. Ke, J. Yang, S. Kitipornchai, Dynamic stability of functionally graded carbon nanotube-reinforced composite beams, Mech. Adv. Mater. Struct. 20 (1) (2013) 28–37. https://doi.org/10.1080/15376494.2011.581412.

[287] Z.X. Lei, K.M. Liew, J.L. Yu, Free vibration analysis of functionally graded carbon nanotube-reinforced composite plates using the element-free kp-Ritz method in thermal environment, Compos. Struct. 106 (2013) 128–138. https://doi.org/10.1016/j.compstruct.2013.06.003.

[288] N. Wattanasakulpong, V. Ungbhakorn, Analytical solutions for bending, buckling and vibration responses of carbon nanotube-reinforced composite beams resting on elastic foundation, Comput. Mater. Sci. 71 (2013) 201–208. https://doi.org/10.1016/j.commatsci.2013.01.028.

[289] R. Ansari, et al., Nonlinear forced vibration analysis of functionally graded carbon nanotube-reinforced composite Timoshenko beams, Compos. Struct. 113 (2014) 316–327. https://doi.org/10.1016/j.compstruct.2014.03.015.

[290] F. Lin, Y. Xiang, Vibration of carbon nanotube reinforced composite beams based on the first and third order beam theories, Appl. Math. Modell. 38 (15) (2014) 3741–3754. https://doi.org/10.1016/j.apm.2014.02.008.

[291] Z.X. Lei, L.W. Zhang, K.M. Liew, Free vibration analysis of laminated FG-CNT reinforced composite rectangular plates using the kp-Ritz method, Compos. Struct. 127 (2015) 245–259. https://doi.org/10.1016/j.compstruct.2015.03.019.

[292] M. Mirzaei, Y. Kiani, Thermal buckling of temperature dependent FG-CNT reinforced composite conical shells, Aerosp. Sci. Technol. 47 (2015) 42–53. https://doi.org/10.1016/j.ast.2015.09.011.

[293] L.W. Zhang, W.C. Cui, K.M. Liew, Vibration analysis of functionally graded carbon nanotube reinforced composite thick plates with elastically restrained edges, Int. J. Mech. Sci. 103 (2015) 9–21. https://doi.org/10.1016/j.ijmecsci.2015.08.021.

[294] L.W. Zhang, Z.X. Lei, K.M. Liew, An element-free IMLS-Ritz framework for buckling analysis of FG–CNT reinforced composite thick plates resting on Winkler foundations, Eng. Anal. Boundary Elem. 58 (2015) 7–17. https://doi.org/10.1016/j.enganabound.2015.03.004.

[295] L.W. Zhang, Z.X. Lei, K.M. Liew, Buckling analysis of FG-CNT reinforced composite thick skew plates using an element-free approach, Compos. Part B: Eng. 75 (2015) 36–46. https://doi.org/10.1016/j.compositesb.2015.01.033.

[296] L.W. Zhang, Z.X. Lei, K.M. Liew, Vibration characteristic of moderately thick functionally graded carbon nanotube reinforced composite skew plates, Compos. Struct. 122 (2015) 172–183. https://doi.org/10.1016/j.compstruct.2014.11.070.

[297] L.W. Zhang, Z.G. Song, K.M. Liew, State-space Levy method for vibration analysis of FG-CNT composite plates subjected to in-plane loads based on higher-order shear deformation theory, Compos. Struct. 134 (2015) 989–1003. https://doi.org/10.1016/j.compstruct.2015.08.138.

[298] R. Ansari, J. Torabi, Numerical study on the buckling and vibration of functionally graded carbon nanotube-reinforced composite conical shells under axial loading, Compos. Part B: Eng. 95 (2016) 196–208. https://doi.org/10.1016/j.compositesb.2016.03.080.

[299] Y. Kiani, Shear buckling of FG-CNT reinforced composite plates using Chebyshev-Ritz method, Compos. Part B: Eng. 105 (2016) 176–187. https://doi.org/10.1016/j.compositesb.2016.09.001.

[300] Y. Kiani, Free vibration of FG-CNT reinforced composite skew plates, Aerosp. Sci. Technol. 58 (2016) 178–188. https://doi.org/10.1016/j.ast.2016.08.018.

[301] Z.X. Lei, L.W. Zhang, K.M. Liew, Analysis of laminated CNT reinforced functionally graded plates using the element-free kp-Ritz method, Compos. Part B: Eng. 84 (2016) 211–221. https://doi.org/10.1016/j.compositesb.2015.08.081.

[302] Z.X. Lei, L.W. Zhang, K.M. Liew, Buckling analysis of CNT reinforced functionally graded laminated composite plates, Compos. Struct. 152 (2016) 62–73. https://doi.org/10.1016/j.compstruct.2016.05.047.

[303] Z.X. Lei, L.W. Zhang, K.M. Liew, Vibration of FG-CNT reinforced composite thick quadrilateral plates resting on Pasternak foundations, Eng. Anal. Boundary Elem. 64 (2016) 1–11. https://doi.org/10.1016/j.enganabound.2015.11.014.

[304] K. Mehar, S.K. Panda, Geometrical nonlinear free vibration analysis of FG-CNT reinforced composite flat panel under uniform thermal field, Compos. Struct. 143 (2016) 336–346. https://doi.org/10.1016/j.compstruct.2016.02.038.

[305] M. Mehri, H. Asadi, Q. Wang, Buckling and vibration analysis of a pressurized CNT reinforced functionally graded truncated conical shell under an axial compression using HDQ method, Comput. Meth. Appl. Mech. Eng. 303 (2016) 75–100. https://doi.org/10.1016/j.cma.2016.01.017.

[306] B.A. Selim, L.W. Zhang, K.M. Liew, Vibration analysis of CNT reinforced functionally graded composite plates in a thermal environment based on Reddy's higher-order shear deformation theory, Compos. Struct. 156 (2016) 276–290. https://doi.org/10.1016/j.compstruct.2015.10.026.

[307] Z.G. Song, L.W. Zhang, K.M. Liew, Vibration analysis of CNT-reinforced functionally graded composite cylindrical shells in thermal environments, Int. J. Mech. Sci. 115-116 (2016) 339–347. https://doi.org/10.1016/j.ijmecsci.2016.06.020.

[308] Z.G. Song, L.W. Zhang, K.M. Liew, Active vibration control of CNT-reinforced composite cylindrical shells via piezoelectric patches, Compos. Struct. 158 (2016) 92–100. https://doi.org/10.1016/j.compstruct.2016.09.031.

[309] Z.G. Song, L.W. Zhang, K.M. Liew, Dynamic responses of CNT reinforced composite plates subjected to impact loading, Compos. Part B: Eng. 99 (2016) 154–161. https://doi.org/10.1016/j.compositesb.2016.06.034.

[310] Z.G. Song, L.W. Zhang, K.M. Liew, Active vibration control of CNT reinforced functionally graded plates based on a higher-order shear deformation theory, Int. J. Mech. Sci. 105 (2016) 90–101. https://doi.org/10.1016/j.ijmecsci.2015.11.019.

[311] M. Wang, Z.-M. Li, P. Qiao, Semi-analytical solutions to buckling and free vibration analysis of carbon nanotube-reinforced composite thin plates, Compos. Struct. 144 (2016) 33–43. https://doi.org/10.1016/j.compstruct.2016.02.025.

[312] L.W. Zhang, et al., Free vibration analysis of triangular CNT-reinforced composite plates subjected to in-plane stresses using FSDT element-free method, Compos. Struct. 149 (2016) 247–260. https://doi.org/10.1016/j.compstruct.2016.04.019.

[313] F. Tornabene, et al., Effect of agglomeration on the natural frequencies of functionally graded carbon nanotube-reinforced laminated composite doubly-curved shells, Compos. Part B: Eng. 89 (2016) 187–218. https://doi.org/10.1016/j.compositesb.2015.11.016.

[314] N. Fantuzzi, et al., Free vibration analysis of arbitrarily shaped functionally graded carbon nanotube-reinforced plates, Compos. Part B: Eng. 115 (2017) 384–408. https://doi.org/10.1016/j.compositesb.2016.09.021.

[315] R. Ansari, J. Torabi, M.F. Shojaei, Buckling and vibration analysis of embedded functionally graded carbon nanotube-reinforced composite annular sector plates under thermal loading, Compos. Part B: Eng. 109 (2017) 197–213. https://doi.org/10.1016/j.compositesb.2016.10.050.

[316] H. Asadi, M. Souri, Q. Wang, A numerical study on flow-induced instabilities of supersonic FG-CNT reinforced composite flat panels in thermal environments, Compos. Struct. 171 (2017) 113–125. https://doi.org/10.1016/j.compstruct.2017.02.003.

[317] F. Ebrahimi, N. Farazmandnia, Thermo-mechanical analysis of carbon nanotube-reinforced composite sandwich beams, Coupled Syst. Mech. 6 (2) (2017) 207–227. https://doi.org/10.12989/csm.2017.6.2.207.

[318] F. Ebrahimi, N. Farazmandnia, Thermo-mechanical vibration analysis of sandwich beams with functionally graded carbon nanotube-reinforced composite face sheets based on a higher-order shear deformation beam theory, Mech. Adv. Mater. Struct. 24 (10) (2017) 820–829. https://doi.org/10.1080/15376494.2016.1196786.

[319] F. Ebrahimi, S. Habibi, Low-velocity impact response of laminated FG-CNT reinforced composite plates in thermal environment, Adv. Nano Res. 5 (2) (2017) 69–97. https://doi.org/10.12989/anr.2017.5.2.069.

[320] M.M. Keleshteri, H. Asadi, Q. Wang, Large amplitude vibration of FG-CNT reinforced composite annular plates with integrated piezoelectric layers on elastic foundation, Thin-Walled Struct. 120 (2017) 203–214. https://doi.org/10.1016/j.tws.2017.08.035.

[321] Y. Kiani, Free vibration of FG-CNT reinforced composite spherical shell panels using Gram-Schmidt shape functions, Compos. Struct. 159 (2017) 368–381. https://doi.org/10.1016/j.compstruct.2016.09.079.

[322] Y. Kiani, Thermal post-buckling of FG-CNT reinforced composite plates, Compos. Struct. 159 (2017) 299–306. https://doi.org/10.1016/j.compstruct.2016.09.084.

[323] P. Kumar, J. Srinivas, Vibration, buckling and bending behavior of functionally graded multi-walled carbon nanotube reinforced polymer composite plates using the

layer-wise formulation, Compos. Struct. 177 (2017) 158–170. https://doi.org/10.1016/j.compstruct.2017.06.055.

[324] K. Mehar, S.K. Panda, T.R. Mahapatra, Thermoelastic nonlinear frequency analysis of CNT reinforced functionally graded sandwich structure, Eur. J. Mech. A. Solids 65 (2017) 384–396. https://doi.org/10.1016/j.euromechsol.2017.05.005.

[325] M.M. Ardestani, L.W. Zhang, K.M. Liew, Isogeometric analysis of the effect of CNT orientation on the static and vibration behaviors of CNT-reinforced skew composite plates, Comput. Meth. Appl. Mech. Eng. 317 (2017) 341–379. https://doi.org/10.1016/j.cma.2016.12.009.

[326] M. Mohammadzadeh-Keleshteri, H. Asadi, M.M. Aghdam, Geometrical nonlinear free vibration responses of FG-CNT reinforced composite annular sector plates integrated with piezoelectric layers, Compos. Struct. 171 (2017) 100–112. https://doi.org/10.1016/j.compstruct.2017.01.048.

[327] B.A. Selim, L.W. Zhang, K.M. Liew, Active vibration control of CNT-reinforced composite plates with piezoelectric layers based on Reddy's higher-order shear deformation theory, Compos. Struct. 163 (2017) 350–364. https://doi.org/10.1016/j.compstruct.2016.11.011.

[328] H.-S. Shen, X.Q. He, D.-Q. Yang, Vibration of thermally postbuckled carbon nanotube-reinforced composite beams resting on elastic foundations, Int. J. Non Linear Mech. 91 (2017) 69–75. https://doi.org/10.1016/j.ijnonlinmec.2017.02.010.

[329] L.W. Zhang, On the study of the effect of in-plane forces on the frequency parameters of CNT-reinforced composite skew plates, Compos. Struct. 160 (2017) 824–837. https://doi.org/10.1016/j.compstruct.2016.10.116.

[330] L.W. Zhang, M.M. Ardestani, K.M. Liew, Isogeometric approach for buckling analysis of CNT-reinforced composite skew plates under optimal CNT-orientation, Compos. Struct. 163 (2017) 365–384. https://doi.org/10.1016/j.compstruct.2016.12.047.

[331] L.W. Zhang, et al., Modeling of dynamic responses of CNT-reinforced composite cylindrical shells under impact loads, Comput. Meth. Appl. Mech. Eng. 313 (2017) 889–903. https://doi.org/10.1016/j.cma.2016.10.020.

[332] A.K. Baltacıoğlu, Ö. Civalek, Vibration analysis of circular cylindrical panels with CNT reinforced and FGM composites, Compos. Struct. 202 (2018) 374–388. https://doi.org/10.1016/j.compstruct.2018.02.024.

[333] F. Ebrahimi, N. Farazmandnia, Vibration analysis of functionally graded carbon nanotube-reinforced composite sandwich beams in thermal environment, Adv. Aircraft Spacecraft Sci. 5 (1) (2018) 107–128. https://doi.org/10.12989/aas.2018.5.1.107.

[334] F. Ebrahimi, N. Farazmandnia, Thermal buckling analysis of functionally graded carbon nanotube-reinforced composite sandwich beams, Steel Compos. Struct. 27 (2) (2018) 149–159. https://doi.org/10.12989/scs.2018.27.2.149.

[335] F. Ebrahimi, P. Rostami, Wave propagation analysis of carbon nanotube reinforced composite beams, Eur. Phys. J. Plus 133 (7) (2018) 285. https://doi.org/10.1140/epjp/i2018-12069-y.

[336] F. Ebrahimi, P. Rostami, Propagation of elastic waves in thermally affected embedded carbon-nanotube-reinforced composite beams via various shear deformation plate theories, Struct. Eng. Mech. 66 (4) (2018) 495–504. https://doi.org/10.12989/sem.2018.66.4.495.

[337] Y. Kiani, R. Dimitri, F. Tornabene, Free vibration of FG-CNT reinforced composite skew cylindrical shells using the Chebyshev-Ritz formulation, Compos. Part B: Eng. 147 (2018) 169–177. https://doi.org/10.1016/j.compositesb.2018.04.028.

[338] R. Ansari, J. Torabi, R. Hassani, A comprehensive study on the free vibration of arbitrary shaped thick functionally graded CNT-reinforced composite plates, Eng. Struct. 181 (2019) 653–669. https://doi.org/10.1016/j.engstruct.2018.12.049.

[339] S. Chakraborty, T. Dey, R. Kumar, Stability and vibration analysis of CNT-reinforced functionally graded laminated composite cylindrical shell panels using semi-analytical approach, Compos. Part B: Eng. 168 (2019) 1–14. https://doi.org/10.1016/j.compositesb.2018.12.051.

[340] M. Di Sciuva, M. Sorrenti, Bending, free vibration and buckling of functionally graded carbon nanotube-reinforced sandwich plates, using the extended refined zigzag theory, Compos. Struct. 227 (2019) 111324. https://doi.org/10.1016/j.compstruct.2019.111324.

[341] F. Ebrahimi, et al., Buckling and vibration characteristics of a carbon nanotube-reinforced spinning cantilever cylindrical 3D shell conveying viscous fluid flow and carrying spring-mass systems under various temperature distributions, Proc. Inst. Mech. Eng. Part C: J. Mech. Eng. Sci. 233 (13) (2019) 4590–4605 https://doi.org/10.1177/0954406219832323.

[342] K. Mehar, et al., Numerical buckling analysis of graded CNT-reinforced composite sandwich shell structure under thermal loading, Compos. Struct. 216 (2019) 406–414. https://doi.org/10.1016/j.compstruct.2019.03.002.

[343] Z. Qin, et al., Free vibration analysis of rotating functionally graded CNT reinforced composite cylindrical shells with arbitrary boundary conditions, Compos. Struct. 220 (2019) 847–860. https://doi.org/10.1016/j.compstruct.2019.04.046.

[344] B. Safaei, et al., Critical buckling temperature and force in porous sandwich plates with CNT-reinforced nanocomposite layers, Aerosp. Sci. Technol. 91 (2019) 175–185. https://doi.org/10.1016/j.ast.2019.05.020.

[345] H. SafarPour, B. Ghanbari, M. Ghadiri, Buckling and free vibration analysis of high speed rotating carbon nanotube reinforced cylindrical piezoelectric shell, Appl. Math. Modell. 65 (2019) 428–442. https://doi.org/10.1016/j.apm.2018.08.028.

[346] J. Torabi, R. Ansari, R. Hassani, Numerical study on the thermal buckling analysis of CNT-reinforced composite plates with different shapes based on the higher-order shear deformation theory, Eur. J. Mech. A. Solids 73 (2019) 144–160. https://doi.org/10.1016/j.euromechsol.2018.07.009.

[347] A. Ghorbanpour Arani, F. Kiani, H. Afshari, Free and forced vibration analysis of laminated functionally graded CNT-reinforced composite cylindrical panels, J. Sandwich Struct. Mater. 23 (1) (2021) 255–278. https://doi.org/10.1177/1099636219830787.

[348] Ö. Civalek, M. Avcar, Free vibration and buckling analyses of CNT reinforced laminated non-rectangular plates by discrete singular convolution method, Eng. Comput. (2020). https://doi.org/10.1007/s00366-020-01168-8.

[349] Ö. Civalek, S. Dastjerdi, B. Akgöz, Buckling and free vibrations of CNT-reinforced cross-ply laminated composite plates, Mech. Based Des. Struct. Mach. (2020) 1–18. https://doi.org/10.1080/15397734.2020.1766494.

[350] M. Mirzaei, Vibrations of FG-CNT reinforced composite cylindrical panels with cutout, Mech. Based Des. Struct. Mach. (2020) 1–21. https://doi.org/10.1080/15397734.2019.1705165.

[351] R. Moradi-Dastjerdi, et al., Buckling behavior of porous CNT-reinforced plates integrated between active piezoelectric layers, Eng. Struct. 222 (2020) 111141. https://doi.org/10.1016/j.engstruct.2020.111141.

[352] R. Moradi-Dastjerdi, et al., Static performance of agglomerated CNT-reinforced porous plates bonded with piezoceramic faces, Int. J. Mech. Sci. 188 (2020) 105966. https://doi.org/10.1016/j.ijmecsci.2020.105966.

[353] K.M. Liew, A. Alibeigloo, Predicting bucking and vibration behaviors of functionally graded carbon nanotube reinforced composite cylindrical panels with three-dimensional flexibilities, Compos. Struct. 256 (2021) 113039. https://doi.org/10.1016/j.compstruct.2020.113039.

[354] E.T. Thostenson, et al., Carbon nanotube/carbon fiber hybrid multiscale composites, J. Appl. Phys. 91 (9) (2002) 6034–6037. https://doi.org/10.1063/1.1466880.

[355] E. Bekyarova, et al., Multiscale carbon nanotube–carbon fiber reinforcement for advanced epoxy composites, Langmuir 23 (7) (2007) 3970–3974. https://doi.org/10.1021/la062743p.
[356] M. Kim, et al., Processing, characterization, and modeling of carbon nanotube-reinforced multiscale composites, Compos. Sci. Technol. 69 (3) (2009) 335–342. https://doi.org/10.1016/j.compscitech.2008.10.019.
[357] M.T. Kim, et al., Property enhancement of a carbon fiber/epoxy composite by using carbon nanotubes, Compos. Part B: Eng. 42 (5) (2011) 1257–1261. https://doi.org/10.1016/j.compositesb.2011.02.005.
[358] S.P. Sharma, S.C. Lakkad, Effect of CNTs growth on carbon fibers on the tensile strength of CNTs grown carbon fiber-reinforced polymer matrix composites, Compos. Part A 42 (1) (2011) 8–15. https://doi.org/10.1016/j.compositesa.2010.09.008.
[359] A.Y. Boroujeni, et al., Hybrid carbon nanotube–carbon fiber composites with improved in-plane mechanical properties, Compos. Part B: Eng. 66 (2014) 475–483. https://doi.org/10.1016/j.compositesb.2014.06.010.
[360] R.L. Zhang, et al., Polyhedral oligomeric silsesquioxanes/carbon nanotube/carbon fiber multiscale composite: influence of a novel hierarchical reinforcement on the interfacial properties, Appl. Surf. Sci. 353 (2015) 224–231. https://doi.org/10.1016/j.apsusc.2015.06.156.
[361] H.W. Zhou, et al., Carbon fiber/carbon nanotube reinforced hierarchical composites: effect of CNT distribution on shearing strength, Compos. Part B: Eng. 88 (2016) 201–211. https://doi.org/10.1016/j.compositesb.2015.10.035.
[362] M.S. Chaudhry, A. Czekanski, Z.H. Zhu, Characterization of carbon nanotube enhanced interlaminar fracture toughness of woven carbon fiber reinforced polymer composites, Int. J. Mech. Sci. 131-132 (2017) 480–489. https://doi.org/10.1016/j.ijmecsci.2017.06.016.
[363] S. Khan, H.S. Bedi, P.K. Agnihotri, Augmenting mode-II fracture toughness of carbon fiber/epoxy composites through carbon nanotube grafting, Eng. Fract. Mech. 204 (2018) 211–220. https://doi.org/10.1016/j.engfracmech.2018.10.014.
[364] S. Wang, J. Qiu, Enhancing thermal conductivity of glass fiber/polymer composites through carbon nanotubes incorporation, Compos. Part B: Eng. 41 (7) (2010) 533–536. https://doi.org/10.1016/j.compositesb.2010.07.002.
[365] P. Agnihotri, S. Basu, K.K. Kar, Effect of carbon nanotube length and density on the properties of carbon nanotube-coated carbon fiber/polyester composites, Carbon 49 (9) (2011) 3098–3106. https://doi.org/10.1016/j.carbon.2011.03.032.
[366] X. Liu, et al., Fabrication and electromagnetic interference shielding effectiveness of carbon nanotube reinforced carbon fiber/pyrolytic carbon composites, Carbon 68 (2014) 501–510. https://doi.org/10.1016/j.carbon.2013.11.027.
[367] J. Zhang, et al., Effect of hierarchical structure on electrical properties and percolation behavior of multiscale composites modified by carbon nanotube coating, Compos. Sci. Technol. 164 (2018) 160–167. https://doi.org/10.1016/j.compscitech.2018.05.037.
[368] M. Rafiee, et al., Geometrically nonlinear free vibration of shear deformable piezoelectric carbon nanotube/fiber/polymer multiscale laminated composite plates, J. Sound Vib. 333 (14) (2014) 3236–3251. https://doi.org/10.1016/j.jsv.2014.02.033.
[369] X.Q. He, et al., Large amplitude vibration of fractionally damped viscoelastic CNTs/fiber/polymer multiscale composite beams, Compos. Struct. 131 (2015) 1111–1123. https://doi.org/10.1016/j.compstruct.2015.06.038.
[370] M. Rafiee, F. Nitzsche, M. Labrosse, Rotating nanocomposite thin-walled beams undergoing large deformation, Compos. Struct. 150 (2016) 191–199. https://doi.org/10.1016/j.compstruct.2016.05.014.
[371] M. Ahmadi, R. Ansari, H. Rouhi, Multi-scale bending, buckling and vibration analyses of carbon fiber/carbon nanotube-reinforced polymer nanocomposite plates with various shapes, Physica E 93 (2017) 17–25. https://doi.org/10.1016/j.physe.2017.05.009.

[372] F. Ebrahimi, S. Habibi, Thermal effects on nonlinear dynamic characteristics of polymer-CNT-fiber multiscale nanocomposite structures, Struct. Eng. Mech. 67 (4) (2018) 403–415. https://doi.org/10.12989/sem.2018.67.4.403.

[373] F. Ebrahimi, S. Habibi, Nonlinear eccentric low-velocity impact response of a polymer-carbon nanotube-fiber multiscale nanocomposite plate resting on elastic foundations in hygrothermal environments, Mech. Adv. Mater. Struct. 25 (5) (2018) 425–438. https://doi.org/10.1080/15376494.2017.1285453.

[374] M. Rafiee, F. Nitzsche, M.R. Labrosse, Modeling and mechanical analysis of multi-scale fiber-reinforced graphene composites: nonlinear bending, thermal post-buckling and large amplitude vibration, Int. J. Non Linear Mech. 103 (2018) 104–112. https://doi.org/10.1016/j.ijnonlinmec.2018.05.004.

[375] A. Dabbagh, A. Rastgoo, F. Ebrahimi, Finite element vibration analysis of multi-scale hybrid nanocomposite beams via a refined beam theory, Thin-Walled Struct. 140 (2019) 304–317. https://doi.org/10.1016/j.tws.2019.03.031.

[376] F. Ebrahimi, A. Dabbagh, Vibration analysis of multi-scale hybrid nanocomposite plates based on a Halpin-Tsai homogenization model, Compos. Part B: Eng. 173 (2019) 106955. https://doi.org/10.1016/j.compositesb.2019.106955.

[377] F. Ebrahimi, A. Dabbagh, An analytical solution for static stability of multi-scale hybrid nanocomposite plates, Eng. Comput. 37 (1) (2021) 545–559. https://doi.org/10.1007/s00366-019-00840-y.

[378] F. Ebrahimi, A. Dabbagh, Vibration analysis of graphene oxide powder-/carbon fiber-reinforced multi-scale porous nanocomposite beams: a finite-element study, Eur. Phys. J. Plus 134 (5) (2019) 225. https://doi.org/10.1140/epjp/i2019-12594-1.

[379] F. Ebrahimi, A. Dabbagh, On thermo-mechanical vibration analysis of multi-scale hybrid composite beams, J. Vib. Control 25 (4) (2019) 933–945. https://doi.org/10.1177/1077546318806800.

[380] F. Ebrahimi, A. Dabbagh, A. Rastgoo, Free vibration analysis of multi-scale hybrid nanocomposite plates with agglomerated nanoparticles, Mech. Based Des. Struct. Mach. 49 (4) (2021) 487–510. https://doi.org/10.1080/15397734.2019.1692665.

[381] F. Ebrahimi, M. Karimiasl, Dynamic modeling of a multi-scale sandwich composite panel containing flexible core and MR smart layer, Eur. Phys. J. Plus 134 (12) (2019) 622. https://doi.org/10.1140/epjp/i2019-12662-6.

[382] F. Ebrahimi, A. Seyfi, Wave propagation response of multi-scale hybrid nanocomposite shell by considering aggregation effect of CNTs, Mech. Based Des. Struct. Mach. 49 (1) (2021) 59–80. https://doi.org/10.1080/15397734.2019.1666722.

[383] F. Ebrahimi, A. Seyfi, A. Dabbagh, Wave dispersion characteristics of agglomerated multi-scale hybrid nanocomposite beams, J. Strain Anal. Eng. Des. 54 (4) (2019) 276–289. https://doi.org/10.1177/0309324719862713.

[384] M. Karimiasl, F. Ebrahimi, Large amplitude vibration of viscoelastically damped multiscale composite doubly curved sandwich shell with flexible core and MR layers, Thin-Walled Struct. 144 (2019) 106128. https://doi.org/10.1016/j.tws.2019.04.020.

[385] M. Karimiasl, F. Ebrahimi, B. Akgöz, Buckling and post-buckling responses of smart doubly curved composite shallow shells embedded in SMA fiber under hygro-thermal loading, Compos. Struct. 223 (2019) 110988. https://doi.org/10.1016/j.compstruct.2019.110988.

[386] M. Karimiasl, F. Ebrahimi, V. Mahesh, Nonlinear free and forced vibration analysis of multiscale composite doubly curved shell embedded in shape-memory alloy fiber under hygrothermal environment, J. Vib. Control 25 (13) (2019) 1945–1957. https://doi.org/10.1177/1077546319842426.

[387] M. Karimiasl, F. Ebrahimi, M. Vinyas, Nonlinear vibration analysis of multiscale doubly curved piezoelectric composite shell in hygrothermal environment, J. Intell. Mater. Syst. Struct. 30 (10) (2019) 1594–1609. https://doi.org/10.1177/1045389X19835956.

[388] A. Dabbagh, A. Rastgoo, F. Ebrahimi, Static stability analysis of agglomerated multi-scale hybrid nanocomposites via a refined theory, Eng. Comput. 37 (3) (2021) 2225–2244. https://doi.org/10.1007/s00366-020-00939-7.

[389] A. Dabbagh, A. Rastgoo, F. Ebrahimi, Thermal buckling analysis of agglomerated multiscale hybrid nanocomposites via a refined beam theory, Mech. Based Des. Struct. Mach. 49 (3) (2021) 403–429. https://doi.org/10.1080/15397734.2019.1692666.

[390] A. Dabbagh, A. Rastgoo, F. Ebrahimi, Post-buckling analysis of imperfect multi-scale hybrid nanocomposite beams rested on a nonlinear stiff substrate, Eng. Comput. (2020). https://doi.org/10.1007/s00366-020-01064-1.

[391] F. Ebrahimi, A. Dabbagh, Vibration analysis of multi-scale hybrid nanocomposite shells by considering nanofillers' aggregation, Waves Random Complex Media (2020) 1–19. https://doi.org/10.1080/17455030.2020.1810363.

[392] F. Ebrahimi, A. Dabbagh, Vibration analysis of fluid-conveying multi-scale hybrid nanocomposite shells with respect to agglomeration of nanofillers, Defence Technol. 17 (1) (2021) 212–225. https://doi.org/10.1016/j.dt.2020.01.007.

[393] F. Ebrahimi, A. Dabbagh, A. Rastgoo, Static stability analysis of multi-scale agglomerated nanocomposite shells, Mech. Based Des. Struct. Mach. (2020) 1–17. https://doi.org/10.1080/15397734.2020.1848585.

[394] F. Ebrahimi, et al., Agglomeration effects on static stability analysis of multi-scale hybrid nanocomposite plates, Comput. Mater. Continua 63 (1) (2020) 41–64. https://doi.org/10.32604/cmc.2020.07947.

[395] F. Ebrahimi, A. Seyfi, Wave propagation response of agglomerated multi-scale hybrid nanocomposite plates, Waves Random Complex Media (2020) 1–25. https://doi.org/10.1080/17455030.2020.1821933.

[396] M. Karimiasl, F. Ebrahimi, V. Mahesh, On nonlinear vibration of sandwiched polymer-CNT/GPL-fiber nanocomposite nanoshells, Thin-Walled Struct. 146 (2020) 106431. https://doi.org/10.1016/j.tws.2019.106431.

[397] M. Karimiasl, F. Ebrahimi, V. Mahesh, Hygrothermal postbuckling analysis of smart multiscale piezoelectric composite shells, Eur. Phys. J. Plus 135 (2) (2020) 242. https://doi.org/10.1140/epjp/s13360-020-00137-w.

[398] M. Safarpour, et al., On the nonlinear dynamics of a multi-scale hybrid nanocomposite disk, Eng. Comput. 37 (3) (2021) 2369–2388. https://doi.org/10.1007/s00366-020-00949-5.

[399] M. Taheri, F. Ebrahimi, Buckling analysis of CFRP plates: a porosity-dependent study considering the GPLs-reinforced interphase between fiber and matrix, Eur. Phys. J. Plus 135 (7) (2020) 549. https://doi.org/10.1140/epjp/s13360-020-00581-8.

[400] H.-S. Shen, Y. Xiang, Y. Fan, Nonlinear vibration of functionally graded graphene-reinforced composite laminated cylindrical shells in thermal environments, Compos. Struct. 182 (2017) 447–456. https://doi.org/10.1016/j.compstruct.2017.09.010.

[401] H.-S. Shen, A comparison of buckling and postbuckling behavior of FGM plates with piezoelectric fiber reinforced composite actuators, Compos. Struct. 91 (3) (2009) 375–384. https://doi.org/10.1016/j.compstruct.2009.06.005.

[402] Y. Han, J. Elliott, Molecular dynamics simulations of the elastic properties of polymer/carbon nanotube composites, Comput. Mater. Sci. 39 (2) (2007) 315–323. https://doi.org/10.1016/j.commatsci.2006.06.011.

[403] L. Wang, H. Hu, Flexural wave propagation in single-walled carbon nanotubes, Phys. Rev. B 71 (19) (2005) 195412. https://doi.org/10.1103/PhysRevB.71.195412.

[404] R. Arasteh, et al., A study on effect of waviness on mechanical properties of multi-walled carbon nanotube/epoxy composites using modified Halpin–Tsai theory, J. Macromol. Sci. Part B 50 (12) (2011) 2464–2480. https://doi.org/10.1080/00222348.2011.579868.

[405] I.A. Kazakov, A.N. Krasnovskii, P.S. Kishuk, The influence of randomly oriented CNTs on the elastic properties of unidirectionally aligned composites, Mech. Mater. 134 (2019) 54–60. https://doi.org/10.1016/j.mechmat.2019.04.002.

[406] F. Ebrahimi, A. Rastgoo, An analytical study on the free vibration of smart circular thin FGM plate based on classical plate theory, Thin-Walled Struct. 46 (12) (2008) 1402–1408. https://doi.org/10.1016/j.tws.2008.03.008.

[407] F. Ebrahimi, A. Rastgoo, Free vibration analysis of smart annular FGM plates integrated with piezoelectric layers, Smart Mater. Struct. 17 (1) (2008) 015044. https://doi.org/10.1088/0964-1726/17/1/015044.

[408] F. Ebrahimi, M.H. Naei, A. Rastgoo, Geometrically nonlinear vibration analysis of piezoelectrically actuated FGM plate with an initial large deformation, J. Mech. Sci. Technol. 23 (8) (2009) 2107–2124. https://doi.org/10.1007/s12206-009-0358-8.

[409] F. Ebrahimi, A. Rastgo, Vibration analysis of thin circular FGM plate coupled with piezoelectric layers, Amirkabir J. Mech. Eng. 41 (1) (2009) 1–8. https://doi.org/10.22060/mej.2009.251.

[410] F. Ebrahimi, A. Rastgoo, Nonlinear vibration of smart circular functionally graded plates coupled with piezoelectric layers, Int. J. Mech. Mater. Des. 5 (2) (2009) 157–165. https://doi.org/10.1007/s10999-008-9091-1.

[411] H. Sepiani, F. Ebrahimi, H. Karimipour, A mathematical model for smart functionally graded beam integrated with shape memory alloy actuators, J. Mech. Sci. Technol. 23 (12) (2009) 3179–3190. https://doi.org/10.1007/s12206-009-0919-x.

[412] F. Ebrahimi, A. Rastgoo, M.N. Bahrami, Investigating the thermal environment effects on geometrically nonlinear vibration of smart functionally graded plates, J. Mech. Sci. Technol. 24 (3) (2010) 775–791. https://doi.org/10.1007/s12206-010-0102-4.

[413] M. Aghelinejad, et al., Nonlinear thermomechanical post-buckling analysis of thin functionally graded annular plates based on Von-Karman's plate theory, Mech. Adv. Mater. Struct. 18 (5) (2011) 319–326. https://doi.org/10.1080/15376494.2010.516880.

[414] F. Ebrahimi, A. Rastgoo, Nonlinear vibration analysis of piezo-thermo-electrically actuated functionally graded circular plates, Arch. Appl. Mech. 81 (3) (2011) 361–383. https://doi.org/10.1007/s00419-010-0415-x.

[415] F. Ebrahimi, Analytical investigation on vibrations and dynamic response of functionally graded plate integrated with piezoelectric layers in thermal environment, Mech. Adv. Mater. Struct. 20 (10) (2013) 854–870. https://doi.org/10.1080/15376494.2012.677098.

[416] F. Ebrahimi, M. Boreiry, Investigating various surface effects on nonlocal vibrational behavior of nanobeams, Appl. Phys. A 121 (3) (2015) 1305–1316. https://doi.org/10.1007/s00339-015-9512-6.

[417] F. Ebrahimi, S. Dashti, Free vibration analysis of a rotating non-uniform functionally graded beam, Steel Compos. Struct. 19 (5) (2015) 1279–1298. https://doi.org/10.12989/scs.2015.19.5.1279.

[418] F. Ebrahimi, et al., Application of the differential transformation method for nonlocal vibration analysis of functionally graded nanobeams, J. Mech. Sci. Technol. 29 (3) (2015) 1207–1215. https://doi.org/10.1007/s12206-015-0234-7.

[419] F. Ebrahimi, E. Salari, Thermo-mechanical vibration analysis of nonlocal temperature-dependent FG nanobeams with various boundary conditions, Compos. Part B: Eng. 78 (2015) 272–290. https://doi.org/10.1016/j.compositesb.2015.03.068.

[420] F. Ebrahimi, E. Salari, Nonlocal thermo-mechanical vibration analysis of functionally graded nanobeams in thermal environment, Acta Astronaut. 113 (2015) 29–50. https://doi.org/10.1016/j.actaastro.2015.03.031.

[421] F. Ebrahimi, E. Salari, Size-dependent free flexural vibrational behavior of functionally graded nanobeams using semi-analytical differential transform method, Compos. Part B: Eng. 79 (2015) 156–169. https://doi.org/10.1016/j.compositesb.2015.04.010.

[422] F. Ebrahimi, E. Salari, A semi-analytical method for vibrational and buckling analysis of functionally graded nanobeams considering the physical neutral axis position, Comput. Model. Eng. Sci. 105 (2) (2015) 151–181. https://doi.org/10.3970/cmes.2015.105.151.

[423] F. Ebrahimi, E. Salari, S.A.H. Hosseini, Thermomechanical vibration behavior of FG nanobeams subjected to linear and non-linear temperature distributions, J. Thermal Stresses 38 (12) (2015) 1360–1386. https://doi.org/10.1080/01495739.2015.1073980.

[424] F. Ebrahimi, M.R. Barati, Magneto-electro-elastic buckling analysis of nonlocal curved nanobeams, Eur. Phys. J. Plus 131 (9) (2016) 346. https://doi.org/10.1140/epjp/i2016-16346-5.

[425] F. Ebrahimi, M.R. Barati, On nonlocal characteristics of curved inhomogeneous Euler–Bernoulli nanobeams under different temperature distributions, Appl. Phys. A 122 (10) (2016) 880. https://doi.org/10.1007/s00339-016-0399-7.

[426] F. Ebrahimi, M.R. Barati, Vibration analysis of viscoelastic inhomogeneous nanobeams incorporating surface and thermal effects, Appl. Phys. A 123 (1) (2016) 5. https://doi.sorg/10.1007/s00339-016-0511-z.

[427] F. Ebrahimi, M.R. Barati, Magnetic field effects on nonlocal wave dispersion characteristics of size-dependent nanobeams, Appl. Phys. A 123 (1) (2016) 81. https://doi.org/10.1007/s00339-016-0646-y.

[428] F. Ebrahimi, M.R. Barati, A. Dabbagh, Wave dispersion characteristics of axially loaded magneto-electro-elastic nanobeams, Appl. Phys. A 122 (11) (2016) 949. https://doi.org/10.1007/s00339-016-0465-1.

[429] F. Ebrahimi, M. Daman, Dynamic modeling of embedded curved nanobeams incorporating surface effects, Coupled Syst. Mech. 5 (3) (2016) 255–268. https://doi.org/10.12989/csm.2016.5.3.255.

[430] F. Ebrahimi, J. Ehyaei, R. Babaei, Thermal buckling of FGM nanoplates subjected to linear and nonlinear varying loads on Pasternak foundation, Adv. Mater. Res. 5 (4) (2016) 245–261. https://doi.org/10.12989/amr.2016.5.4.245.

[431] F. Ebrahimi, F. Ghasemi, E. Salari, Investigating thermal effects on vibration behavior of temperature-dependent compositionally graded Euler beams with porosities, Meccanica 51 (1) (2016) 223–249. https://doi.org/10.1007/s11012-015-0208-y.

[432] F. Ebrahimi, M. Hashemi, On vibration behavior of rotating functionally graded double-tapered beam with the effect of porosities, Proc. Inst. Mech. Eng. Part G J. Aerosp. Eng. 230 (10) (2016) 1903–1916. https://doi.org/10.1177/0954410015619647.

[433] F. Ebrahimi, S.H.S. Hosseini, Thermal effects on nonlinear vibration behavior of viscoelastic nanosize plates, J. Thermal Stresses 39 (5) (2016) 606–625. https://doi.org/10.1080/01495739.2016.1160684.

[434] F. Ebrahimi, S.H.S. Hosseini, Double nanoplate-based NEMS under hydrostatic and electrostatic actuations, Eur. Phys. J. Plus 131 (5) (2016) 160. https://doi.org/10.1140/epjp/i2016-16160-1.

[435] F. Ebrahimi, E. Salari, Analytical modeling of dynamic behavior of piezo-thermo-electrically affected sigmoid and power-law graded nanoscale beams, Appl. Phys. A 122 (9) (2016) 793. https://doi.org/10.1007/s00339-016-0273-7.

[436] F. Ebrahimi, E. Salari, S.A.H. Hosseini, In-plane thermal loading effects on vibrational characteristics of functionally graded nanobeams, Meccanica 51 (4) (2016) 951–977. https://doi.org/10.1007/s11012-015-0248-3.

[437] F. Ebrahimi, G.R. Shaghaghi, Thermal effects on nonlocal vibrational characteristics of nanobeams with non-ideal boundary conditions, Smart Struct. Syst. 18 (6) (2016) 1087–1109. https://doi.org/10.12989/sss.2016.18.6.1087.

[438] F. Ebrahimi, G.R. Shaghaghi, M. Boreiry, A semi-analytical evaluation of surface and nonlocal effects on buckling and vibrational characteristics of nanotubes with various boundary conditions, Int. J. Struct. Stab. Dyn. 16 (06) (2016) 1550023. https://doi.org/10.1142/S0219455415500236.

[439] F. Ebrahimi, G.R. Shaghaghi, M. Boreiry, An investigation into the influence of thermal loading and surface effects on mechanical characteristics of nanotubes, Struct. Eng. Mech. 57 (1) (2016) 179–200. https://doi.org/10.12989/sem.2016.57.1.179.

[440] F. Ebrahimy, S.H.S. Hosseini, Nonlinear electroelastic vibration analysis of NEMS consisting of double-viscoelastic nanoplates, Appl. Phys. A 122 (10) (2016) 922. https://doi.org/10.1007/s00339-016-0452-6.

[441] J. Ehyaei, F. Ebrahimi, E. Salari, Nonlocal vibration analysis of FG nano beams with different boundary conditions, Adv. Nano Res. 4 (2) (2016) 85–111. https://doi.org/10.12989/anr.2016.4.2.085.

[442] F. Ebrahimi, M.R. Barati, Surface effects on the vibration behavior of flexoelectric nanobeams based on nonlocal elasticity theory, Eur. Phys. J. Plus 132 (1) (2017) 19. https://doi.org/10.1140/epjp/i2017-11320-5.

[443] F. Ebrahimi, M.R. Barati, Small-scale effects on hygro-thermo-mechanical vibration of temperature-dependent nonhomogeneous nanoscale beams, Mech. Adv. Mater. Struct. 24 (11) (2017) 924–936. https://doi.org/10.1080/15376494.2016.1196795.

[444] F. Ebrahimi, M.R. Barati, Through-the-length temperature distribution effects on thermal vibration analysis of nonlocal strain-gradient axially graded nanobeams subjected to nonuniform magnetic field, J. Thermal Stresses 40 (5) (2017) 548–563. https://doi.org/10.1080/01495739.2016.1254076.

[445] F. Ebrahimi, M.R. Barati, Damping vibration analysis of smart piezoelectric polymeric nanoplates on viscoelastic substrate based on nonlocal strain gradient theory, Smart Mater. Struct. 26 (6) (2017) 065018. https://doi.org/10.1088/1361-665X/aa6eec.

[446] F. Ebrahimi, M.R. Barati, Flexural wave propagation analysis of embedded S-FGM nanobeams under longitudinal magnetic field based on nonlocal strain gradient theory, Arab. J. Sci. Eng. 42 (5) (2017) 1715–1726. https://doi.org/10.1007/s13369-016-2266-4.

[447] F. Ebrahimi, M.R. Barati, Vibration analysis of viscoelastic inhomogeneous nanobeams resting on a viscoelastic foundation based on nonlocal strain gradient theory incorporating surface and thermal effects, Acta Mech. 228 (3) (2017) 1197–1210. https://doi.org/10.1007/s00707-016-1755-6.

[448] F. Ebrahimi, M.R. Barati, Size-dependent dynamic modeling of inhomogeneous curved nanobeams embedded in elastic medium based on nonlocal strain gradient theory, Proc. Inst. Mech. Eng. Part C: J. Mech. Eng. Sci. 231 (23) (2017) 4457–4469. https://doi.org/10.1177/0954406216668912.

[449] F. Ebrahimi, M.R. Barati, Static stability analysis of embedded flexoelectric nanoplates considering surface effects, Appl. Phys. A 123 (10) (2017) 666. https://doi.org/10.1007/s00339-017-1265-y.

[450] F. Ebrahimi, M.R. Barati, Dynamic modeling of preloaded size-dependent nanocrystalline nano-structures, Appl. Math. Mech. 38 (12) (2017) 1753–1772. https://doi.org/10.1007/s10483-017-2291-8.

[451] F. Ebrahimi, M.R. Barati, Modeling of smart magnetically affected flexoelectric/piezoelectric nanostructures incorporating surface effects, Nanomater. Nanotechnol. 7 (2017) 1847980417713106. https://doi.org/10.1177/1847980417713106.

[452] F. Ebrahimi, M.R. Barati, P. Haghi, Thermal effects on wave propagation characteristics of rotating strain gradient temperature-dependent functionally graded nanoscale beams, J. Thermal Stresses 40 (5) (2017) 535–547. https://doi.org/10.1080/01495739.2016.1230483.

[453] F. Ebrahimi, A. Dabbagh, Wave propagation analysis of smart rotating porous heterogeneous piezo-electric nanobeams, Eur. Phys. J. Plus 132 (4) (2017) 153. https://doi.org/10.1140/epjp/i2017-11366-3.

[454] F. Ebrahimi, A. Dabbagh, Wave propagation analysis of embedded nanoplates based on a nonlocal strain gradient-based surface piezoelectricity theory, Eur. Phys. J. Plus 132 (11) (2017) 449. https://doi.org/10.1140/epjp/i2017-11694-2.

[455] F. Ebrahimi, M. Daman, Analytical investigation of the surface effects on nonlocal vibration behavior of nanosize curved beams, Adv. Nano Res. 5 (1) (2017) 35–47. https://doi.org/10.12989/anr.2017.5.1.035.

[456] F. Ebrahimi, M. Daman, R.E. Fardshad, Surface effects on vibration and buckling behavior of embedded nanoarches, Struct. Eng. Mech. 64 (1) (2017) 1–10. https://doi.org/10.12989/sem.2017.64.1.001.

[457] F. Ebrahimi, M. Hashemi, Vibration analysis of non-uniform imperfect functionally graded beams with porosities in thermal environment, J. Mech. 33 (6) (2017) 739–757. https://doi.org/10.1017/jmech.2017.81.

[458] F. Ebrahimi, S.H.S. Hosseini, Surface effects on nonlinear dynamics of NEMS consisting of double-layered viscoelastic nanoplates, Eur. Phys. J. Plus 132 (4) (2017) 172. https://doi.org/10.1140/epjp/i2017-11400-6.

[459] F. Ebrahimi, S.H.S. Hosseini, Effect of temperature on pull-in voltage and nonlinear vibration behavior of nanoplate-based NEMS under hydrostatic and electrostatic actuations, Acta Mech. Solida Sin. 30 (2) (2017) 174–189. https://doi.org/10.1016/j.camss.2017.02.001.

[460] F. Ebrahimi, et al., Thermo-mechanical vibration analysis of rotating nonlocal nanoplates applying generalized differential quadrature method, Mech. Adv. Mater. Struct. 24 (15) (2017) 1257–1273. https://doi.org/10.1080/15376494.2016.1227499.

[461] M. Dehghan, F. Ebrahimi, On wave dispersion characteristics of magneto-electro-elastic nanotubes considering the shell model based on the nonlocal strain gradient elasticity theory, Eur. Phys. J. Plus 133 (11) (2018) 466. https://doi.org/10.1140/epjp/i2018-12304-7.

[462] F. Ebrahimi, R. Babaei, G.R. Shaghaghi, Nonlocal buckling characteristics of heterogeneous plates subjected to various loadings, Adv. Aircraft Spacecraft Sci. 5 (5) (2018) 515–531. https://doi.org/10.12989/aas.2018.5.5.515.

[463] F. Ebrahimi, M.R. Barati, Vibration analysis of piezoelectrically actuated curved nanosize FG beams via a nonlocal strain-electric field gradient theory, Mech. Adv. Mater. Struct. 25 (4) (2018) 350–359. https://doi.org/10.1080/15376494.2016.1255830.

[464] F. Ebrahimi, M.R. Barati, Vibration analysis of size-dependent flexoelectric nanoplates incorporating surface and thermal effects, Mech. Adv. Mater. Struct. 25 (7) (2018) 611–621. https://doi.org/10.1080/15376494.2017.1285464.

[465] F. Ebrahimi, M.R. Barati, Buckling analysis of nonlocal strain gradient axially functionally graded nanobeams resting on variable elastic medium, Proc. Inst. Mech. Eng. Part C J. Mech. Eng. Sci. 232 (11) (2018) 2067–2078. https://doi.org/10.1177/0954406217713518.

[466] F. Ebrahimi, M.R. Barati, Scale-dependent effects on wave propagation in magnetically affected single/double-layered compositionally graded nanosize beams, Waves Random Complex Media 28 (2) (2018) 326–342. https://doi.org/10.1080/17455030.2017.1346331.

[467] F. Ebrahimi, M.R. Barati, Longitudinal varying elastic foundation effects on vibration behavior of axially graded nanobeams via nonlocal strain gradient elasticity theory, Mech. Adv. Mater. Struct. 25 (11) (2018) 953–963. https://doi.org/10.1080/15376494.2017.1329467.

[468] F. Ebrahimi, M.R. Barati, Free vibration analysis of couple stress rotating nanobeams with surface effect under in-plane axial magnetic field, J. Vib. Control 24 (21) (2018) 5097–5107. https://doi.org/10.1177/1077546317744719.

[469] F. Ebrahimi, M.R. Barati, Magnetic field effects on buckling characteristics of smart flexoelectrically actuated piezoelectric nanobeams based on nonlocal and surface elasticity theories, Microsyst. Technol. 24 (5) (2018) 2147–2157. https://doi.org/10.1007/s00542-017-3652-x.

[470] F. Ebrahimi, M.R. Barati, Nonlocal and surface effects on vibration behavior of axially loaded flexoelectric nanobeams subjected to in-plane magnetic field, Arab. J. Sci. Eng. 43 (3) (2018) 1423–1433. https://doi.org/10.1007/s13369-017-2943-y.

[471] F. Ebrahimi, M.R. Barati, Axial magnetic field effects on dynamic characteristics of embedded multiphase nanocrystalline nanobeams, Microsyst. Technol. 24 (8) (2018) 3521–3536. https://doi.org/10.1007/s00542-018-3771-z.

[472] F. Ebrahimi, M.R. Barati, Surface and flexoelectricity effects on size-dependent thermal stability analysis of smart piezoelectric nanoplates, Struct. Eng. Mech. 67 (2) (2018) 143–153. https://doi.org/10.12989/sem.2018.67.2.143.

[473] F. Ebrahimi, M.R. Barati, Wave propagation analysis of smart strain gradient piezo-magneto-elastic nonlocal beams, Struct. Eng. Mech. 66 (2) (2018) 237–248. https://doi.org/10.12989/sem.2018.66.2.237.

[474] F. Ebrahimi, M.R. Barati, Thermo-mechanical vibration analysis of nonlocal flexoelectric/piezoelectric beams incorporating surface effects, Struct. Eng. Mech. 65 (4) (2018) 435–445. https://doi.org/10.12989/sem.2018.65.4.435.

[475] F. Ebrahimi, M.R. Barati, P. Haghi, Wave propagation analysis of size-dependent rotating inhomogeneous nanobeams based on nonlocal elasticity theory, J. Vib. Control 24 (17) (2018) 3809–3818. https://doi.org/10.1177/1077546317711537.

[476] F. Ebrahimi, M. Boreiry, G.R. Shaghaghi, Nonlinear vibration analysis of electro-hygro-thermally actuated embedded nanobeams with various boundary conditions, Microsyst. Technol. 24 (12) (2018) 5037–5054. https://doi.org/10.1007/s00542-018-3924-0.

[477] F. Ebrahimi, A. Dabbagh, NSGT-based acoustical wave dispersion characteristics of thermo-magnetically actuated double-nanobeam systems, Struct. Eng. Mech. 68 (6) (2018) 701–711. https://doi.org/10.12989/sem.2018.68.6.701.

[478] F. Ebrahimi, A. Dabbagh, Wave dispersion characteristics of nonlocal strain gradient double-layered graphene sheets in hygro-thermal environments, Struct. Eng. Mech. 65 (6) (2018) 645–656. https://doi.org/10.12989/sem.2018.65.6.645.

[479] F. Ebrahimi, A. Dabbagh, Wave dispersion characteristics of embedded graphene platelets-reinforced composite microplates, Eur. Phys. J. Plus 133 (4) (2018) 151. https://doi.org/10.1140/epjp/i2018-11956-5.

[480] F. Ebrahimi, A. Dabbagh, On modeling wave dispersion characteristics of protein lipid nanotubules, J. Biomech. 77 (2018) 1–7. https://doi.org/10.1016/j.jbiomech.2018.05.038.

[481] F. Ebrahimi, A. Dabbagh, Thermo-magnetic field effects on the wave propagation behavior of smart magnetostrictive sandwich nanoplates, Eur. Phys. J. Plus 133 (3) (2018) 97. https://doi.org/10.1140/epjp/i2018-11910-7.

[482] F. Ebrahimi, A. Dabbagh, On wave dispersion characteristics of double-layered graphene sheets in thermal environments, J. Electromagn. Waves Applic. 32 (15) (2018) 1869–1888. https://doi.org/10.1080/09205071.2017.1417918.

[483] F. Ebrahimi, A. Dabbagh, Wave propagation analysis of magnetostrictive sandwich composite nanoplates via nonlocal strain gradient theory, Proc. Inst. Mech. Eng. Part C: J. Mech. Eng. Sci. 232 (22) (2018) 4180–4192. https://doi.org/10.1177/0954406217748687.

[484] F. Ebrahimi, A. Dabbagh, Wave dispersion characteristics of orthotropic double-nanoplate-system subjected to a longitudinal magnetic field, Microsyst. Technol. 24 (7) (2018) 2929–2939. https://doi.org/10.1007/s00542-018-3738-0.

[485] F. Ebrahimi, P. Haghi, Elastic wave dispersion modelling within rotating functionally graded nanobeams in thermal environment, Adv. Nano Res. 6 (3) (2018) 201–217. https://doi.org/10.12989/anr.2018.6.3.201.

[486] F. Ebrahimi, P. Haghi, Wave dispersion analysis of rotating heterogeneous nanobeams in thermal environment, Adv. Nano Res. 6 (1) (2018) 21–37. https://doi.org/10.12989/anr.2018.6.1.021.

[487] F. Ebrahimi, P. Haghi, A. Dabbagh, Analytical wave dispersion modeling in advanced piezoelectric double-layered nanobeam systems, Struct. Eng. Mech. 67 (2) (2018) 175–183. https://doi.org/10.12989/sem.2018.67.2.175.

[488] F. Ebrahimi, P. Haghi, A.M. Zenkour, Modelling of thermally affected elastic wave propagation within rotating Mori–Tanaka-based heterogeneous nanostructures, Microsyst. Technol. 24 (6) (2018) 2683–2693. https://doi.org/10.1007/s00542-018-3800-y.

[489] F. Ebrahimi, E. Heidari, Thermo-elastic analysis of rotating functionally graded microdiscs incorporating surface and nonlocal effects, Adv. Aircraft Spacecraft Sci. 5 (3) (2018) 295–318. https://doi.org/10.12989/aas.2018.5.3.295.

[490] F. Ebrahimi, M. Karimiasl, Nonlocal and surface effects on the buckling behavior of flexoelectric sandwich nanobeams, Mech. Adv. Mater. Struct. 25 (11) (2018) 943–952. https://doi.org/10.1080/15376494.2017.1329468.

[491] F. Ebrahimi, F. Mahmoodi, Vibration analysis of carbon nanotubes with multiple cracks in thermal environment, Adv. Nano Res. 6 (1) (2018) 57–80. https://doi.org/10.12989/anr.2018.6.1.057.

[492] F. Ebrahimi, H. Safarpour, Vibration analysis of inhomogeneous nonlocal beams via a modified couple stress theory incorporating surface effects, Wind Struct. 27 (6) (2018) 431–438. https://doi.org/10.12989/was.2018.27.6.431.

[493] M. Dehghan, F. Ebrahimi, M. Vinyas, Wave dispersion analysis of magnetic-electrically affected fluid-conveying nanotubes in thermal environment, Proc. Inst. Mech. Eng. Part C: J. Mech. Eng. Sci. 233 (19-20) (2019) 7116–7131. https://doi.org/10.1177/0954406219869752.

[494] F. Ebrahimi, M.R. Barati, Damping vibration behavior of viscoelastic porous nanocrystalline nanobeams incorporating nonlocal–couple stress and surface energy effects, Iran. J. Sci. Technol. Trans. Mech. Eng. 43 (2) (2019) 187–203. https://doi.org/10.1007/s40997-017-0127-8.

[495] F. Ebrahimi, M.R. Barati, Dynamic modeling of embedded nanoplate systems incorporating flexoelectricity and surface effects, Microsyst. Technol. 25 (1) (2019) 175–187. https://doi.org/10.1007/s00542-018-3946-7.

[496] F. Ebrahimi, M.R. Barati, V. Mahesh, Dynamic modeling of smart magneto-electro-elastic curved nanobeams, Adv. Nano Res. 7 (3) (2019) 145–155. https://doi.org/10.12989/anr.2019.7.3.145.

[497] F. Ebrahimi, M.R. Barati, F. Tornabene, Mechanics of nonlocal advanced magneto-electro-viscoelastic plates, Struct. Eng. Mech. 71 (3) (2019) 257–269. https://doi.org/10.12989/sem.2019.71.3.257.

[498] F. Ebrahimi, A. Dabbagh, Wave dispersion characteristics of heterogeneous nanoscale beams via a novel porosity-based homogenization scheme, Eur. Phys. J. Plus 134 (4) (2019) 157. https://doi.org/10.1140/epjp/i2019-12510-9.

[499] F. Ebrahimi, A. Dabbagh, A novel porosity-based homogenization scheme for propagation of waves in axially-excited FG nanobeams, Adv. Nano Res. 7 (6) (2019) 379–390. https://doi.org/10.12989/anr.2019.7.6.379.

[500] F. Ebrahimi, A. Dabbagh, Wave propagation responses of double-layered graphene sheets in hygrothermal environment, Handbook of Graphene, 8, Wiley, New Jersey, NJ, 2019, pp. 289–307. https://doi.org/10.1002/9781119468455.ch132.

[501] F. Ebrahimi, et al., Analysis of propagation characteristics of elastic waves in heterogeneous nanobeams employing a new two-step porosity-dependent homogenization scheme, Adv. Nano Res. 7 (2) (2019) 135–143. https://doi.org/10.12989/anr.2019.7.2.135.

[502] F. Ebrahimi, et al., Hygro-thermal effects on wave dispersion responses of magnetostrictive sandwich nanoplates, Adv. Nano Res. 7 (3) (2019) 157–167. https://doi.org/10.12989/anr.2019.7.3.157.

[503] F. Ebrahimi, M. Dehghan, A. Seyfi, Eringen's nonlocal elasticity theory for wave propagation analysis of magneto-electro-elastic nanotubes, Adv. Nano Res. 7 (1) (2019) 1–11. https://doi.org/10.12989/anr.2019.7.1.001.

[504] F. Ebrahimi, R.E. Fardshad, V. Mahesh, Frequency response analysis of curved embedded magneto-electro-viscoelastic functionally graded nanobeams, Adv. Nano Res. 7 (6) (2019) 391–403. https://doi.org/10.12989/anr.2019.7.6.391.

[505] F. Ebrahimi, S.H.S. Hosseini, Nonlinear vibration analysis of prestressed double layered nanoscale viscoelastic plates, Int. J. Acoust. Vibr. 24 (3) (2019) 394–407. https://doi.org/10.20855/ijav.2019.24.31115.

[506] F. Ebrahimi, S.H.S. Hosseini, S. Sedighi Bayrami, Nonlinear forced vibration of prestressed graphene sheets subjected to a mechanical shock: an analytical study, Thin-Walled Struct. 141 (2019) 293–307 https://doi.org/10.1016/j.tws.2019.04.038.

[507] F. Ebrahimi, et al., Surface effects on scale-dependent vibration behavior of flexoelectric sandwich nanobeams, Adv. Nano Res. 7 (2) (2019) 77–88. https://doi.org/10.12989/anr.2019.7.2.077.

[508] F. Ebrahimi, F. Zokaee, V. Mahesh, Analysis of the size-dependent wave propagation of a single lamellae based on the nonlocal strain gradient theory, Biomater. Biomed. Eng. 4 (1) (2019) 45–58. https://doi.org/10.12989/bme.2019.4.1.045.

[509] R. Selvamani, F. Ebrahimi, Axisymmetric vibration in a submerged piezoelectric rod coated with thin film, Applied Mathematics and Scientific Computing, Birkhäuser, Cham, Switzerland, 2019, pp. 203–211. https://doi.org/10.1007/978-3-030-01123-9_21.

[510] R. Asrari, F. Ebrahimi, M.M. Kheirikhah, On post-buckling characteristics of functionally graded smart magneto-electro-elastic nanoscale shells, Struct. Eng. Mech. 9 (1) (2020) 33–45. https://doi.org/10.12989/anr.2020.9.1.033.

[511] M. Dehghan, F. Ebrahimi, M. Vinyas, Wave dispersion characteristics of fluid-conveying magneto-electro-elastic nanotubes, Eng. Comput. 36 (4) (2020) 1687–1703. https://doi.org/10.1007/s00366-019-00790-5.

[512] F. Ebrahimi, M.R. Barati, Propagation of waves in nonlocal porous multi-phase nanocrystalline nanobeams under longitudinal magnetic field, Waves Random Complex Media 30 (2) (2020) 308–327. https://doi.org/10.1080/17455030.2018.1506596.

[513] F. Ebrahimi, S. Bayrami Sedighi, Wave propagation analysis of a rectangular sandwich composite plate with tunable magneto-rheological fluid core, J. Vib. Control 27 (11–12) (2021) 1231–1239. https://doi.org/10.1177/1077546320938189.

[514] F. Ebrahimi, S. Bayrami Sedighi, Wave dispersion characteristics of a rectangular sandwich composite plate with tunable magneto-rheological fluid core rested on a visco-Pasternak foundation, Mech. Based Des. Struct. Mach. (2020) 1–14. https://doi.org/10.1080/15397734.2020.1716244.

[515] F. Ebrahimi, A. Dabbagh, Viscoelastic wave propagation analysis of axially motivated double-layered graphene sheets via nonlocal strain gradient theory, Waves Random Complex Media 30 (1) (2020) 157–176. https://doi.org/10.1080/17455030.2018.1490505.

[516] F. Ebrahimi, S.H.S. Hosseini, Parametrically excited nonlinear dynamics and instability of double-walled nanobeams under thermo-magneto-mechanical loads, Microsyst. Technol. 26 (4) (2020) 1121–1132. https://doi.org/10.1007/s00542-019-04638-2.

[517] F. Ebrahimi, S.H.S. Hosseini, Double harmonically excited nonlinear vibration of viscoelastic piezoelectric nanoplates subjected to thermo-electro-mechanical forces, J. Vib. Control 26 (7-8) (2020) 430–446. https://doi.org/10.1177/1077546319889785.

[518] F. Ebrahimi, S.H.S. Hosseini, Nonlinear dynamic modeling of smart graphene/piezoelectric composite nanoplates subjected to dual frequency excitation, Eng. Res. Express 2 (2) (2020) 025019. https://doi.org/10.1088/2631-8695/ab3916.

[519] F. Ebrahimi, S.H.S. Hosseini, Effect of residual surface stress on parametrically excited nonlinear dynamics and instability of double-walled nanobeams: an analytical study, Eng. Comput. 37 (3) (2021) 1219–1230. https://doi.org/10.1007/s00366-019-00879-x.

[520] F. Ebrahimi, S.H.S. Hosseini, Nonlinear dynamics and stability of viscoelastic nanoplates considering residual surface stress and surface elasticity effects: a parametric excitation analysis, Eng. Comput. 37 (3) (2021) 1709–1722. https://doi.org/10.1007/s00366-019-00906-x.

[521] F. Ebrahimi, S.H.S. Hosseini, R. Selvamani, Thermo-electro-elastic nonlinear stability analysis of viscoelastic double-piezo nanoplates under magnetic field, Struct. Eng. Mech. 73 (5) (2020) 565–584. https://doi.org/10.12989/sem.2020.73.5.565.

[522] F. Ebrahimi, et al., Dynamic characteristics of hygro-magneto-thermo-electrical nanobeam with non-ideal boundary conditions, Adv. Nano Res. 8 (2) (2020) 169–182. https://doi.org/10.12989/anr.2020.8.2.169.

[523] F. Ebrahimi, A. Seyfi, Wave propagation analysis of smart inhomogeneous piezoelectric nanosize beams rested on an elastic medium, Waves Random Complex Media (2020) 1–20. https://doi.org/10.1080/17455030.2020.1817625.

[524] F. Ebrahimi, F. Zokaee, Modeling free vibration analysis of osteon as bone unite, Biomater. Biomech. Bioeng. 5 (1) (2020) 1–10. https://doi.org/10.12989/bme.2020.5.1.001.

[525] R. Selvamani, et al., Nonlinear magneto-thermo-elastic vibration of mass sensor armchair carbon nanotube resting on an elastic substrate, Curved Layered Struct. 7 (1) (2020) 153–165. https://doi.org/10.1515/cls-2020-0012.

[526] R. Selvamani, M.M.S. Jayan, F. Ebrahimi, Nonlinear ultrasonic waves in a magneto-flexo-thermally actuated single walled armchair carbon nanotube embedded on polymer matrix, World J. Eng. 18 (1) (2020) 1–13. https://doi.org/10.1108/WJE-02-2020-0066.

[527] A. Shariati, et al., Investigation of microstructure and surface effects on vibrational characteristics of nanobeams based on nonlocal couple stress theory, Adv. Nano Res. 8 (3) (2020) 191–202. http://dx.doi.org/10.12989/anr.2020.8.3.191.

[528] A. Shariati, et al., On the nonlinear dynamics of viscoelastic graphene sheets conveying nanoflow: parametric excitation analysis, Mech. Based Des. Struct. Mach. (2020) 1–18. https://doi.org/10.1080/15397734.2020.1728544.

[529] A. Shariati, et al., Nonlinear dynamics and vibration of reinforced piezoelectric scale-dependent plates as a class of nonlinear Mathieu–Hill systems: parametric excitation analysis, Eng. Comput. 37 (3) (2021) 2285–2301. https://doi.org/10.1007/s00366-020-00942-y.

[530] A. Shariati, et al., Effect of residual surface stress on parametrically excited nonlinear dynamics and instability of viscoelastic piezoelectric nanoelectromechanical resonators, Eng. Comput. 37 (3) (2021) 1835–1850. https://doi.org/10.1007/s00366-019-00916-9.

[531] A. Shariati, et al., Wave propagation analysis of electro-rheological fluid-filled sandwich composite beam, Mech. Based Des. Struct. Mach., (2020) 1–10. https://doi.org/10.1080/15397734.2020.1745646.

[532] F. Ebrahimi and A. Dabbagh, Magnetic field effects on thermally affected propagation of acoustical waves in rotary double-nanobeam systems. Waves Random Complex Media, 2021. **31**(1): p. 25-45. https://doi.org/10.1080/17455030.2018.1558308.

[533] F. Ebrahimi, A. Dabbagh, T. Rabczuk, On wave dispersion characteristics of magnetostrictive sandwich nanoplates in thermal environments, Eur. J. Mech. A Solids 85 (2021) 104130. https://doi.org/10.1016/j.euromechsol.2020.104130.

[534] F. Ebrahimi, S.H.S. Hosseini, Nonlinear vibration and dynamic instability analysis nanobeams under thermo-magneto-mechanical loads: a parametric excitation study, Eng. Comput. 37 (1) (2021) 395–408. https://doi.org/10.1007/s00366-019-00830-0.

[535] R. Selvamani, M.M.S. Jayan, F. Ebrahimi, Ultrasonic waves in a single walled armchair carbon nanotube resting on nonlinear foundation subjected to thermal and in plane magnetic fields, Coupled Syst. Mech. 10 (1) (2021) 39–60. https://doi.org/10.12989/csm.2021.10.1.039.

[536] R. Selvamani, M.M.S. Jayan, F. Ebrahimi, Thermomagnetic field effects on stability analysis of a single-walled fluid-conveying carbon nanotube rested on polymer matrix, Nanosci. Technol.: Int. J. 12 (2) (2021) 31–57. https://doi.org/10.1615/NanoSciTechnolIntJ.2021033910.

[537] A. Rastgoo, F. Ebrahimi, A.F. Dizaji, On the existence of periodic solution for equation of motion of thick beams having arbitrary cross section with tip mass under harmonic support motion, Int. J. Mech. Mater. Des. 3 (1) (2006) 29–38. https://doi.org/10.1007/s10999-006-9011-1.

[538] F. Ebrahimi, A. Rastgoo, On the application of Mindlin's plate theory to free vibration analysis of piezoelectric coupled circular FGM plate, in: ASME 2008 Pressure Vessels and Piping Conference, Chicago, IL, USA, 2008.

[539] F. Ebrahimi, A. Rastgoo, M.H. Kargarnovin, Analytical investigation on axisymmetric free vibrations of moderately thick circular functionally graded plate integrated with piezoelectric layers, J. Mech. Sci. Technol. 22 (6) (2008) 1058–1072. https://doi.org/10.1007/s12206-008-0303-2.

[540] F. Ebrahimi, A. Rastgoo, FSDPT based study for vibration analysis of piezoelectric coupled annular FGM plate, J. Mech. Sci. Technol. 23 (8) (2009) 2157–2168. https://doi.org/10.1007/s12206-009-0433-1.

[541] F. Ebrahimi, A. Rastgoo, A.A. Atai, A theoretical analysis of smart moderately thick shear deformable annular functionally graded plate, Eur. J. Mech. A Solids 28 (5) (2009) 962–973. https://doi.org/10.1016/j.euromechsol.2008.12.008.

[542] F. Ebrahimi, H. Sepiani, Transverse shear and rotary inertia effects on the stability analysis of functionally graded shells under combined static and periodic axial loadings, J. Mech. Sci. Technol. 24 (12) (2010) 2359–2366. https://doi.org/10.1007/s12206-010-0924-0.

[543] F. Ebrahimi, H.A. Sepiani, An investigation on the influence of transverse shear and rotary inertia on vibration and buckling of functionally graded cylindrical shells, Mech. Adv. Mater. Struct. 17 (3) (2010) 176–182. https://doi.org/10.1080/15376490903243845.

[544] F. Ebrahimi, H.A. Sepiani, Vibration and buckling analysis of cylindrical shells made of functionally graded materials under combined static and periodic axial forces, Adv. Compos. Lett. 19 (2) (2010) 096369351001900202, doi:10.1177/096369351001900202.

[545] H.A. Sepiani, et al., Vibration and buckling analysis of two-layered functionally graded cylindrical shell, considering the effects of transverse shear and rotary inertia, Mater. Des. 31 (3) (2010) 1063–1069. https://doi.org/10.1016/j.matdes.2009.09.052.

[546] F. Ebrahimi, M. Mokhtari, Transverse vibration analysis of rotating porous beam with functionally graded microstructure using the differential transform method, J. Braz. Soc. Mech. Sci. Eng. 37 (4) (2015) 1435–1444. https://doi.org/10.1007/s40430-014-0255-7.

[547] F. Ebrahimi, M. Mokhtari, Vibration analysis of spinning exponentially functionally graded Timoshenko beams based on differential transform method, Proc. Inst. Mech. Eng. Part G J. Aerosp. Eng. 229 (14) (2015) 2559–2571. https://doi.org/10.1177/0954410015580801.

[548] F. Ebrahimi, P. Nasirzadeh, A nonlocal Timoshenko beam theory for vibration analysis of thick nanobeams using differential transform method, J. Theor. Appl. Mech. 53 (4) (2015) 1041–1052. https://doi.org/10.15632/jtam-pl.53.4.1041.

[549] F. Ebrahimi, P. Nasirzadeh, Small-scale effects on transverse vibrational behavior of single-walled carbon nanotubes with arbitrary boundary conditions, Eng. Solid Mech. 3 (2) (2015) 131–144. https://doi.org/10.5267/j.esm.2015.1.002.

[550] F. Ebrahimi, E. Salari, Thermal buckling and free vibration analysis of size dependent Timoshenko FG nanobeams in thermal environments, Compos. Struct. 128 (2015) 363–380. https://doi.org/10.1016/j.compstruct.2015.03.023.

[551] F. Ebrahimi, E. Salari, Size-dependent thermo-electrical buckling analysis of functionally graded piezoelectric nanobeams, Smart Mater. Struct. 24 (12) (2015) 125007. https://doi.org/10.1088/0964-1726/24/12/125007.

[552] F. Ebrahimi, M. Zia, Large amplitude nonlinear vibration analysis of functionally graded Timoshenko beams with porosities, Acta Astronaut. 116 (2015) 117–125. https://doi.org/10.1016/j.actaastro.2015.06.014.

[553] F. Ebrahimi, M. Daman, Investigating surface effects on thermomechanical behavior of embedded circular curved nanosize beams, J. Eng. 2016 (2016) 9848343. https://doi.org/10.1155/2016/9848343.

[554] F. Ebrahimi, A. Jafari, Thermo-mechanical vibration analysis of temperature-dependent porous FG beams based on Timoshenko beam theory, Struct. Eng. Mech. 59 (2) (2016) 343–371. https://doi.org/10.12989/sem.2016.59.2.343.

[555] F. Ebrahimi, M. Mokhtari, Free vibration analysis of a rotating Mori–Tanaka-based functionally graded beam via differential transformation method, Arab. J. Sci. Eng. 41 (2) (2016) 577–590. https://doi.org/10.1007/s13369-015-1689-7.

[556] F. Ebrahimi, E. Salari, Effect of various thermal loadings on buckling and vibrational characteristics of nonlocal temperature-dependent functionally graded nanobeams, Mech. Adv. Mater. Struct. 23 (12) (2016) 1379–1397. https://doi.org/10.1080/15376494.2015.1091524.

[557] F. Ebrahimi, E. Salari, Thermal loading effects on electro-mechanical vibration behavior of piezoelectrically actuated inhomogeneous size-dependent Timoshenko nanobeams, Adv. Nano Res. 4 (3) (2016) 197–228. https://doi.org/10.12989/anr.2016.4.3.197.

[558] F. Ebrahimi, N. Shafiei, Application of Eringen's nonlocal elasticity theory for vibration analysis of rotating functionally graded nanobeams, Smart Struct. Syst. 17 (5) (2016) 837–857. https://doi.org/10.12989/sss.2016.17.5.837.

[559] F. Ebrahimi, M. Daman, Dynamic characteristics of curved inhomogeneous nonlocal porous beams in thermal environment, Struct. Eng. Mech. 64 (1) (2017) 121–133. https://doi.org/10.12989/sem.2017.64.1.121.

[560] F. Ebrahimi, M. Daman, Nonlocal thermo-electro-mechanical vibration analysis of smart curved FG piezoelectric Timoshenko nanobeam, Smart Struct. Syst. 20 (3) (2017) 351–368. https://doi.org/10.12989/sss.2017.20.3.351.

[561] F. Ebrahimi, E. Salari, Semi-analytical vibration analysis of functionally graded size-dependent nanobeams with various boundary conditions, Smart Struct. Syst. 19 (3) (2017) 243–257. https://doi.org/10.12989/sss.2017.19.3.243.

[562] F. Ebrahimi, R. Babaei, G.R. Shaghaghi, Vibration analysis thermally affected viscoelastic nanosensors subjected to linear varying loads, Adv. Nano Res. 6 (4) (2018) 399–422. https://doi.org/10.12989/anr.2018.6.4.399.

[563] F. Ebrahimi, R.E. Fardshad, Dynamic modeling of nonlocal compositionally graded temperature-dependent beams, Adv. Aircraft Spacecraft Sci. 5 (1) (2018) 141–164. https://doi.org/10.12989/aas.2018.5.1.141.

[564] F. Ebrahimi, R.E. Fardshad, Analytical solution for scale-dependent static stability analysis of temperature-dependent nanobeams subjected to uniform temperature distributions, Wind Struct. 26 (4) (2018) 205–214. https://doi.org/10.12989/was.2018.26.4.205.

[565] F. Ebrahimi, E. Salari, Effect of non-uniform temperature distributions on nonlocal vibration and buckling of inhomogeneous size-dependent beams, Adv. Nano Res. 6 (4) (2018) 377–397. https://doi.org/10.12989/anr.2018.6.4.377.

[566] F. Ebrahimi, A. Dabbagh, A. Rastgoo, Vibration analysis of porous metal foam shells rested on an elastic substrate, J. Strain Anal. Eng. Des. 54 (3) (2019) 199–208. https://doi.org/10.1177/0309324719852555.

[567] F. Ebrahimi, M. Daman, V. Mahesh, Thermo-mechanical vibration analysis of curved imperfect nano-beams based on nonlocal strain gradient theory, Adv. Nano Res. 7 (4) (2019) 249–263. https://doi.org/10.12989/anr.2019.7.4.249.

[568] F. Ebrahimi, M. Habibi, H. Safarpour, On modeling of wave propagation in a thermally affected GNP-reinforced imperfect nanocomposite shell, Eng. Comput. 35 (4) (2019) 1375–1389. https://doi.org/10.1007/s00366-018-0669-4.

[569] F. Ebrahimi, et al., Wave propagation analysis of a spinning porous graphene nanoplatelet-reinforced nanoshell, Waves Random Complex Media (2019) 1–27. 10.1080/17455030.2019.1694729.

[570] F. Ebrahimi, et al., Wave dispersion characteristics of porous graphene platelet-reinforced composite shells, Struct. Eng. Mech. 71 (1) (2019) 99–107. https://doi.org/10.12989/sem.2019.71.1.099.

[571] M.S.H. Al-Furjan, et al., A coupled thermomechanics approach for frequency information of electrically composite microshell using heat-transfer continuum problem, Eur. Phys. J. Plus 135 (10) (2020) 837. https://doi.org/10.1140/epjp/s13360-020-00764-3.

[572] M.S.H. Al-Furjan, et al., Wave dispersion characteristics of high-speed-rotating laminated nanocomposite cylindrical shells based on four continuum mechanics theories, Waves Random Complex Media (2020) 1–27. https://doi.org/10.1080/17455030.2020.1831099.

[573] R. Asrari, F. Ebrahimi, M.M. Kheirikhah, On scale-dependent stability analysis of functionally graded magneto-electro-thermo-elastic cylindrical nanoshells, Struct. Eng. Mech. 75 (6) (2020) 659–674. https://doi.org/10.12989/sem.2020.75.6.659.

[574] R. Asrari, et al., Buckling analysis of heterogeneous magneto-electro-thermo-elastic cylindrical nanoshells based on nonlocal strain gradient elasticity theory, Mech. Based Des. Struct. Mach. (2020) 1–24. https://doi.org/10.1080/15397734.2020.1728545.

[575] F. Ebrahimi, et al., Thermal buckling and forced vibration characteristics of a porous GNP reinforced nanocomposite cylindrical shell, Microsyst. Technol. 26 (2) (2020) 461–473. https://doi.org/10.1007/s00542-019-04542-9.

[576] F. Ebrahimi, S.H.S. Hosseini, Investigation of flexoelectric effect on nonlinear forced vibration of piezoelectric/functionally graded porous nanocomposite resting on viscoelastic foundation, J. Strain Anal. Eng. Des. 55 (1-2) (2020) 53–68. https://doi.org/10.1177/0309324719890868.

[577] F. Ebrahimi, S.H.S. Hosseini, Resonance analysis on nonlinear vibration of piezoelectric/FG porous nanocomposite subjected to moving load, Eur. Phys. J. Plus 135 (2) (2020) 215. https://doi.org/10.1140/epjp/s13360-019-00011-4.

[578] F. Ebrahimi, A. Seyfi, Propagation of flexural waves in anisotropic fluid-conveying cylindrical shells, Symmetry 12 (6) (2020) 901. https://doi.org/10.3390/sym12060901.

[579] F. Ebrahimi, A. Seyfi, Studying propagation of wave in metal foam cylindrical shells with graded porosities resting on variable elastic substrate, Eng. Comput. (2020). https://doi.org/10.1007/s00366-020-01069-w.

[580] K. Mercan, F. Ebrahimi, Ö. Civalek, Vibration of angle-ply laminated composite circular and annular plates, Steel Compos. Struct. 34 (1) (2020) 141–154. https://doi.org/10.12989/scs.2020.34.1.141.

[581] H. Moayedi, et al., Application of nonlocal strain–stress gradient theory and GDQEM for thermo-vibration responses of a laminated composite nanoshell, Eng. Comput. (2020) 1–16. https://doi.org/10.1007/s00366-020-01002-1.

[582] A. Shokrgozar, et al., Viscoelastic dynamics and static responses of a graphene nanoplatelets-reinforced composite cylindrical microshell, Mech. Based Des. Struct. Mach. (2020) 1–28. https://doi.org/10.1080/15397734.2020.1719509.

[583] E. Yarali, et al., Magnetorheological elastomer composites: modeling and dynamic finite element analysis, Compos. Struct. 254 (2020) 112881. https://doi.org/10.1016/j.compstruct.2020.112881.

[584] R. Zare, et al., Influence of imperfection on the smart control frequency characteristics of a cylindrical sensor-actuator GPLRC cylindrical shell using a proportional-derivative smart controller, Smart Struct. Syst. 26 (4) (2020) 469–480. https://doi.org/10.12989/sss.2020.26.4.469.

[585] M.S.H. Al-Furjan, et al., Enhancing vibration performance of a spinning smart nanocomposite reinforced microstructure conveying fluid flow, Eng. Comput. (2021) 1–16. https://doi.org/10.1007/s00366-020-01255-w.

[586] F. Ebrahimi, P. Hafezi, A. Dabbagh, Buckling analysis of embedded graphene oxide powder-reinforced nanocomposite shells, Defence Technol. 17 (1) (2021) 226–233. https://doi.org/10.1016/j.dt.2020.02.010.

[587] F. Ebrahimi, A. Seyfi, Wave propagation response of multi-scale hybrid nanocomposite shell by considering aggregation effect of CNTs, Mech. Based Des. Struct. Mach. 49 (1) (2021) 59–80. https://doi.org/10.1080/15397734.2019.1666722.

[588] F. Ebrahimi, E. Salari, Thermo-mechanical vibration analysis of a single-walled carbon nanotube embedded in an elastic medium based on higher-order shear deformation beam theory, J. Mech. Sci. Technol. 29 (9) (2015) 3797–3803. https://doi.org/10.1007/s12206-015-0826-2.

[589] M. Ghadiri, et al., Electro-thermo-mechanical vibration analysis of embedded single-walled boron nitride nanotubes based on nonlocal third-order beam theory, Int. J. Multiscale Comput. Eng. 13 (5) (2015) 443–461. https://doi.org/10.1615/IntJMultCompEng.2015013784.

[590] F. Ebrahimi, M.R. Barati, A nonlocal higher-order refined magneto-electro-viscoelastic beam model for dynamic analysis of smart nanostructures, Int. J. Eng. Sci. 107 (2016) 183–196. https://doi.org/10.1016/j.ijengsci.2016.08.001.

[591] F. Ebrahimi, M.R. Barati, Hygrothermal buckling analysis of magnetically actuated embedded higher order functionally graded nanoscale beams considering the neutral surface position, J. Thermal Stresses 39 (10) (2016) 1210–1229. https://doi.org/10.1080/01495739.2016.1215726.

[592] F. Ebrahimi, M.R. Barati, Static stability analysis of smart magneto-electro-elastic heterogeneous nanoplates embedded in an elastic medium based on a four-variable refined plate theory, Smart Mater. Struct. 25 (10) (2016) 105014. https://doi.org/10.1088/0964-1726/25/10/105014.

[593] F. Ebrahimi, M.R. Barati, A unified formulation for dynamic analysis of nonlocal heterogeneous nanobeams in hygro-thermal environment, Appl. Phys. A 122 (9) (2016) 792. https://doi.org/10.1007/s00339-016-0322-2.

[594] F. Ebrahimi, M.R. Barati, Dynamic modeling of smart shear-deformable heterogeneous piezoelectric nanobeams resting on Winkler–Pasternak foundation, Appl. Phys. A 122 (11) (2016) 952. https://doi.org/10.1007/s00339-016-0466-0.

[595] F. Ebrahimi, M.R. Barati, Magnetic field effects on buckling behavior of smart size-dependent graded nanoscale beams, Eur. Phys. J. Plus 131 (7) (2016) 238. https://doi.org/10.1140/epjp/i2016-16238-8.

[596] F. Ebrahimi, M.R. Barati, An exact solution for buckling analysis of embedded piezo-electro-magnetically actuated nanoscale beams, Adv. Nano Res. 4 (2) (2016) 65–84. https://doi.org/10.12989/anr.2016.4.2.065.

[597] F. Ebrahimi, M.R. Barati, Dynamic modeling of a thermo-piezo-electrically actuated nanosize beam subjected to a magnetic field, Appl. Phys. A 122 (4) (2016) 451. https://doi.org/10.1007/s00339-016-0001-3.

[598] F. Ebrahimi, M.R. Barati, Size-dependent thermal stability analysis of graded piezomagnetic nanoplates on elastic medium subjected to various thermal environments, Appl. Phys. A 122 (10) (2016) 910, doi:10.1007/s00339-016-0441-9.

[599] F. Ebrahimi, M.R. Barati, A nonlocal higher-order shear deformation beam theory for vibration analysis of size-dependent functionally graded nanobeams, Arab. J. Sci. Eng. 41 (5) (2016) 1679–1690. https://doi.org/10.1007/s13369-015-1930-4.

[600] F. Ebrahimi, M.R. Barati, Vibration analysis of nonlocal beams made of functionally graded material in thermal environment, Eur. Phys. J. Plus 131 (8) (2016) 279. https://doi.org/10.1140/epjp/i2016-16279-y.

[601] F. Ebrahimi, M.R. Barati, Wave propagation analysis of quasi-3D FG nanobeams in thermal environment based on nonlocal strain gradient theory, Appl. Phys. A 122 (9) (2016) 843. https://doi.org/10.1007/s00339-016-0368-1.

[602] F. Ebrahimi, M.R. Barati, An exact solution for buckling analysis of embedded piezoelectro-magnetically actuated nanoscale beams, Adv. Nano Res. 4 (2) (2016) 65–84. https://doi.org/10.12989/anr.2016.4.2.065.

[603] F. Ebrahimi, M.R. Barati, Electromechanical buckling behavior of smart piezoelectrically actuated higher-order size-dependent graded nanoscale beams in thermal environment, Int. J. Smart Nano Mater. 7 (2) (2016) 69–90. https://doi.org/10.1080/19475411.2016.1191556.

[604] F. Ebrahimi, M.R. Barati, Temperature distribution effects on buckling behavior of smart heterogeneous nanosize plates based on nonlocal four-variable refined plate theory, Int. J. Smart Nano Mater. 7 (3) (2016) 119–143. https://doi.org/10.1080/19475411.2016.1223203.

[605] F. Ebrahimi, M.R. Barati, Thermal environment effects on wave dispersion behavior of inhomogeneous strain gradient nanobeams based on higher order refined beam theory, J. Thermal Stresses 39 (12) (2016) 1560–1571. https://doi.org/10.1080/01495739.2016.1219243.

[606] F. Ebrahimi, M.R. Barati, Thermal buckling analysis of size-dependent FG nanobeams based on the third-order shear deformation beam theory, Acta Mech. Solida Sin. 29 (5) (2016) 547–554. https://doi.org/10.1016/S0894-9166(16)30272-5.

[607] F. Ebrahimi, M.R. Barati, Nonlocal thermal buckling analysis of embedded magneto-electro-thermo-elastic nonhomogeneous nanoplates, Iran. J. Sci. Technol. Trans. Mech. Eng. 40 (4) (2016) 243–264. https://doi.org/10.1007/s40997-016-0029-1.

[608] F. Ebrahimi, M.R. Barati, Analytical solution for nonlocal buckling characteristics of higher-order inhomogeneous nanosize beams embedded in elastic medium, Adv. Nano Res. 4 (3) (2016) 229–249. https://doi.org/10.12989/anr.2016.4.3.229.

[609] F. Ebrahimi, M.R. Barati, A. Dabbagh, A nonlocal strain gradient theory for wave propagation analysis in temperature-dependent inhomogeneous nanoplates, Int. J. Eng. Sci. 107 (2016) 169–182. https://doi.org/10.1016/j.ijengsci.2016.07.008.

[610] F. Ebrahimi, M.R. Barati, P. Haghi, Nonlocal thermo-elastic wave propagation in temperature-dependent embedded small-scaled nonhomogeneous beams, Eur. Phys. J. Plus 131 (11) (2016) 383. https://doi.org/10.1140/epjp/i2016-16383-0.

[611] F. Ebrahimi, A. Dabbagh, M.R. Barati, Wave propagation analysis of a size-dependent magneto-electro-elastic heterogeneous nanoplate, Eur. Phys. J. Plus 131 (12) (2016) 433. https://doi.org/10.1140/epjp/i2016-16433-7.

[612] F. Ebrahimi, S. Habibi, Deflection and vibration analysis of higher-order shear deformable compositionally graded porous plate, Steel Compos. Struct. 20 (1) (2016) 205–225. https://doi.org/10.12989/scs.2016.20.1.205.

[613] F. Ebrahimi, A. Jafari, A higher-order thermomechanical vibration analysis of temperature-dependent FGM beams with porosities, J. Eng. 2016 (2016) 9561504. https://doi.org/10.1155/2016/9561504.

[614] F. Ebrahimi, A. Jafari, Buckling behavior of smart MEE-FG porous plate with various boundary conditions based on refined theory, Adv. Mater. Res. 5 (4) (2016) 279–298. https://doi.org/10.12989/amr.2016.5.4.279.

[615] F. Ebrahimi, M.R. Barati, A nonlocal strain gradient refined beam model for buckling analysis of size-dependent shear-deformable curved FG nanobeams, Compos. Struct. 159 (2017) 174–182. https://doi.org/10.1016/j.compstruct.2016.09.058.

[616] F. Ebrahimi, M.R. Barati, Hygrothermal effects on vibration characteristics of viscoelastic FG nanobeams based on nonlocal strain gradient theory, Compos. Struct. 159 (2017) 433–444. https://doi.org/10.1016/j.compstruct.2016.09.092.

[617] F. Ebrahimi, M.R. Barati, Size-dependent vibration analysis of viscoelastic nanocrystalline silicon nanobeams with porosities based on a higher order refined beam theory, Compos. Struct. 166 (2017) 256–267. https://doi.org/10.1016/j.compstruct.2017.01.036.

[618] F. Ebrahimi, M.R. Barati, Dynamic modeling of magneto-electrically actuated compositionally graded nanosize plates lying on elastic foundation, Arab. J. Sci. Eng. 42 (5) (2017) 1977–1997. https://doi.org/10.1007/s13369-017-2413-6.

[619] F. Ebrahimi, M.R. Barati, Investigating physical field effects on the size-dependent dynamic behavior of inhomogeneous nanoscale plates, Eur. Phys. J. Plus 132 (2) (2017) 88. https://doi.org/10.1140/epjp/i2017-11357-4.

[620] F. Ebrahimi, M.R. Barati, A general higher-order nonlocal couple stress based beam model for vibration analysis of porous nanocrystalline nanobeams, Superlattices Microstruct. 112 (2017) 64–78. https://doi.org/10.1016/j.spmi.2017.09.010.

[621] F. Ebrahimi, M.R. Barati, Buckling analysis of piezoelectrically actuated smart nanoscale plates subjected to magnetic field, J. Intell. Mater. Syst. Struct. 28 (11) (2017) 1472–1490. https://doi.org/10.1177/1045389X16672569.

[622] F. Ebrahimi, M.R. Barati, Electro-magnetic effects on nonlocal dynamic behavior of embedded piezoelectric nanoscale beams, J. Intell. Mater. Syst. Struct. 28 (15) (2017) 2007–2022. https://doi.org/10.1177/1045389X16682850.

[623] F. Ebrahimi, M.R. Barati, Buckling analysis of nonlocal third-order shear deformable functionally graded piezoelectric nanobeams embedded in elastic medium, J. Braz. Soc. Mech. Sci. Eng. 39 (3) (2017) 937–952. https://doi.org/10.1007/s40430-016-0551-5.

[624] F. Ebrahimi, M.R. Barati, Buckling analysis of smart size-dependent higher order magneto-electro-thermo-elastic functionally graded nanosize beams, J. Mech. 33 (1) (2017) 23–33. https://doi.org/10.1017/jmech.2016.46.

[625] F. Ebrahimi, M.R. Barati, Magnetic field effects on dynamic behavior of inhomogeneous thermo-piezo-electrically actuated nanoplates, J. Braz. Soc. Mech. Sci. Eng. 39 (6) (2017) 2203–2223. https://doi.org/10.1007/s40430-016-0646-z.

[626] F. Ebrahimi, M.R. Barati, Porosity-dependent vibration analysis of piezo-magnetically actuated heterogeneous nanobeams, Mech. Syst. Sig. Process. 93 (2017) 445–459. https://doi.org/10.1016/j.ymssp.2017.02.021.

[627] F. Ebrahimi, M.R. Barati, Vibration analysis of embedded size dependent FG nanobeams based on third-order shear deformation beam theory, Struct. Eng. Mech. 61 (6) (2017) 721–736. https://doi.org/10.12989/sem.2017.61.6.721.

[628] F. Ebrahimi, M.R. Barati, A third-order parabolic shear deformation beam theory for nonlocal vibration analysis of magneto-electro-elastic nanobeams embedded in two-parameter elastic foundation, Adv. Nano Res. 5 (4) (2017) 313–336. https://doi.org/10.12989/anr.2017.5.4.313.

[629] F. Ebrahimi, M.R. Barati, Thermal-induced nonlocal vibration characteristics of heterogeneous beams, Adv. Mater. Res. 6 (2) (2017) 93–128. https://doi.org/10.12989/amr.2017.6.2.093.

[630] F. Ebrahimi, M.R. Barati, Vibration analysis of heterogeneous nonlocal beams in thermal environment, Coupled Syst. Mech. 6 (3) (2017) 251–272. https://doi.org/10.12989/csm.2017.6.3.251.

[631] F. Ebrahimi, M.R. Barati, A.M. Zenkour, Vibration analysis of smart embedded shear deformable nonhomogeneous piezoelectric nanoscale beams based on nonlocal

elasticity theory, Int. J. Aeronaut. Space Sci. 18 (2) (2017) 255–269. https://doi.org/10.5139/IJASS.2017.18.2.255.

[632] F. Ebrahimi, A. Dabbagh, Nonlocal strain gradient based wave dispersion behavior of smart rotating magneto-electro-elastic nanoplates, Mater. Res. Express 4 (2) (2017) 025003. https://doi.org/10.1088/2053-1591/aa55b5.

[633] F. Ebrahimi, A. Dabbagh, On flexural wave propagation responses of smart FG magneto-electro-elastic nanoplates via nonlocal strain gradient theory, Compos. Struct. 162 (2017) 281–293. https://doi.org/10.1016/j.compstruct.2016.11.058.

[634] F. Ebrahimi, M. Daman, A. Jafari, Nonlocal strain gradient-based vibration analysis of embedded curved porous piezoelectric nano-beams in thermal environment, Smart Struct. Syst. 20 (6) (2017) 709–728. https://doi.org/10.12989/sss.2017.20.6.709.

[635] F. Ebrahimi, P. Haghi, Wave propagation analysis of rotating thermoelastically-actuated nanobeams based on nonlocal strain gradient theory, Acta Mech. Solida Sin. 30 (6) (2017) 647–657. https://doi.org/10.1016/j.camss.2017.09.007.

[636] F. Ebrahimi, A. Jafari, Investigating vibration behavior of smart imperfect functionally graded beam subjected to magnetic-electric fields based on refined shear deformation theory, Adv. Nano Res. 5 (4) (2017) 281–301. https://doi.org/10.12989/anr.2017.5.4.281.

[637] F. Ebrahimi, A. Jafari, M.R. Barati, Dynamic modeling of porous heterogeneous micro/nanobeams, Eur. Phys. J. Plus 132 (12) (2017) 521. https://doi.org/10.1140/epjp/i2017-11754-7.

[638] F. Ebrahimi, A. Jafari, M.R. Barati, Vibration analysis of magneto-electro-elastic heterogeneous porous material plates resting on elastic foundations, Thin-Walled Struct. 119 (2017) 33–46. https://doi.org/10.1016/j.tws.2017.04.002.

[639] F. Ebrahimi, A. Jafari, M.R. Barati, Free vibration analysis of smart porous plates subjected to various physical fields considering neutral surface position, Arab. J. Sci. Eng. 42 (5) (2017) 1865–1881. https://doi.org/10.1007/s13369-016-2348-3.

[640] F. Ebrahimi, F. Mahmoodi, M.R. Barati, Thermo-mechanical vibration analysis of functionally graded micro/nanoscale beams with porosities based on modified couple stress theory, Adv. Mater. Res. 6 (3) (2017) 279–301. https://doi.org/10.12989/amr.2017.6.3.279.

[641] F. Ebrahimi, N. Shafiei, Influence of initial shear stress on the vibration behavior of single-layered graphene sheets embedded in an elastic medium based on Reddy's higher-order shear deformation plate theory, Mech. Adv. Mater. Struct. 24 (9) (2017) 761–772. https://doi.org/10.1080/15376494.2016.1196781.

[642] F. Ebrahimi, M.R. Barati, Damping vibration analysis of graphene sheets on viscoelastic medium incorporating hygro-thermal effects employing nonlocal strain gradient theory, Compos. Struct. 185 (2018) 241–253. https://doi.org/10.1016/j.compstruct.2017.10.021.

[643] F. Ebrahimi, M.R. Barati, A modified nonlocal couple stress-based beam model for vibration analysis of higher-order FG nanobeams, Mech. Adv. Mater. Struct. 25 (13) (2018) 1121–1132. https://doi.org/10.1080/15376494.2017.1365979.

[644] F. Ebrahimi, M.R. Barati, Effect of three-parameter viscoelastic medium on vibration behavior of temperature-dependent non-homogeneous viscoelastic nanobeams in a hygro-thermal environment, Mech. Adv. Mater. Struct. 25 (5) (2018) 361–374. https://doi.org/10.1080/15376494.2016.1255831.

[645] F. Ebrahimi, M.R. Barati, Vibration analysis of nonlocal strain gradient embedded single-layer graphene sheets under nonuniform in-plane loads, J. Vib. Control 24 (20) (2018) 4751–4763. https://doi.org/10.1177/1077546317734083.

[646] F. Ebrahimi, M.R. Barati, Influence of neutral surface position on dynamic characteristics of in-homogeneous piezo-magnetically actuated nanoscale plates, Proc. Inst. Mech. Eng. Part C: J. Mech. Eng. Sci. 232 (17) (2018) 3125–3143. https://doi.org/10.1177/0954406217728977.

[647] F. Ebrahimi, M.R. Barati, Static stability analysis of double-layer graphene sheet system in hygro-thermal environment, Microsyst. Technol. 24 (9) (2018) 3713–3727. https://doi.org/10.1007/s00542-018-3827-0.

[648] F. Ebrahimi, M.R. Barati, Vibration analysis of embedded biaxially loaded magneto-electrically actuated inhomogeneous nanoscale plates, J. Vib. Control 24 (16) (2018) 3587–3607. https://doi.org/10.1177/1077546317708105.

[649] F. Ebrahimi, M.R. Barati, Vibration analysis of graphene sheets resting on the orthotropic elastic medium subjected to hygro-thermal and in-plane magnetic fields based on the nonlocal strain gradient theory, Proc. Inst. Mech. Eng. Part C: J. Mech. Eng. Sci. 232 (13) (2018) 2469–2481. https://doi.org/10.1177/0954406217720232.

[650] F. Ebrahimi, M.R. Barati, Nonlocal strain gradient theory for damping vibration analysis of viscoelastic inhomogeneous nano-scale beams embedded in visco-Pasternak foundation, J. Vib. Control 24 (10) (2018) 2080–2095. https://doi.org/10.1177/1077546316678511.

[651] F. Ebrahimi, M.R. Barati, Damping vibration behavior of visco-elastically coupled double-layered graphene sheets based on nonlocal strain gradient theory, Microsyst. Technol. 24 (3) (2018) 1643–1658. https://doi.org/10.1007/s00542-017-3529-z.

[652] F. Ebrahimi, M.R. Barati, A nonlocal strain gradient refined plate model for thermal vibration analysis of embedded graphene sheets via DQM, Struct. Eng. Mech. 66 (6) (2018) 693–701. https://doi.org/10.12989/sem.2018.66.6.693.

[653] F. Ebrahimi, M.R. Barati, Vibration analysis of parabolic shear-deformable piezoelectrically actuated nanoscale beams incorporating thermal effects, Mech. Adv. Mater. Struct. 25 (11) (2018) 917–929. https://doi.org/10.1080/15376494.2017.1323141.

[654] F. Ebrahimi, M.R. Barati, Stability analysis of porous multi-phase nanocrystalline nonlocal beams based on a general higher-order couple-stress beam model, Struct. Eng. Mech. 65 (4) (2018) 465–476. https://doi.org/10.12989/sem.2018.65.4.465.

[655] F. Ebrahimi, M.R. Barati, Stability analysis of functionally graded heterogeneous piezoelectric nanobeams based on nonlocal elasticity theory, Adv. Nano Res. 6 (2) (2018) 93–112. https://doi.org/10.12989/anr.2018.6.2.093.

[656] F. Ebrahimi, M.R. Barati, Vibration analysis of smart piezoelectrically actuated nanobeams subjected to magneto-electrical field in thermal environment, J. Vib. Control 24 (3) (2018) 549–564. https://doi.org/10.1177/1077546316646239.

[657] F. Ebrahimi, M.R. Barati, Hygro-thermal vibration analysis of bilayer graphene sheet system via nonlocal strain gradient plate theory, J. Braz. Soc. Mech. Sci. Eng. 40 (9) (2018) 428. https://doi.org/10.1007/s40430-018-1350-y.

[658] F. Ebrahimi, M.R. Barati, A unified formulation for modeling of inhomogeneous nonlocal beams, Struct. Eng. Mech. 66 (3) (2018) 369–377. https://doi.org/10.12989/sem.2018.66.3.369.

[659] F. Ebrahimi, M.R. Barati, Size-dependent thermally affected wave propagation analysis in nonlocal strain gradient functionally graded nanoplates via a quasi-3D plate theory, Proc. Inst. Mech. Eng. Part C: J. Mech. Eng. Sci. 232 (1) (2018) 162–173. https://doi.org/10.1177/0954406216674243.

[660] F. Ebrahimi, M.R. Barati, A. Dabbagh, Wave propagation in embedded inhomogeneous nanoscale plates incorporating thermal effects, Waves Random Complex Media 28 (2) (2018) 215–235. https://doi.org/10.1080/17455030.2017.1337281.

[661] F. Ebrahimi, M.R. Barati, A.M. Zenkour, A new nonlocal elasticity theory with graded nonlocality for thermo-mechanical vibration of FG nanobeams via a nonlocal third-order shear deformation theory, Mech. Adv. Mater. Struct. 25 (6) (2018) 512–522. https://doi.org/10.1080/15376494.2017.1285458.

[662] F. Ebrahimi, A. Dabbagh, Wave dispersion characteristics of rotating heterogeneous magneto-electro-elastic nanobeams based on nonlocal strain gradient elasticity theory, J. Electromagn. Waves Applic. 32 (2) (2018) 138–169. https://doi.org/10.1080/09205071.2017.1369903.

[663] F. Ebrahimi, A. Dabbagh, Effect of humid-thermal environment on wave dispersion characteristics of single-layered graphene sheets, Appl. Phys. A 124 (4) (2018) 301. https://doi.org/10.1007/s00339-018-1734-y.

[664] F. Ebrahimi, R.E. Fardshad, Modeling the size effect on vibration characteristics of functionally graded piezoelectric nanobeams based on Reddy's shear deformation beam theory, Adv. Nano Res. 6 (2) (2018) 113–133. https://doi.org/10.12989/anr.2018.6.2.113.

[665] F. Ebrahimi, S. Habibi, Nonlinear dynamic response analysis of carbon fiber reinforced polymer enhanced with carbon nanotubes on elastic foundations in thermal environments, Amirkabir J. Mech. Eng. 50 (1) (2018) 73–90. https://doi.org/10.22060/mej.2016.774.

[666] F. Ebrahimi, P. Haghi, A nonlocal strain gradient theory for scale-dependent wave dispersion analysis of rotating nanobeams considering physical field effects, Coupled Syst. Mech. 7 (4) (2018) 373–393. https://doi.org/10.12989/csm.2018.7.4.373.

[667] F. Ebrahimi, E. Heidari, Vibration characteristics of advanced nanoplates in humid-thermal environment incorporating surface elasticity effects via differential quadrature method, Struct. Eng. Mech. 68 (1) (2018) 131–157. https://doi.org/10.12989/sem.2018.68.1.131.

[668] F. Ebrahimi, E. Heidari, Surface effects on nonlinear vibration and buckling analysis of embedded FG nanoplates via refined HOSDPT in hygrothermal environment considering physical neutral surface position, Adv. Aircraft Spacecraft Sci. 5 (6) (2018) 691–729. https://doi.org/10.12989/aas.2018.5.6.691.

[669] F. Ebrahimi, A. Jafari, A four-variable refined shear-deformation beam theory for thermo-mechanical vibration analysis of temperature-dependent FGM beams with porosities, Mech. Adv. Mater. Struct. 25 (3) (2018) 212–224. https://doi.org/10.1080/15376494.2016.1255820.

[670] F. Ebrahimi, M.R. Barati, Hygrothermal effects on static stability of embedded single-layer graphene sheets based on nonlocal strain gradient elasticity theory, J. Thermal Stresses 42 (12) (2019) 1535–1550. https://doi.org/10.1080/01495739.2019.1662352.

[671] F. Ebrahimi, M.R. Barati, Vibration analysis of biaxially compressed double-layered graphene sheets based on nonlocal strain gradient theory, Mech. Adv. Mater. Struct. 26 (10) (2019) 854–865. https://doi.org/10.1080/15376494.2018.1430267.

[672] F. Ebrahimi, M.R. Barati, A nonlocal strain gradient mass sensor based on vibrating hygro-thermally affected graphene nanosheets, Iran. J. Sci. Technol. Trans. Mech. Eng. 43 (2) (2019) 205–220. https://doi.org/10.1007/s40997-017-0131-z.

[673] F. Ebrahimi, M.R. Barati, On static stability of electro-magnetically affected smart magneto-electro-elastic nanoplates, Adv. Nano Res. 7 (1) (2019) 63–75. https://doi.org/10.12989/anr.2019.7.1.063.

[674] F. Ebrahimi, M.R. Barati, Buckling characteristics of bilayer graphene sheets subjected to humid thermomechanical loading, Handbook of Graphene, 8, Wiley, New Jersey, NJ, 2019, pp. 433–454. https://doi.org/10.1002/9781119468455.ch138.

[675] F. Ebrahimi, A. Dabbagh, Application of the nonlocal strain gradient elasticity on the wave dispersion behaviors of inhomogeneous nanosize beams, Eur. Phys. J. Plus 134 (3) (2019) 112. https://doi.org/10.1140/epjp/i2019-12464-x.

[676] F. Ebrahimi, A. Dabbagh, Thermo-mechanical wave dispersion analysis of nonlocal strain gradient single-layered graphene sheet rested on elastic medium, Microsyst. Technol. 25 (2) (2019) 587–597. https://doi.org/10.1007/s00542-018-3972-5.

[677] F. Ebrahimi, A. Dabbagh, Ö. Civalek, Vibration analysis of magnetically affected graphene oxide-reinforced nanocomposite beams, J. Vib. Control 25 (23-24) (2019) 2837–2849. https://doi.org/10.1177/1077546319861002.

[678] F. Ebrahimi, E. Heidari, Surface effects on nonlinear vibration of embedded functionally graded nanoplates via higher order shear deformation plate theory, Mech. Adv. Mater. Struct. 26 (8) (2019) 671–699. https://doi.org/10.1080/15376494.2017.1410908.

[679] F. Ebrahimi, A. Jafari, Thermo-mechanical vibration analysis of imperfect inhomogeneous beams based on a four-variable refined shear deformation beam theory considering neutral surface position, Int. J. Acoust. Vibr. 24 (3) (2019) 426–439. https://doi.org/10.20855/ijav.2019.24.31237.

[680] F. Ebrahimi, A. Jafari, V. Mahesh, Assessment of porosity influence on dynamic characteristics of smart heterogeneous magneto-electro-elastic plates, Struct. Eng. Mech. 72 (1) (2019) 113–129. https://doi.org/10.12989/sem.2019.72.1.113.

[681] F. Ebrahimi, M. Karimiasl, V. Mahesh, Vibration analysis of magneto-flexo-electrically actuated porous rotary nanobeams considering thermal effects via nonlocal strain gradient elasticity theory, Adv. Nano Res. 7 (4) (2019) 223–231. https://doi.org/10.12989/anr.2019.7.4.223.

[682] F. Ebrahimi, F. Mahmoodi, A modified couple stress theory for buckling analysis of higher order inhomogeneous microbeams with porosities, Proc. Inst. Mech. Eng. Part C: J. Mech. Eng. Sci. 233 (8) (2019) 2855–2866. https://doi.org/10.1177/0954406218791642.

[683] F. Ebrahimi, et al., Buckling analysis of graphene oxide powder-reinforced nanocomposite beams subjected to non-uniform magnetic field, Struct. Eng. Mech. 71 (4) (2019) 351–361. https://doi.org/10.12989/sem.2019.71.4.351.

[684] F. Ebrahimi, et al., Thermal buckling analysis of embedded graphene-oxide powder-reinforced nanocomposite plates, Adv. Nano Res. 7 (5) (2019) 293–310. https://doi.org/10.12989/anr.2019.7.5.293.

[685] F. Ebrahimi, S. Qaderi, Stability analysis of embedded graphene platelets reinforced composite plates in thermal environment, Eur. Phys. J. Plus 134 (7) (2019) 349. https://doi.org/10.1140/epjp/i2019-12581-6.

[686] F. Ebrahimi, A. Seyfi, A. Dabbagh, A novel porosity-dependent homogenization procedure for wave dispersion in nonlocal strain gradient inhomogeneous nanobeams, Eur. Phys. J. Plus 134 (5) (2019) 226. https://doi.org/10.1140/epjp/i2019-12547-8.

[687] F. Ebrahimi, A. Seyfi, A. Dabbagh, Dispersion of waves in FG porous nanoscale plates based on NSGT in thermal environment, Adv. Nano Res. 7 (5) (2019) 325–335. https://doi.org/10.12989/anr.2019.7.5.325.

[688] R.E. Fardshad, Y. Mohammadi, F. Ebrahimi, Modeling wave propagation in graphene sheets influenced by magnetic field via a refined trigonometric two-variable plate theory, Struct. Eng. Mech. 72 (3) (2019) 329–338. https://doi.org/10.12989/sem.2019.72.3.329.

[689] M. Karimiasl, F. Ebrahimi, V. Mahesh, Nonlinear forced vibration of smart multiscale sandwich composite doubly curved porous shell, Thin-Walled Struct. 143 (2019) 106152. https://doi.org/10.1016/j.tws.2019.04.044.

[690] M. Karimiasl, K. Kargarfard, F. Ebrahimi, Buckling of magneto-electro-hygro-thermal piezoelectric nanoplates system embedded in a visco-Pasternak medium based on nonlocal theory, Microsyst. Technol. 25 (3) (2019) 1031–1042. https://doi.org/10.1007/s00542-018-4082-0.

[691] S. Qaderi, F. Ebrahimi, V. Mahesh, Free vibration analysis of graphene platelets-reinforced composites plates in thermal environment based on higher-order shear deformation plate theory, Int. J. Aeronaut. Space Sci. 20 (4) (2019) 902–912. https://doi.org/10.1007/s42405-019-00184-3.

[692] S. Qaderi, F. Ebrahimi, A. Seyfi, An investigation of the vibration of multi-layer composite beams reinforced by graphene platelets resting on two parameter viscoelastic foundation, SN Appl. Sci. 1 (5) (2019) 399. https://doi.org/10.1007/s42452-019-0252-7.

[693] S. Qaderi, F. Ebrahimi, M. Vinyas, Dynamic analysis of multi-layered composite beams reinforced with graphene platelets resting on two-parameter viscoelastic foundation, Eur. Phys. J. Plus 134 (7) (2019) 339. https://doi.org/10.1140/epjp/i2019-12739-2.

[694] M. Vinyas, et al., Numerical analysis of the vibration response of skew magneto-electro-elastic plates based on the higher-order shear deformation theory, Compos. Struct. 214 (2019) 132–142. https://doi.org/10.1016/j.compstruct.2019.02.010.

[695] M. Vinyas, et al., A finite element-based assessment of free vibration behaviour of circular and annular magneto-electro-elastic plates using higher order shear deformation theory, J. Intell. Mater. Syst. Struct. 30 (16) (2019) 2478–2501. https://doi.org/10.1177/1045389X19862386.

[696] M.A. Amani, et al., A machine learning-based model for the estimation of the temperature-dependent moduli of graphene oxide reinforced nanocomposites and its application in a thermally affected buckling analysis, Eng. Comput. 37 (3) (2021) 2245–2255. https://doi.org/10.1007/s00366-020-00945-9.

[697] F. Ebrahimi, M.R. Barati, Ö. Civalek, Application of Chebyshev–Ritz method for static stability and vibration analysis of nonlocal microstructure-dependent nanostructures, Eng. Comput. 36 (3) (2020) 953–964. https://doi.org/10.1007/s00366-019-00742-z.

[698] F. Ebrahimi, A. Dabbagh, M. Taheri, Vibration analysis of porous metal foam plates rested on viscoelastic substrate, Eng. Comput. (2020). https://doi.org/10.1007/s00366-020-01031-w.

[699] F. Ebrahimi, A. Jafari, R. Selvamani, Thermal buckling analysis of magneto-electro-elastic porous FG beam in thermal environment, Adv. Nano Res. 8 (1) (2020) 83–94. https://doi.org/10.12989/anr.2020.8.1.083.

[700] F. Ebrahimi, M. Karimiasl, R. Selvamani, Bending analysis of magneto-electro piezoelectric nanobeams system under hygro-thermal loading, Adv. Nano Res. 8 (3) (2020) 203–214. https://doi.org/10.12989/anr.2020.8.3.203.

[701] F. Ebrahimi, M. Nouraei, A. Dabbagh, Thermal vibration analysis of embedded graphene oxide powder-reinforced nanocomposite plates, Eng. Comput. 36 (3) (2020) 879–895. https://doi.org/10.1007/s00366-019-00737-w.

[702] F. Ebrahimi, M. Nouraei, A. Dabbagh, Modeling vibration behavior of embedded graphene-oxide powder-reinforced nanocomposite plates in thermal environment, Mech. Based Des. Struct. Mach. 48 (2) (2020) 217–240. https://doi.org/10.1080/15397734.2019.1660185.

[703] F. Ebrahimi, M. Nouraei, A. Seyfi, Wave dispersion characteristics of thermally excited graphene oxide powder-reinforced nanocomposite plates, Waves Random Complex Media (2020) 1–29. https://doi.org/10.1080/17455030.2020.1767829.

[704] F. Ebrahimi, A. Seyfi, Studying propagation of wave of metal foam rectangular plates with graded porosities resting on Kerr substrate in thermal environment via analytical method, Waves Random Complex Media (2020) 1–24. https://doi.org/10.1080/17455030.2020.1802531.

[705] S. Qaderi, F. Ebrahimi, Vibration analysis of polymer composite plates reinforced with graphene platelets resting on two-parameter viscoelastic foundation, Eng. Comput. (2020) 1–17. https://doi.org/10.1007/s00366-020-01066-z.

[706] M. Safarpour, et al., Frequency characteristics of FG-GPLRC viscoelastic thick annular plate with the aid of GDQM, Thin-Walled Struct. 150 (2020) 106683. https://doi.org/10.1016/j.tws.2020.106683.

[707] R. Selvamani, M. Mahaveer Sreejayan, F. Ebrahimi, Static stability analysis of mass sensors consisting of hygro-thermally activated graphene sheets using a nonlocal strain gradient theory, Eng. Trans. 68 (3) (2020) 269–295. https://doi.org/10.24423/engtrans.1187.20200904.

[708] E. Shamsaddini Lori, et al., The critical voltage of a GPL-reinforced composite microdisk covered with piezoelectric layer, Eng. Comput. (2020). https://doi.org/10.1007/s00366-020-01004-z.

[709] E. Shamsaddini Lori, et al., Frequency characteristics of a GPL-reinforced composite microdisk coupled with a piezoelectric layer, Eur. Phys. J. Plus 135 (2) (2020) 144. https://doi.org/10.1140/epjp/s13360-020-00217-x.

[710] A. Shariati, et al., Investigating vibrational behavior of graphene sheets under linearly varying in-plane bending load based on the nonlocal strain gradient theory, Adv. Nano Res. 8 (4) (2020) 265–276. https://doi.org/10.12989/anr.2020.8.4.265.

[711] A. Shariati, et al., On bending characteristics of smart magneto-electro-piezoelectric nanobeams system, Adv. Nano Res. 9 (3) (2020) 183–191. https://doi.org/10.12989/anr.2020.9.3.183.

[712] A. Shariati, et al., On transient hygrothermal vibration of embedded viscoelastic flexoelectric/piezoelectric nanobeams under magnetic loading, Adv. Nano Res. 8 (1) (2020) 49–58. https://doi.org/10.12989/anr.2020.8.1.049.

[713] A. Shariati, et al., On buckling characteristics of polymer composite plates reinforced with graphene platelets, Eng. Comput. (2020) 1–12. https://doi.org/10.1007/s00366-020-00992-2.

[714] F. Ebrahimi, et al., Vibration analysis of porous magneto-electro-elastically actuated carbon nanotube-reinforced composite sandwich plate based on a refined plate theory, Eng. Comput. 37 (2) (2021) 921–936. https://doi.org/10.1007/s00366-019-00864-4.

[715] F. Ebrahimi, M. Karimiasl, V. Mahesh, Chaotic dynamics and forced harmonic vibration analysis of magneto-electro-viscoelastic multiscale composite nanobeam, Eng. Comput. 37 (2) (2021) 937–950. https://doi.org/10.1007/s00366-019-00865-3.

[716] F. Ebrahimi, M. Karimiasl, A. Singhal, Magneto-electro-elastic analysis of piezoelectric–flexoelectric nanobeams rested on silica aerogel foundation, Eng. Comput. 37 (2) (2021) 1007–1014. https://doi.org/10.1007/s00366-019-00869-z.

[717] F. Ebrahimi, R. Nopour, A. Dabbagh, Effect of viscoelastic properties of polymer and wavy shape of the CNTs on the vibrational behaviors of CNT/glass fiber/polymer plates, Eng. Comput. (2021) 1–14. https://doi.org/10.1007/s00366-021-01387-7.

[718] F. Ebrahimi, A. Seyfi, Wave dispersion analysis of embedded MWCNTs-reinforced nanocomposite beams by considering waviness and agglomeration factors, Waves Random Complex Media (2021) 1–20. https://doi.org/10.1080/17455030.2021.1883148.

[719] F. Ebrahimi, A. Seyfi, A wave propagation study for porous metal foam beams resting on an elastic foundation, Waves Random Complex Media (2021) 1–15. https://doi.org/10.1080/17455030.2021.1905909.

[720] F. Ebrahimi, et al., Influence of magnetic field on the wave propagation response of functionally graded (FG) beam lying on elastic foundation in thermal environment, Waves Random Complex Media (2021) 1–19. https://doi.org/10.1080/17455030.2020.1847359.

[721] M. Karimiasl, F. Ebrahimi, V. Mahesh, Postbuckling analysis of piezoelectric multiscale sandwich composite doubly curved porous shallow shells via homotopy perturbation method, Eng. Comput. 37 (1) (2021) 561–577. https://doi.org/10.1007/s00366-019-00841-x.

[722] A.H. Nayfeh, D.T. Mook, Nonlinear Oscillations, Wiley, New York, NY, 1995. https://doi.org/10.1002/9783527617586.

[723] S.H. Strogatz, Nonlinear Dynamics and Chaos, second ed., CRC Press, Boca Raton, FL, 2015. https://doi.org/10.1201/9780429492563.

[724] F. Ebrahimi, A. Dabbagh, Wave Propagation Analysis of Smart Nanostructures, first ed., CRC Press, Boca Raton, FL, 2019. https://doi.org/10.1201/9780429279225.

[725] N.S. Nise, Control Systems Engineering, eighth ed., Wiley, New Jersey, NJ, 2019.

[726] F.Z. Zaoui, D. Ouinas, A. Tounsi, New 2D and quasi-3D shear deformation theories for free vibration of functionally graded plates on elastic foundations, Compos. Part B: Eng. 159 (2019) 231–247. https://doi.org/10.1016/j.compositesb.2018.09.051.

[727] H.-T. Thai, T.P. Vo, Bending and free vibration of functionally graded beams using various higher-order shear deformation beam theories, Int. J. Mech. Sci. 62 (1) (2012) 57–66. https://doi.org/10.1016/j.ijmecsci.2012.05.014.

[728] M. Mohammadi, A. Rastgoo, Primary and secondary resonance analysis of FG/lipid nanoplate with considering porosity distribution based on a nonlinear elastic medium, Mech. Adv. Mater. Struct. 27 (20) (2020) 1709–1730. https://doi.org/10.1080/15376494.2018.1525453.

[729] H.-T. Thai, D.-H. Choi, Analytical solutions of refined plate theory for bending, buckling and vibration analyses of thick plates, Appl. Math. Modell. 37 (18) (2013) 8310–8323. https://doi.org/10.1016/j.apm.2013.03.038.

[730] H. Zeighampour, Y.T. Beni, Analysis of conical shells in the framework of coupled stresses theory, Int. J. Eng. Sci. 81 (2014) 107–122. https://doi.org/10.1016/j.ijengsci.2014.04.008.

[731] X.-F. Li, B.-L. Wang, J.-C. Han, A higher-order theory for static and dynamic analyses of functionally graded beams, Arch. Appl. Mech. 80 (10) (2010) 1197–1212. https://doi.org/10.1007/s00419-010-0435-6.

[732] N.R. Raravikar, et al., Temperature dependence of radial breathing mode Raman frequency of single-walled carbon nanotubes, Phys. Rev. B 66 (23) (2002) 235424. https://doi.org/10.1103/PhysRevB.66.235424.

[733] P.K. Schelling, P. Keblinski, Thermal expansion of carbon structures, Phys. Rev. B 68 (3) (2003) 035425. https://doi.org/10.1103/PhysRevB.68.035425.

[734] J. Yang, H. Wu, S. Kitipornchai, Buckling and postbuckling of functionally graded multilayer graphene platelet-reinforced composite beams, Compos. Struct. 161 (2017) 111–118. https://doi.org/10.1016/j.compstruct.2016.11.048.

[735] Y. Wang, et al., Eigenvalue buckling of functionally graded cylindrical shells reinforced with graphene platelets (GPL), Compos. Struct. 202 (2018) 38–46. https://doi.org/10.1016/j.compstruct.2017.10.005.

[736] N. Wattanasakulpong, A. Chaikittiratana, Exact solutions for static and dynamic analyses of carbon nanotube-reinforced composite plates with Pasternak elastic foundation, Appl. Math. Modell. 39 (18) (2015) 5459–5472. https://doi.org/10.1016/j.apm.2014.12.058.

[737] R. Ansari, R. Gholami, Nonlinear primary resonance of third-order shear deformable functionally graded nanocomposite rectangular plates reinforced by carbon nanotubes, Compos. Struct. 154 (2016) 707–723. https://doi.org/10.1016/j.compstruct.2016.07.023.

[738] Y. Heydarpour, M.M. Aghdam, P. Malekzadeh, Free vibration analysis of rotating functionally graded carbon nanotube-reinforced composite truncated conical shells, Compos. Struct. 117 (2014) 187–200. https://doi.org/10.1016/j.compstruct.2014.06.023.

[739] R. Ansari, E. Hasrati, J. Torabi, Nonlinear vibration response of higher-order shear deformable FG-CNTRC conical shells, Compos. Struct. 222 (2019) 110906. https://doi.org/10.1016/j.compstruct.2019.110906.

[740] Y.-G. Hu, et al., Nonlocal shell model for elastic wave propagation in single- and double-walled carbon nanotubes, J. Mech. Phys. Solids 56 (12) (2008) 3475–3485. https://doi.org/10.1016/j.jmps.2008.08.010.

Index

Page numbers followed by "*f*" and "*t*" indicate, figures and tables respectively.

A

Agglomeration
　parameters, 234, 283, 300, 303, 97
　phenomenon, 312, 27
　and waviness phenomena, 297
Amplitude–response curves, 275, 301
Automotive industry, composite materials in, 3
　automobile devices, design of, 4
　carbon fiber (CF), 3
　CF-reinforced polymer (CFRP), 3
　FRPs, energy absorption mechanism in, 4
　FRPs, stress–strain behavior of, 4
　glass fibers (GFs), 4
　green composites terminology, 6
　metal-based composites, 6
　metal-matrix composites, 6
　thermoplastics, 4

B

Beam's free vibrations, 231
Biomedical applications/composites in, 26
　anterior cruciate ligament/ ACL, 27
　correction of refractive errors of the human eyes, 26
　ionic polymeric metal composites, 26
Bode and Root Locus diagrams, 278
Bode diagram, 249, 256, 285
Bode stability criterion, 126
Boundary condition (BC), 6, 111
Buckling, of shells, 17
Buckling analysis, of multiscale hybrid polymer nanocomposite materials, 3
Buckling load matrix, 132

C

Carbon nanotube (CNT)
　curvy shape of, 90
　mass fraction of, 82
　reinforced polymer nanocomposite, 81
　　composition of, 84*f*
　　dimensionless buckling load of, 70
　　homogenization of, 81
　　3D Mori–Tanaka method, 101
　　Eshelby–Mori–Tanaka method, 94
　　Halpin–Tsai method, 85
　　modified Halpin–Tsai method, 87
　　Mori–Tanaka method, 92
　　rule of the mixture, 82
　　material properties of, 83
　　Poisson's ratio, 82, 85
　　Young's and shear modules values, 82
　　Young's modulus of, 100*f*, 85
　　volume fraction of, 87*f*, 82, 85
Carbon nanotube (CNT)-reinforced nanocomposites, 29, 29
　agglomeration, 31
　double-walled, 30
　functionalized, 31
　local deformations, 32
　multi-walled, 30
　nonstraight, 32
　single-walled, 30
　thermal conductivity, 33
　thermo-electro-mechanical properties, 33
　waviness phenomenon, 32
Chebyshev polynomials, 165
Classical
　beam theory, 108
　shell theory, 161
Coefficient of thermal expansions (CTE), 12

D

Detuning parameter, 118
Dilute tensor, 92
3D Mori–Tanaka method, 6, 101
　in-plane shear modulus G_{12}, 101
　longitudinal Young's modulus E_{11}, 101
　out-of-plane shear modulus G_{23}, 101
　plane-strain bulk modulus K_{23}, 101
　transverse Young's modulus E_{22}, 101
Duffing equation, 142

E

Eigen function, 131
Elastic constants of Hill, 93

359

Electric arc discharge, 37, 37
Eshelby–Mori–Tanaka method, 94
Eshelby tensor, 92
Euler–Bernoulli beam
  hypothesis, 247
  theory
    nonlinear forced vibration problem, 112
    postbuckling problem, 109
    wave propagation problem, 121
Euler–Bernoulli beam theory, 12, 108, 108
Euler–Lagrange equations, of beam, 128

F
Fiber-reinforced composites, 2
First-order shear deformation shell theory (FSDT), 167, 107
  buckling problem, 168
  nonlinear forced vibration problem, 175 187–175 187
  wave propagation problem, 172
Free vibration analysis, 262, 290
Free vibration problem, 132, 132
Frequency–response curve, 276
Fusion applications, composites in, 22
  artificial composite muscles, 26
  chemical vapor infiltration method, 23
  CMCs of silicon carbide-matrix/silicon carbide-fiber, 22
  graphite fiber, 22
  SiC fibers synthetized with chemical vapor deposition, 23

G
Galerkin's method, 131, 135, 160, 153
Glass fiber-reinforced (GFR) composites, 4
Gradient index, 243, 249, 264
Graphene, lattice constant of, 86
Grobman theorem, 239

H
Halpin–Tsai method, 85
  modified, 87
Hamilton's principle, 107, 144
Hartman–Grobman theorem, 269, 272, 121, 185
Higher-order shear deformation theory (HSDT), 231, 107, 126
Hill's elastic constants, 93
Homogenization, of CNTR PNC, 81
  3D Mori–Tanaka method, 101
  Eshelby–Mori–Tanaka method, 94
  Halpin–Tsai method, 85

  modified Halpin–Tsai method, 87
  Mori–Tanaka method, 92
  rule of the mixture, 82
Hook's law, 122, 164

J
Jacobi matrix, 239, 120, 185

K
Kinematic relations, 107
  of beams, 108
    bending problem, 128
    buckling problem, 130
    Euler–Bernoulli beam theory, 108
    free vibration problem, 132
    nonlinear forced vibration problem, 112
    postbuckling problem, 109
    refined higher-order beam theory, 126
    wave propagation problem, 121
  of plates, 136
    buckling problem, 152
    free vibration problem, 155
    Kirchhoff–Love plate theory, 136
    nonlinear forced vibration problem, 137
    refined higher-order plate theory, 145
    wave propagation problem, 142
  of shells, 161
    buckling problem, 168
    classical shell theory, 161
    first-order shear deformation shell theory, 167
    nonlinear forced vibration problem, 175 187–175 187
    wave propagation problem, 172
Kirchhoff–Love plate theory, 136
  nonlinear forced vibration problem, 137
  wave propagation problem, 142
Kirchhoff's circuit law, 125

L
Lame's constant, of matrix, 103
Laser ablation, 38, 38
Linearization
  procedure, 243
  technique, 272

M
Mass moments of inertia, 156
Micromechanical homogenization, 81
  homogenization of CNTR PNC, 81
    3D Mori–Tanaka method, 101
    Eshelby–Mori–Tanaka method, 94

Halpin–Tsai method, 85
  modified Halpin–Tsai method, 87
  Mori–Tanaka method, 92
  rule of the mixture, 82
homogenization of MSH PNC, 104
Molecular dynamics (MD) simulations, 82, 83
Mori–Tanaka method, 92
Multiscale hybrid polymer nanocomposite, 231, 1, 81
  bending analysis of, 1
  buckling analysis of, 3
  buckling load of, 70, 33
  deflection of, 14, 20
  Euler–Bernoulli beam model, 76
  FRP composites, 74
  homogenization of, 104
  Kirchhoff-Love plate theorem, 76
  maximum deflection of, 28 29f
  postbuckling analysis of, 3
  postbuckling path of, 77
  Rayleigh-Ritz FE solution, 76
  silane-functionalized MWCNTs, 74

N
Natural fiber-reinforced composites, 6
Naval and marine applications, composite materials in, 8
  corvettes, 8
  patrol boats and corvettes of -US navy, 8
  mine-countermeasure vessels, 8
  project YS-2000, 8
  polymeric composites, 10
  fire-retardant composites, 10
Navier's analytical solution, 180
Nonlinear forced vibration
  analysis, 239, 297
  problem, 112, 175 187,
    see also (Euler–Bernoulli beam theory; First–order shear deformation shell theory (FSDT))

O
Oscillatory carbon nanotube-reinforced PNC beam, 231

P
Particulate composites, 2
Pasternak spring, 35, 109, 168
Percent of overshoot (PO), 125
Plates, bending and buckling of, 8
  bending analysis of MSH PNC plates, 10

buckling analysis of MSH PNC plates, 13
point load, 26
sinusoidal load, 26
uniform load, 26
Poisson's ratio, of CNTR PNC, 82
Polymer nanocomposite (PNC) materials, 81
  partial agglomeration in, 97
  shear modules of, 99
Postbuckling analysis, of multiscale hybrid polymer nanocomposite, 3
Postbuckling problem, 109,
    see also Euler–Bernoulli beam theory

R
Refined higher-order beam theory, 126
  bending problem, 128
  buckling problem, 130
  free vibration problem, 132
Refined higher-order plate theory, 145
  bending problem, 147
  buckling problem, 152
  free vibration problem, 155
Root Locus diagram, 249

S
Scattered waves, 249
Shells, buckling of, 17
Single-walled CNT (SWCNT), 83
  Hill's elastic constants of, 94
Space and aerospace applications, composite materials in, 10
  Boron-aluminum, 10
  discontinuous reinforced aluminum, 11
  graphite-reinforced aluminum-based, 10
  metal-matrix composites, 10
  polymer-matrix composites, 12
  rubber-based, 12
  shape memory polymer matrix/SMP composite, 12
  SMPCs, shape memory feature of, 12
S–S thin-walled beams, 239
Stiffness-enhancement mechanism, 231
Stiffness matrix, 132

T
Taylor expansion, 89
Thermal environments, composite materials in, 13
  ceramic-matrix composites, 13
  light-emitting diode, 16
  silicone, 16
Thin-walled plates, 269

Transducers, sensors, and actuators,
  composites in, 17
  energy conversion, 17
  hydrostatic stress, 18
  polymeric matrices with piezoelectric
    ceramics, 18
  transducers, 18

V
Visco-Pasternak medium, 249

W
Wave propagation, 262, 290
  analysis, 247, 277, 303
  problem, 121, 142, 172,
    see also (Euler–Bernoulli beam theory;
    First–order shear deformation shell
    theory (FSDT); Kirchhoff-Love plate
    theory)
Waviness coefficient, 89
Waviness phenomenon, 301, 5
Winkler–Pasternak elastic seat, 12, 130
Winkler spring, 109, 168